STUD MANAGERS' HANDBOOK
VOLUME 18

International Stockmen's School Handbooks

Stud Managers' Handbook
Volume 18

edited by Frank H. Baker

The 1983 *International Stockmen's School Handbooks* include more than 200 technical papers presented at this year's Stockmen's School—sponsored by Winrock International—by outstanding animal scientists, agribusiness leaders, and livestock producers expert in animal technology, animal management, and general fields relevant to animal agriculture.

The *Handbooks* represent advanced technology in a problem-oriented form readily accessible to livestock producers, operators of family farms, managers of agribusinesses, scholars, and students of animal agriculture. The *Beef Cattle Science Handbook*, the *Dairy Science Handbook*, the *Sheep and Goat Handbook*, and the *Stud Managers' Handbook* each include papers on such general topics as genetics and selection; general anatomy and physiology; reproduction; behavior and animal welfare; feeds and nutrition; pastures, ranges, and forests; health, diseases, and parasites; buildings, equipment, and environment; animal management; marketing and economics (including product processing, when relevant); farm and ranch business management and economics; computer use in animal enterprises; and production systems. The four *Handbooks* also contain papers specifically related to the type of animal considered.

Frank H. Baker is director of the International Stockmen's School at Winrock International, where he is also program officer of the National Program. An animal production and nutrition specialist, Dr. Baker has served as dean of the School of Agriculture at Oklahoma State University, president of the American Society of Animal Science, president of the Council on Agricultural Science and Technology, and executive secretary of the National Beef Improvement Federation.

A Winrock International Project

Serving People Through Animal Agriculture

This handbook is composed of papers presented at the International Stockmen's School January 2–6, 1983, San Antonio, Texas sponsored by Winrock International

A worldwide need exists to more productively exploit animal agriculture in the efficient utilization of natural and human resources. It is in filling this need and carrying out the public service aspirations of the late Winthrop Rockefeller, Governor of Arkansas, that Winrock International bases its mission to advance agriculture for the benefit of people. Winrock's focus is to help generate income, supply employment, and provide food through the use of animals.

INTERNATIONAL STOCKMEN'S SCHOOL HANDBOOKS

STUD MANAGERS' HANDBOOK VOLUME 18

edited by Frank H. Baker

LONDON AND NEW YORK

First published 1983 by Westview Press

Published 2019 by Routledge
52 Vanderbilt Avenue, New York, NY 10017
2 Park Square, Milton Park, Abingdon, Oxon OX14 4RN

Routledge is an imprint of the Taylor & Francis Group, an informa business

ISBN 13: 978-0-367-28903-4(hbk)
ISBN 13: 978-0-367-30449-2(pbk)

CONTENTS

Part 4. ANATOMY, PHYSIOLOGY, AND REPRODUCTION

Part 5. BEHAVIOR AND TRAINING

Part 6. ANIMAL PSYCHOLOGY, ENVIRONMENT, AND WELFARE

Part 7. FEEDS AND NUTRITION

Part 8. PASTURE, FORAGE, AND RANGE

Part 9. HEALTH, DISEASE, AND PARASITES

PREFACE

The Stud Managers' Handbook includes presentations made at the International Stockmen's School, January 2-6, 1983. The faculty members of the School who authored this eighteenth volume of the Handbook, along with books on Dairy Cattle, Beef Cattle, and Sheep and Goats, are scholars, stockmen, and agribusiness leaders with national and international reputations. The papers are a mixture of tried and true technology and practices with new concepts from the latest research results of experiments in all parts of the world. Relevant information and concepts from many related disciplines are included.

The School has been held annually since 1963 under Agriservices Foundation sponsorship; before that it was held for 20 years at Washington State University. Dr. M. E. Ensminger, the School's founder, is now Chairman Emeritus. Transfer of the School to sponsorship by Winrock International with Dr. Frank H. Baker as Director occurred late in 1981. The 1983 School is the first under Winrock International's sponsorship after a one-year hiatus to transfer sponsorship from one organization to the other.

The five basic aims of the School are to:

1. address needs identified by commercial livestock producers and industries of the United States and other countries,
2. serve as an educational bridge between the livestock industry and its technical base in the universities,
3. mobilize and interact with the livestock industry's best minds and most experienced workers,
4. incorporate new livestock industry audiences into the technology transfer process on a continuing basis, and
5. improve the teaching of animal science technology.

Wide dissemination of the technology to livestock producers throughout the world is an important purpose of the Handbooks and the School. Improvement of animal production and management is vital to the ultimate solution of hunger problems of many nations. The subject matter, the style of presentation, and opinions expressed in the papers are those of the authors and do not necessarily reflect the opinions of Winrock International.

This handbook is copyrighted in the name of Winrock International and Winrock International encourages its use and the use of its contents. Permission of Winrock International and the authors should be requested for reproduction of the material. In the case of papers with individual copyrights or of illustrations reproduced by permission of other publishers (indicated on the copyright page), contact those authors or publishers.

ACKNOWLEDGMENTS

Winrock International expresses special appreciation to the individual authors, staff members, and all others who contributed to the preparation of the _Stud Managers' Handbook_. Each of the papers (lectures) was prepared by the individual authors. The following editorial, word processing, and secretarial staff of Winrock International assisted the School Director in reading and editing the papers for delivery to the publishers.

Editorial Assistance

Jim Bemis, Production Editor
Essie Raun
Betty Stonaker

Word Processing and Secretarial Assistance

Patty Allison, General Coordinator
Shirley Zimmerman, Coordinator of Word Processing
Darlene Galloway
Tammy Chism
Jamie Whittington
Venetta Vaughn
Kerri Alexander, Computing Specialist Assistant
Ramona Jolly, Assistant to the School Director
Natalie Young, Secretary for the School

Part 1

THE HORSE INDUSTRY AND ITS HERITAGE

1

CENTAURS, CAVALRY, AND COWBOYS

Richard L. Willham

- "A Horse! a horse! my kingdom for a horse!"
King Richard III, Act V, Sc 4,7. Shakespeare.

To those mounted on horses, the world is somehow different! Being astride a horse above other men, and having a sense of mobile power between legs, has always transformed man both in physical ability and psychic stance. Observing a stallion dominate and control his band of mares on the vast steppes of Eurasia must have introduced transhumant man to the potential of the horse as a partner so that man could dominate and control the hunt, his grazing livestock, and other men.

Some cultures were mounted even as the curtains of antiquity parted; earth-bound man, the cultivator of soil, already knew fear of the mounted as reflected in the CENTAUR of mythology. The livestock of the civilizations developing on the fertile flood plains were subservient to agriculture. They contributed to food production as a source of power, meat, fat, and clothing. The horse did not. The horses of defense and conquest cost the nations. What the Egyptians and their slaves ate when Ramses II stabled 70,000 chariot horses along the Nile remains a mystery. The pastoral nomads of the Eurasian steppes astride their horses and armed with bows and arrows posed a constant threat to developing civilizations. The history of the fertile crescent is one of successive waves of horsemen conquering complacent cultures and in succession being assimilated; the rich civilizations of the ancient world were the result.

The purpose of this paper is to explore the role of the horse in the cultural evolution of man. Primary emphasis will be on mounted man. Three aspects of the role are represented in the title: CENTAURS, CALVARY, and COWBOYS. The mythological centaur represents the role of the horse in enriching the humanities; cavalry represents the involvement of the horse in the wars of man; and the word cowboys in the title represent the contribution of the horse in herding pastoral stock through the ages. The roles of the horse as food, as transport (especially during the railway age), as a late source of power for agriculture, and as a chariot animal will receive only brief attention. The presentation is roughly chronological to better capture the connections inherent in the web of history. The paper is written for lovers of horseflesh and leather.

CHRONOLOGY

In some 50 million years, the terrier-sized forest mammal with multitoes evolved primarily on the North American continent into the single-toed massive beast of Europe and the swift exploiter of the vast steppes of Eurasia. The rapidly expanding savannas of some 4 million years ago enticed man from the forests and induced a more omnivorous diet. Meat eating from the clan hunt gave man more time for social interaction, and the control of fire some half a million years ago allowed man to survive the last, long glacial period. The refuse piles of the cave dwellers contained the disjointed bones of many horses. Man depicted the horse in his cave art of some 20 thousand years ago, signaling that he would become a shaper of his world rather than just another ill-equipped mammal of an ecosystem.

After the last glaciation, many clans of man became transhumant, following the herds of grass eaters north in the summer and south in the winter. It was during this time, along with the domestication of the Asiatic wolf as a hunting companion and several cloven-hooved ruminants, that man developed his relationship with the steppe horse. Transport and the natural herding instinct made the horse useful, as did the increased mobility they provided for man. Cultural exchange was possible, but so was war! Mounted archers, in times of hardship on the steppes, were to be reckoned with by people developing a settled agriculture.

CIVILIZATIONS

The Scythians rode into the history of the Near East around 6000 B.C. The Hyksos of Syria conquered the stable Egyptian civilization in 1730 B.C., but gave the Egyptians their chariot horses. The entourage of nations of the Fertile Crescent represented one horse culture after another conquering and then being assimilated, each adding new flavor to the cultural heritage. The earliest bits and bridles were retained for centuries. The yoking of onagers to chariots was adopted for the horse only to make the chariot an ineffective war machine except on level ground. Choking-down was not solved for centuries, but the chariot age gave Europe the Palfrey that paced, the gait that could propel a chariot at a good speed without choking down the team.

The mounted bowman of the steppes gave man one of his oldest games. It was a pure Mongol war game, later called "buz kashi". The horsemen gathered. An end point and a distant flag were set. A headless calf was dropped among the milling horsemen. The objective for each horseman was to pick up the calf, carry it around the flag, and return it to the endpoint. Whatever tactics that were needed were used. From the game came polo, which was imported from Bengal by British soldiers in India.

The Scythians and later the Parthians not only gave civilization the chariot horse but their threatening presence also gave the cities their giant walls. The Assyrians and later the Persians cemented their empires together with a pony express system. The horses of these early civilizations contributed to the imperial conquests necessitated by the continual need for new land as yields under irrigation went down.

MEDITERRANEAN BASIN

The Phoenicians by 1000 B.C. were the horse traders of the Mediterranean basin. The common pastoral nomad heritage of the basin peoples has been preserved for us in Greek mythology. The great winged horse, Pegasus, was caught and tamed by Minerva in all her wisdom. When Bellerophon set out to kill Chemera, a dragon, Minerva gave him a golden bridle and Bellerophon became the first man to ride and to conduct mounted warfare; he probably started the romance of dragon slaying in the Middle Ages. Actually the Surmatians of 300 B.C. used scale armor made from horses' hooves for themselves and their horse. They probably looked dragon-like.

Thrace was the legendary source of the mythological CENTAUR, which may have been the origin of bull fighting by man on horseback. The account of the battle between the Centaurs and the Lapithae of Thessaly reflects settled man's fear of the horseman. The man and horse were one. All of the bad and fearsome traits of man were given to the Centaurs by the Greeks.

Neptune and Athena disputed as to who could give the greatest benefit to man. Athena gave the olive tree while Neptune gave man the horse; Neptune's waves still resemble the shapes and movements of the horse's mane and tail. Greek sea navigation terms were the same used in equitation. The bridle was the anchor. Homer called ships horses of the sea. The Trojan horse may really have been a troop ship.

Four-mile horse races were held in the 23rd Olympiad of 684 B.C. The Parthenon frieze of the Acropolis in Athens completed in 440 B.C. captures the horse in all his majesty. Off-side mounting was the rule since vaulting on with a spear required that it be retained in the throwing hand.

Horse blocks for mounting were maintained along the roadsides. Horses were even taught to kneel. Such was riding without the stirrup. Hippocrates in 400 B.C. noted the swelling in the legs of the Scythians.

Philip of Macedonia is credited with the first use of massed cavalry charges against the Greeks in 340 B.C. His son, Alexander the Great, used his cavalry to great advantage. In the battle of Granicus in 334 B.C. he lost his horse Bucephalus. The town around his burial site was named in the horse's honor.

Roman riding, called singularis to differentiate it from team driving with the chariot, was copied from the Greeks. The Greek book on horsemanship called the Hippike, written in 365 B.C. by Xenophone, influenced all future cultures of the Western world, including the Romans. Romulus in 725 B.C. began the order of Equites consisting of 300 mounted men. This privileged group wore red tunics with two narrow stripes of purple. "Being in purple" was a sought-after honor during the time of Roman empire. A cavalry legion consisted of 300 men, as compared to an infantry legion of 6000. This ratio of 1 to 20 possibly represented the effectiveness of cavalry. Although the Roman phalanx won the empire and maintained the period of stability, increasing pressures from the mounted Parthians and barbaric hordes forced more legions on the frontier of the empire to become cavalry. Back in Rome, the chariot races at the Circus Maximus were splendid.

DARK AGES

The Western Roman Empire fell from within; however, the continual pressure from the mounted bowmen of the steppes helped. The Empire held the Gothic tide, who themselves were pushed West by the Huns, until 408 A.D. when Alarie began the invasion of Italy. In 410 A.D. Rome was sacked; the Christian Goths fanatically destroyed the Greek and Roman statues, symbols to them of pagan worship. The Goths were mounted; supplies were carried in nets behind the pad saddle.

By 435 A.D. the European barbarians and the Romans were living in temporary harmony. This was upset by the Huns moving even farther West under Attila, "the scourge of God". His war and butchering were halted by a united Europe near Orleans in northern France where he met a decisive defeat in the battle of Chalons. Only by catching the Huns in camp without full use of their cavalry was Europe saved from destruction. Rome was saved only when Pope Leo the Great took advantage of the Hun's fear of religious curses. The Huns used saddles with stirrups.

The eastern Roman Empire survived the horsemen and developed into the stoic Byzantine empire under Constantine and Justinian. They buffered Europe so that Clovis, who became Christian in 496 A.D., could consolidate parts of

France. Near Poitiers in 507 A.D., the Goths were driven from the country. The Teutonic hordes of Clovis were equally divided between those mounted and those on foot, since the foot soldier was paired with the mounted soldier, taking his place if necessary. All supplies were carried on the course horses.

Mohammad united the Arab world around 500 A.D. using their Arabian horses and scimitars, he began a series of Holy Wars that by 700 A.D. saw them across North Africa. After some fateful intrigue, Tarik the Moor landed at Tarifa on the Iberian coast with some 12 thousand light cavalry. Roderick the Gaul, who ruled Spain, massed some 90 thousand for the battle of Xeres fought on the plains near Cadiz. The outcome was of monumental importance to the Western World. Many of the Gauls were soldiers of fortune. The Moslem army was disciplined and fanatical, besides that Tarik had burned their ships. In 711 A.D. Spain was Moorish and remained so for several hundred years. The Moors contributed their Barb horses, horsemanship, knowledge of the classics, weaving, feral husbandry of livestock, and citrus production to the Iberian culture. Leather was produced in Cordoba and Cordovan remains the best.

The Franks under Charlemagne united Western Europe. Many of his 53 campaigns were retaliations for misdeeds against the Pope in Rome. Riders of the era used saddles with stirrups. The Germanic involvement came in 919 A.D. with the reign of Henry of Saxony. To stop the Huns he made a nine-year truce and began to fortify cities and build a cavalry force using all the European knowledge from as far away as Spain. His "tribute" nine years later was a mangy dog. As a result, the Huns invaded but were driven from the Germanic countries forever. To keep the cavalry arm in condition, Henry started tournaments that developed into a Medieval tradition.

MEDIEVAL EUROPE

Two battles mark the beginning and the eventual end of the armored, lance-carrying knight on his massive charger--now a weapon as well as a means of mobility. Europe's answer to the mounted bowmen of the steppes evolved over centuries.

The first battle changed the government of England. On Friday, October 13, 1066, Harold, King of England, occupied a rise overlooking Hastings. His army had marched 270 miles in 10 days after battling the Danes at Stamford bridge. They numbered 5000 at the most. Duke William of Normandy had landed 16 days earlier with 1000 archers, 4000 men at arms, and 3000 cavalry. William offered terms, but they were not accepted by Harold whose conservative style of fighting behind a wall of shields did not allow combat on horseback. At 9 a.m. the Normans advanced on the 800 yard-long wall of shields. They were beaten back repeatedly. As the Normans moved back again, the Saxons made their mistake and followed them down. The Norman cavalry turned and standing in their stirrups cut the Saxon ranks to pieces beneath their horses' feet. They were shock troops using the lance. Twenty rode to the English standard on the hill and the last four alive hacked Harold to pieces. At 5 p.m., England was Norman. The Bayeux tapestry records the battle of Hastings. The kite shields, the lance in the right hand, and the stirrup ushered in shock troop cavalry and the era of the invincible knight.

The manor, in fact the feudal system of Europe, developed as a ranch to raise the necessary horses and to protect and defend groups, especially against the Viking raids. The age of chivalry grew out of the feudal system of responsibilities. Mounted warfare demanded a host of ancillary horses. The sumpter was the pack horse. The courser was for messengers or heralds and became the race horse. The rouncy was for the squires and became the carriage horse. The massive charger was now a weapon and later became the work animal. The palfrey was a legacy of the chariot age and, because of its gaits, was used for pleasure riding and hawking, and later became a work animal. The Andalusian of Iberia was the knights' charger for ceremony. With all the horses, riding skills were necessary for the nobility. The church induced the soldier nobles to become soldiers of the cross. An order of knighthood was established, with the ceremonies of knighthood kept vividly alive in King Arthur.

The crusades have no equal in fanaticism, massacre, and terrorism. Horses were paramount to both the Christian and the Moslem. The first four of the nine crusades were full-scale wars of religious fervor. Most often associated with the crusades Richard the Lion Heart, King of England, whose exploits in the third crusade inspired poets and writers. The white tunic over his coat of mail with a prominent red cross on his chest was respected even by the Moslems. The heavier horses of Europe were matched with the swift Arabian and Barbs of the Moslems. Such flagrant waste of lives and resources appears now to be truly the sport of the excess nobility of Europe.

ASIAN HORDES

While European nobles coated themselves with armor and fought on heavy horses, the steppes of Eurasia resounded to the hoofbeats of thousands of mounted warriors. The Huns had invaded China as long ago as 1600 B.C. The Great Wall of China was completed in 218 B.C. for the purpose of protecting China from horsemen. It is said Ch'in-Ti had the Wall built by following a magical white horse. The Wall followed his trail except when a great dust storm engulfed the area. When the weather cleared, they found the horse far to the south. They abandoned the wall built in the storm and began following the horse. The off-shoot of the wall exists today. However, the Lsin dynasty in 202 B.C. was taken with cavalry. India, as well, was invaded by the Ayrians early in its history. The horse cultures of the Asian steppes had a continual influence on most civilizations of Asia.

Genghis Khan assumed command of his father's 40 Mongol clans living north of the Great Wall in 1168. After much intrigue and being intoxicated by conquest of all Mongolia, his horde scaled the Great Wall in 1211. Persia, India, and southern Russia were plundered by hordes of horsemen. Imagine a horde of horsemen 10 miles wide and 20 miles long, each with 18 horses. Mobility was the issue. Mare's milk, horse blood, and horse steak carried under the saddle for a day until they were salted constituted the commissary. This coupled with the spoils of war took the hordes over most of the known world. Each officer was in charge of 10 men or 10 units. The stout, walrus-mustached, bow-legged, horseman of the steppes feared only the smell of salt air and mountain breezes. The Khans' conquest of China over a period of 150 years reduced the population form 100 million to 59 million. His son, Kubla Khan, ruled China during the visit of Marco Polo. Horse conquest was not rule; the Mongols were absorbed into the cultures of China, India, Persia, and Russia. The Eastern horsemen spread terror before them, but in the end they carried the pollen that fertilized civilizations over the world. Samarkand of Tamerlanes' Tartars was a tribute to this era of history.

MODERN EUROPE

Meanwhile in Europe, knights were moving toward their demise. In the late summer of 1415 A.D., technology use was to close an era as it had opened it. Henry V of England had invaded France to lay claim to the throne. The battle of Agincourt was the beginning of the end of the mounted, fully-armored knight, although the heavy horse remained as a mobile command post during the reign of Henry VIII. On October 25, both armies had been in position all night. The French had 25,000 knights in a half-mile gap between two woods. At dawn they were in no condition to fight since it had rained all night, and the knights had spent the night in the saddle to keep their armor movable. The English were no better off. Since landing they had spent 17 days traveling some 270 miles. They were outnumbered 4 to 1. The French had 15,000 mounted knights in three ranks. The English were dismounted with wedges of archers between their three groups. By 11 a.m. the French were jostling and hurling insults. Henry moved within 300 yards and set spiked stakes in the sea of mud; his archers fired, wounding and galling the French on their horses. The French then charged. Henry's longbow archers shot their horses and many knights suffocated in the mud or were knifed through the joints in their armor. In a half hour it was all over. The English lost 500; the French lost 10,000. The myth of the invincible knight was shattered by the Welch longbow introduced by Edward I.

The plague, following the little Ice Age of Europe, killed half the population of Europe. This enhanced the value of the common man (as a laborer) as did the historic plays of Shakespeare that gave the English common man his respect. The triumph of the longbow in the hand of the commoner was eulogized. However, the belligerent royal houses of Europe had a new method of destruction--gunpowder.

After Agincourt and before several items of horse equipment moved the charge of war into other avenues of work, the 8th century mouldboard plow let European agriculture feed more people. The iron horseshoe and the simple Chinese horse collar allowed the horse to compete with the ox for farm power and dray work. No longer did the horse choke down. Transportation, speed, and mobility increased dramatically; oats became a part of crop rotation.

The mule of state ridden by Catholic clergy met its demise with the Reformation in Europe. Mules were ridden instead of horses perhaps because David rode only mules. Cardinal Wolsey had a superb pair of mules before Henry VIII fired him.

NEW WORLD

The discovery of the new world by Columbus had monumental consequences for the horse. The horse, in service to man, was returned to the continent where he evolved eons before. The flora of the new world (corn, potatoes, tomatoes, and sugar) revolutionized the old world, but the colonials brought their livestock and let them adapt to the new territory. The voyage for the horse, even in a sling, was not easy; the "horse latitudes" are named for the many horses cast overboard in calms.

The conquistadors with their war horses conquered the American civilizations, repeating again the history of horse conquest. To the Spanish, a gentleman was a caballero or horseman. Spanish law commanded caballeros to ride only stallions; Columbus was granted a dispensation to ride a mule in his old age. Consequently most Cordoban horses used by the conquistadors were stallions. The Spanish had to import mares purposefully.

One conquistador wrote that horses are the most necessary things in the new world because they frighten the enemy most and, after God, to them belongs victory. The glories of the Inca and Aztec civilizations were no match for the mounted men with Morion helmets, steel breast plates, Cordovan boots, and guns in saddle holsters. The cannibalism of natives upset the Spaniards, but in the name of Christianity they reduced the population of the new world from some 25 million to one million in a span of 30 years. Horse heads were piked along with hose of the unlucky conquistadors. Morzillo, the black stallion of Cortez, died in the hands of the Indians; they carved a statue in memory of Morzillo and worshipped it for years.

By 1500 A.D., the crown had a ranch on Hispanola with 60 mares. In less than 10 years, many horse breeding ranches were operating; soon Iberian horses were in less demand than those of the island. The Spanish in the West Indies prospered more from livestock production than from the gold seeking conquistadors they outfitted. The ranchers developed an aristocracy that became the envy of all Europe, but gold or silver horseshoes were not as good as those of iron. Cortez became the first Mexican rancher in 1519. In a short time ranchers had exploited concentric circles out from Mexico City with their feral stock. Coronado, in 1540, was the first trail driver in the Southwest on his search for the seven cities of Cibola. His 228 horsemen took over 1600 horses and mules along with sizable herds of cattle, sheep, and swine.

The French and English came to the new world later and colonized a harsher environment than the Spanish. The English brought horses of many types, primarily from the points of embarcation. Elizabeth I, a palfrey-riding falcon hunter, later disposed of the Spanish Armada and the colonies prospered on the eastern seaboard of the new world.

EUROPE

Gunpowder in Western Europe revolutionized warfare. Sieges of walled cities were shorter and the artillery, made famous later by Napoleon, became a significant arm of military service. Yet the cavalry remained until the machine gun. Knights were replaced by swashbuckling cavaliers with plumed chapeaus, jack boots, and elegant spurs. They helped kings hold their power in the age when peasants became people. European military, especially the Prussians, became infatuated with the Oriental stallion in remount service. State studs were set up over Europe; most stock was imported through Turkey. Equestrian skills rose to heights in the many riding schools. The European wars of the 17th, 18th, and 19th century resulted in Europe being "overhorsed". England continued to codify the sport of kings with its jockey club and practiced fox hunting. Europe used the surplus cavalry horses to go from the horse borne era of Elizabeth I to the age of coaching that lasted until the advent of reliable rail service. No horse person is complete without reading Swift's "A Voyage to the Country of the Houyhnhnms" where horses are the dominant species and man-like creatures the subservient one.

CIVIL WAR

The cavalry arm in the Civil War became the eyes and ears of the armies; mobility instead of shock was the issue. Initially the Southern aristocracy was adept at cavalry tactics and was well horsed. Military leadership was an honored profession, as Lincoln soon found out. The exploits of the dashing cavalier, J.E.B. Stuart on Highfly, were magnificent. Even solid "Stonewall" Jackson had his Old Sorrel. Robert E. Lee on Traveler was the image that the tattered Confederate army fought for until the end. From Sherman's march to the sea on Lexington and Sheridan's ride through the Shennandoah Valley on Winchester, war became, not a gentleman's game, but total economic warfare. But the noble chargers of the generals remained to inspire their men. Beginning with the first large equestrian bronze of Jackson in 1848, government structures became filled with statues of leaders on horseback. The boots reversed in the stirrups of the riderless mounts alongside the lumbering caisson symbolized the importance of the relationship between a man and his horse even today.

INDIANS

The saga of the American Indian and the horse defies imagination. This transition from the Stone Age culture having only the dog, to one with the mobility of the horse was dramatic. The Spanish recognized the military value of their mounts; but, as they took up ranching, the need for vaqueros caused the laws to be broken--and the Indians to learn to ride. The tribes of the South acquired horses early, and the Cherokee horse became part of the Virginia mix of horseflesh. But it was the Plains Indians that quickly developed horse cultures--Comanches, Pueblos, Navajos, and Apaches. The Comanches "God Dog" was appropriated from the Spanish and rapidly transferred to other tribes--wealth, war, coups, and hunting were enhanced by the horse. When Onate moved to Santa Fe in 1598, he brought large herds of horses. The first hanging occurred over horse stealing. The Cheyennes were called the painted horse people. Indians excelled in the art of camouflage.

The horses that became feral in the West multiplied rapidly and began a three-pronged migration north. In 1805, Lewis and Clark met intermountain Indians of the north that had horses. Little horse breeding was done by the Indian, except for the Nez Perce and their Appaloosa. To acquire horses by coup was so much more rewarding. The mustang of the West is primarily Spanish in origin but shows adaptation to the wild. When Indian legends are probed deeply enough, references to horse gods may actually reflect interaction with the earlier horse of the Americas. The Indians developed their own style with the horse. The white man was not copied; he was bettered. The Indian in war paint and feathers and deadly bow repeated again the Centaur effect of the riders of antiquity.

COWBOYS

The golden age of the cowboy lasted around 20 years, yet he has become an American folk hero of gigantic proportion--thanks to the silver screen and the cowboy's supposed Centaur role. He is part California caballero, Mexican vaquero, Rocky Mountain trapper, Dixieland planter, Plains Indian, and eastern cavalryman. His equipment comes from and was named by the Spanish of the Southwest. His herding of feral cattle comes from a long line of ranching tradition going back to the steppes. His part in the transition of the Great Plains of North America from buffalo to the Hereford is truly monumental. The rodeo is now his performing stage. The Book _Smoky_ by Will James speaks of the rugged individual while the art of Russell and Remington clearly illustrates the cowpuncher in action.

RAILWAY AGE

The Iron Horse of the railway age, introduced in the early 1800s, failed to replace the horse as a mover of goods from the "team" tracks of the yards to the inner city. The teamsters' unions of today have a heavy horse heritage. Neither was the steam engine able to displace the horse as power for pulling and turning the ground wheel fast enough to articulate the rapidly evolving, labor saving farm implements. The book _Black Beauty_ is a lucid account of trials of horses during this transition era when some 300,000 horses worked in New York City alone. In Europe, the haymarket streets of the cities reflect the massive requirements of horses of the Victorian age. The crossing sweepers began the manure and bedding removal that was monumental.

ANOTHER "HENRY"

Another "Henry" of a much different line produced the demise of the draft horse. Henry Ford combined the internal combustion engine with mass-produced carriages to usher in our oil-based economy. The bonds of the horse with the railroad age were slowly severed. Teams were unhitched and trucks responded to man's seeming need for speed. Tractors began to replace Tom and Jerry on the farm in the 1920s. Mewes became garages. But horses have not vanished completely. In fact, there are a tremendous number used and loved by people in all walks of life. Their major use is for pleasure, either vicariously or as participants. Racing is big business. The Quarter Horse Association has the largest registry in the world. However, the horse still works as a partner of man in the herding of livestock that feed upon sparse vegetation on the steppes of the world.

The horse may have gone full circle in his use by man. As the sun rises over the mesa and the cowboys take their assignments for the roundup and "step up" into their work saddle, the whole saga of history flashes back across time --when Bellerophon first rode Pegasus, the Centaurs humiliated mere man, the Mongols lathered their mounts in conquest, the knight drove his lance through the infidel, the cavalier dashed over Europe, and the beaten Southern cavalryman helped create the legend of the cowboy. The humanities would be less brilliant without the spirited horse to drive the creative urge of man.

SUMMARY

To the mounted, the world is somehow different! The mythological CENTAUR represents the role of the horse in enriching our humanities, CAVALRY represents the involvement of the horse in the wars of man, and COWBOYS represent the contribution of the horse in herding pastoral stock through the ages. Transhumant man became mounted before most civilizations developed on the fertile flood plains of the world. The mounted bowman from the vast steppes of Eurasia conquered, became assimilated, and mixed cultures into rich tapestries. The Mediterranean basin resounded first to the mythology of the Centaur followed by the Roman Empire, the mounted barbarian hordes, and the classical chariot races of the Byzantine Empire. The feudal system of Europe began with nobility's need of the horse for survival. The Arab world rode into Spain to crossfertilize Western society and the Mongol hordes swept over China, India, Russia, and Persia, only to enrich the evolving cultures after the devastation was forgotten. The battle of Hastings ushered in the knights of Medieval Europe and the battle of Agincourt demonstrated that the knight was obsolete. The conquistadores rode to conquest in the New World, but the Plains Indians took to the horse with their own style and developed horse cultures. The cavalry of Europe perfected Oriental introductions. The industrial revolution powered by steam relied on the horse for town transport and farm work. The cowboy, folk hero of the Americas, relives the herding role of mounted man. The humanities would be deficient without the shaggy horse in cave art or our modern quarter horse paintings and all that lies between.

2
A NATIONAL VIEW OF HORSE PROGRAMS

Michael J. Nolan

For more than two decades the horse world has demonstrated remarkable growth. Both recreational activity and commercial investment in horses have enjoyed dramatic long-term increases. Yet a number of industry observers question whether this degree of growth and profitability can be sustained indefinitely.

On looking at the horse industry from a national-trade-association perspective and in trying to provide some idea of where the business may be going, two major factors should be examined: the projected infusion of new capital and the costs of doing business in the future.

Before going into the specifics of revenue projections, it may be useful to break the industry into segments, to divide racing from showing, and further to subdivide those into select horses, moderate quality horses, and those animals with lesser breeding or conformation.

The sale prices of top quality, select racing and show horses have established new records year after year for the past 7 years--and, so far, enough new investors have come along to provide the fresh capital to fuel yet higher prices. However, there are signs this situation may be changing. The prices paid for top quality fillies and mares, the production resources of the future, have begun to weaken at the Thoroughbred and Standardbred sales. Whether prices resume a steady up-swing if the national economy revives remains to be seen.

Similarly, prices of moderately bred race and show animals have been lagging behind the increases at the top of the market. Horses selling at smaller local or regional sales that do not attract the international buyers or the free-spending newcomers are often bringing less than the cost of breeding and raising them.

Two factors contributing to any possible weakness in the market for horses of investment caliber are oversupply and lack of a sound marketing effort. The oversupply problem has been building for several years, with the number of new foals in the major racing and show breeds up 156% over the past 8 years. The 1981 Economic Recovery Tax Act further exascerbated the trend by substantially shortening the depreciation period for breeding stock while increasing the depreciation period for race horses.

The rapidly increasing supply of horses at the upper levels can be absorbed only if enough new horse owners are created to purchase them. However, there is no industry-wide marketing program, and individual selling often is aimed at short-term sales rather than designed to create satisfied long-term investors. Too many newcomers are subjected to high-pressure sales and hoopla, have bought horses at inflated prices and then have found no one interested in buying those horses or their off-spring. This treatment results in early disillusionment and early departure from the business.

While supply has been growing faster than demand, other factors have begun to squeeze breeders. Raising horses is a labor-intensive industry, and the 1981 tax bill was not helpful to those concerns that are labor intensive, having been designed to help capital-intensive industries. High interest rates and stagnant real estate prices have hurt some horse owners. In addition, there has been little effort in the industry to use new technology to reduce costs.

Horses used for commercial purposes account for less than 20% of the population. The 8.3 million horses in the U.S. have an estimated value of $9 billion, but 1% of those horses are worth $2 billion, more than 20% of the value of all horses. It is the commercial horses that make headlines, and they are the ones that are attractive to investors, but the other 80% are also important. Here, too, there is both good and bad news, and for similar reasons.

A number of surveys have found a high degree of interest in equestrial sports. According to data from both the Gallup organization and Opinion Research Corp., the potential market for horses is equal to or larger than the number of current horse owners. However, the industry as a whole is doing little to sell to that market, and economic and social forces may erode this potential unless a unified industry effort is made to ensure people that they will be able to afford the cost of equestrial recreation.

For over 30 years the horse industry languished while tremendous advances were made in scientific research on both health and management of other livestock. Renewed research efforts have made progress, but horse people have been slow to adapt to new methods.

To close on a positive note, there are new marketing programs being used by state breeders' associations, and the racing programs of both Standardbreds and Thoroughbreds are being modified to increase the potential for profitable racing. New techniques are becoming available to owners and breeders to improve productive efficiency and to reduce production costs. The showing world has led the way in finding sponsors to help defray exhibitor costs, and racing is following that lead. New forms of ownership that reduce the individual risks and permit the neophyte to participate in ownership are gaining wide acceptance. These progressive steps may head off the negative effects of overproduction in the commercial sector. If the recreational market can be strengthened, the 1980s may be even more successful for the horse world than the 1970s.

3

PRESENT DAY USAGE AND FUTURE OPPORTUNITIES FOR DRAFT HORSES AND MULES

Maurice Telleen

There are certain stereotypes that are often associated with draft horses and mules...and with the people who own them. Most of the horses don't weigh a ton and not all of us are Amish...nor do we work for Budweiser.

These stereotypes can get in the way of thinking. So I believe the first order of business might be to deal with them. In the process, maybe we can start looking at the draft horse and mule for what they really are...and what they really offer.

The first stereotype is that anyone in the horse business is a horse lover. Looking over the cast in this production, I recognize the names of many eminent sheep specialists, swine specialists, nutrition specialists, etc. Tell me, do you consider yourselves as "ewe lovers," "pig lovers", and "feed lovers?" Sounds ridiculous doesn't it? The fact that a common expression such as "horse lover" even exists tells us much about what our relationship with the horse has become. It has become a romantic notion.

The truth of the matter is that the heavy horse and mule business is full of people who, while very fond of their animals, are neither hopelessly sentimental about them, nor do they have an incurable itch to show off with them in public. There are thousands of us in this heavy horse and mule business who hold to the notion that a horse, or mule, should work for his owner, not the other way around.

Perhaps that is the reason that most of us regard the relatively high slaughter market that has existed in recent years as an absolute necessity in maintaining stability in the draft-horse market. The fact that it isn't one of our favorite parts of the business does not make it any less important.

One of the largest semi-annual draft horse sales in the nation is held in our home town, Waverly, Iowa. Normally about 500 to 600 head of drafters are sold, both spring and fall. At one of the sales last year, about 70 head went to Japanese buyers. Not all of them were headed for Tokyo restaurants, but some were certainly destined for that. Included in the group were some very serviceable teams of young geldings--horses with years of work still in them. It did not please me to see that kind go, but what displeased me most was that those teams did not have an opportunity to perform several more years of useful service for someone who could have used them.

A couple of years ago the magazine EQUUS published a release entitled "The Exploding Horsemeat Market." The figures astonished me, and I think they might have the same effect on you. In 1972, the U.S. shipped approximately 6.7 million pounds of fresh horsemeat abroad for human consumption. By 1978, according to the USDA statistics, that figure rose to nearly 120 million pounds--a growth of over 1700%.

According to EQUUS, European markets were taking 90% of the U.S. horsemeat abroad for human consumption at that time. Horses costing $.50/lb on the hoof here were selling for over $6.00/lb to the French or Belgian housewife. I don't know whether this trade has grown, shrunk, or stabilized since those 1978 figures were released. The price of slaughter horses has remained relatively high. In a protein-hungry world, I would expect that to continue, whatever cultural taboos we might have on eating horsemeat in our nation. Horsemeat offers nearly as much protein per pound as beef and pork, with far less fat and at a somewhat lower cost. This trade has aroused some public resentment.

I can understand the attitudes of those who would ban or restrict the export of horsemeat. If more of them were in the horse business themselves, it would be a different story. Old feature writers are entitled to their pensions, and so are some old horses...we've got a couple ourselves--horses, that is. While I believe we can safely pension off <u>all</u> the old writers, to do so with <u>all</u> the old horses would create an intolerable economic burden on the people who produce, keep, care for, and use horses. Nor are all the killers (for slaughter) old. Some are simply not much good --misfits and cripples. No serious proposals are ever advanced to deprive the dairyman of his canner-cow market. The trip from Foster Mother of the Human Race to Corned Beef on Rye has come to millions of cows...and it comes without public commentary.

The breeding of Belgian and Percheron horses in their native countries is, and has been for some time, geared to the butcher's block as much as to the harness and collar. The fate of most of the stud colts of our two most popular breeds, in their native countries, is no different from that of the average Charolais bull calf.

Salvage value is an important part of any livestock enterprise. Less so, perhaps, with draft horses because they are a working animal with a longer life span than other domestic species, but still a consideration. It has always struck me as strange that there is more public outcry these days over a Frenchman eating our surplus horses and paying a respectable price to do so than there ever was over Fido eating our surplus horses that sold for $.06/lb during the great liquidation following World War II. I suppose if more Americans kept a Frenchman for a pet there would be less hue and cry about it.

There have been instances of horrendous abuse of horses being transported to slaughter plants and I can offer nothing but contempt for that. Commerce, of any kind, always seems to both need and invite regulation, and I'm certainly for it. But that is more a commentary on human brutality and greed than on the legitimate trade in horsemeat for export.

To get on with the business of draft horse and mule stereotypes, I am not big on nostalgia for its own sake. I do not look back with an uncritical fondness at the "good old days." Most of them weren't so hot. I don't think our country is being farmed very wisely or very well these days, but I doubt that it ever was. So, if you are looking for a draft horse editor who would like nothing better than "going back"--whatever that means--to the "good old days," whenever they were, I am here to disappoint you.

To illustrate how firmly wedded the image of the draft horse and mule is to this "good old days" myth, I wish to refer at some length to a widely quoted speech of a few years ago. It was delivered by Earle E. Gavett with the National Resources Economic Division at the Northeastern Agricultural Economic Council Meeting in Orono, Maine, in June of 1975. The title was "Can We De-Mechanize Agriculture?" It was later published in The Farm Index, an economic research service of the Department of Agriculture and widely quoted. Seldom has a more carelessly crafted speech been treated with such respect. It dealt almost entirely with the public stereotypes regarding both draft horses and mules and draft horse and mule men.

There was one exception to this chorus of approval, and it might not surprise you to learn that it came from my own paper, The Draft Horse Journal. I'd like to quote a bit from our Winter 1976 issue:

"He (Mr. Gavett) starts out by stating that it would take 61 million horses and mules and 31 million farm workers to do the job today. He then states that, 'To some critics of today's farming practices, this whimsical scene is the solution to many problems that have plagued America recently'." From there we go to the inevitable reference to the good old days.

I claim a fair knowledge of the critics of today's agriculture and his description of them describes no one that I know. Most of us do feel that there were some important tools (including horses and mules), attitudes, techniques, and community structures that were discarded in haste, and that these things could and should have been built upon and cannot be easily reconstructed, some of them not at all. This picture, of simple and single-minded critics mounting an antitechnological revolution, is an invention. Horses in themselves are no more antitechnological than legs on people. It is more of the old either-or business, with no middle ground and little room for common sense.

The way he set the stage was not, however, the most interesting part of his speech. The painting, in words, of an idyllic Currier & Ives scene and the subsequent mockery of it, the song of praise to the productivity of modern agribusiness (by now one farmer must be feeding at least 120 people plus playing 9 holes of golf a day), and the title designed to shock you into reading it, WANTED, 61 MILLION HORSES AND 31 MILLION FARM WORKERS...they were all predictable. They are shopworn techniques but still serviceable.

His method of arriving at his conclusions is the real giveaway. Actually his conclusions preceded his methods, which is what stereotyped thinking is all about. Mr. Gavett used the 1918 and 1974 crops, with 1967 as the index year. That is, the 1918 crop was 48% as large as the 1967 crop and the 1974 crop was 109% of the 1967 crop. Thus, production was more than two times greater in our era. He simply multiplied the 26.7 million horses and mules in 1918 by 2.27 because the 1974 crop was 2.27 times larger than the 1918 crop. Now, if you are looking for a truly simplistic approach--there is one for you.

There was no consideration given to total acreage involved, just total production. There was no consideration that millions of those horses and mules in 1918 were not employed in farming. They were still doing the work of trucks and automobiles by the hundreds of thousands in 1918, and as many were recreational animals as we have now.

There was no consideration given to increased yields through hybridization, commercial fertilizer, the use of chemicals, and all of the other practices that came about during that interval...in addition to mechanization. He did acknowledge that improvements and technology in addition to mechanization had accounted for much of this increase but he had been assured by agricultural economists that the point of projection was valid.

I found this to be ABSOLUTELY INCREDIBLE! Have any of these people ever taken a look at the productivity of places such as Lancaster County, Pennsylvania; Elkhart and Goshen Counties in Indiana; Holmes County in Ohio where a very significant part of the land is farmed with draft horses and mules? To ask the question is to answer it. Of course they haven't! Believe me, if those Amish farmers were getting 1918 yields, they would have been out of business years ago. Instead, their communities grow, their young men become farmers, and they don't seem to be going broke.

I am not urging Amishness on this country. I am just saying it is a pity that a society and an industry that prides itself on being so pragmatic, bottom line, hard-nosed, and all the rest of that litany is so blinded by the quaintness of the Amish that they are unable to either acknowledge or examine the success of the Amish. Judging from much of what I read and hear in farming circles these days, we equate survival with success. Believe me, the Amish are survivors! It seems more than a fair and reasonable assumption that their use of live horse and mule power has something to do with their staying power. Nor do we have to lean altogether on the Amish for our examples.

I have cultivated enough corn to know that my horses move just as rapidly (or slowly, depending on your point of view) through 100-bushel-to-the-acre corn as they do through 50-bushel corn.

I have worked with enough old machinery to know that if even a modest fraction of the engineering that went into tractor machinery were to go into horse drawn machinery, fewer horses could do more work easier, better, and more pleasantly for both horse and horseman. I have seen many examples of ingenious adaptations of the most modern machinery to horses and mules.

By conveniently excluding all such considerations and examples, Mr. Gavett arrived at 61,000,000 horses and mules. It was easy, just isolate a couple figures that will get you where you were already determined to go, and multiply. It is pretty obvious that 61,000,000 horses and mules will eat us out of house and home.

I don't know how many horses and mules it would take to farm this country using horses and mules entirely, and neither does M. Gavett. I also know that no one, including the draft horse and mule people themselves, are thinking in those terms. Mr. Gavett, and all the people who quoted him, were simply beating a straw man to death. There are no "across the board" answers to the problems confronting us.

Armed with such figures, it is hardly surprising that Mr. Gavett came up with an estimate that it would take 180 million acres of "prime farmland" (his expression) to feed this inflated army of horses and mules that his one dimension approach had given birth to. This figure proves to be just as faulty as his others.

In the 1930s, the Pioneer Seed Company in Iowa used a lot of draft horses in their operation. They were developing many different inbred lines involving small plots. It was a highly specialized farming operation that the horses lent themselves to very well. The farm manger, Mr. J. Newlin, was a great horseman. It was his rule of thumb that a draft horse required the energy equivalent of 70 bushels of corn per year, and Mr. Newlin, and his company, were very busy working at making a yield of 70 bushels/A on prime farmland as outmoded as the open-pollinated varieties they were replacing. Add to that the fact that horses and mules, when idle, can get along on crop residue (cornstalks and stubble and, like other livestock, can utilize the roughage grown on rough and untillable land. Mr. Gavett's figures on the acreage requirements were just as unreliable as his figures on the animals themselves. In both cases, these figures were accepted, repeated, and quoted with all the solemnity usually reserved for papal bulls. I suppose it should have been surprising...but it wasn't.

The speech contained other interesting information, such as "Legumes are not a cash crop." Last winter was prolonged and severe in northeast Iowa. I have some dairymen friends who had to buy a lot of hay. They would be pleased to know that it was a noncash crop they were paying so handsomely for.

He stated that there were three options for weed control: chemicals, tractor till, and hand weed. Thank God animals can't read. If my horses and sheep ever find that out, life will be more difficult for me and even easier for the weeds.

This leaves only his people figure..the 31 million it would take to operate our farms. Frankly, I have forgotten how he arrived at that figure. With so much thorough research on the first two figures, you can, I'm sure, understand that I would regard that number suspect anyhow.

It is perfectly obvious that it would take a good many more people. What is not obvious to me is that having more people involved in food production is such a bad deal. Frankly, great sections of our country, both inner city and rural, resemble conquered provinces right now. They look like colonies rather than mother countries and that, of course, is exactly what they have become to some degree. Recent figures published by the Des Moines Register showing the percentage of farm tenancy in this state confirm this. I recognize that the ownership is not as remote as the Hudson Bay Company was to the beaver, but it is a far cry, and getting further every year, from the dreams and hopes of the nation that gave birth to the Homestead Act.

I am here as a draft horse editor and breeder, not as a rural sociologist, and this is a livestock meeting, so I will not pursue this much further. The plain fact of the matter is that the draft horse and mule back themselves into a discussion of rural communities without much encouragement. The community-killing type of monoculture that dominates today's agriculture simply wasn't possible with horses and mules. They are limited beasts and they imposed those limits on agriculture itself for decades. This is proof enough for some that they should all go and the sooner the better (except, maybe, the Budweiser hitch) because man should be able to overcome all his restraints. That's one point of view. Given Man's track record, I find myself quite grateful for some of the restraints and limits imposed on us. Take a good look at the language of present day agricultural advertising. Its metaphors are those of conquest in case after case. No wonder so much of the country looks like a colony plantation. It's been conquered, sure enough.

At this point some of you might well raise the question, "What has all this to do with animal agriculture?" My reply is, "A great deal." One of the earmarks of specialization is the tendency to shift the costs and dislocations stemming from its narrowly defined success onto other segments of society. Agriculture has certainly been guilty of this and the result is that the belated concern we hear for rural communities now comes mainly from the churches and social action groups, not from mainstream agriculture. It is safely out of the marketplace with the churches and foundations and can be ignored. Reminds me a bit of Joseph Stalin's rhetorical question, "How many divisions does the Pope have?"

So I'm perfectly content with the knowledge that any significant return of draft horses and mules will reflect itself in more people on the farms and in rural areas. I would regard this as a net gain for our country in a great many ways. The mass exodus from the land has not been voluntary. I find it "curiouser and curiouser" that any industry would take constant pride in the great number of its casualties. The point I am making is that departure of the horse and mule in great numbers from our farms was not an isolated event. The end result was most certainly not what farmers had in mind when they sold the "last team." To illustrate the end result of the direction we are taking, I wish to present table 1 from a recent publication called "The Family Farm: Can It Be Saved?" This booklet was developed as a key resource in the United Methodist Church's effort to combat worldwide hunger. This particular table, concerning two different square miles in California is drawn from the final report issued by California's Small Farm Viability Project.

Table 1 shows the vital statistics for two very different square miles of farmland in the San Joaquin Valley--both in Fresno County. One of the square miles is near Selma in the center of the county; the other is near Huron in the heart of the Westlands Water District. Selma is surrounded by small farms while Huron is surrounded by very large farms. (Reprinted from the June 1978 issue of National Land for People).

Which section of land looks more stable and attractive to you? Which section would be most likely to include a few horses and mules in its power mix? So, when we consider horses and mules, we are really talking about the future shape of agriculture. We aren't talking about "going back" at all.

Harold Breimyer, Agricultural Economist at the University of Missouri, put it very well in a recent presentation. Horses and mules were not on his mind at the time, but they were on mine when I read the following:

"What is at stake is whether anyone cares what kind of agriculture is to prevail. If the preference be for family farming, it is a matter also of willingness to take courageous steps that would truly be effective. More rhetoric in favor of the family farm, put into laws that have an opposite effect, serves no useful purpose.

During the last 20 years, I have learned that the forces being brought to bear on the structure of agriculture are not intrinsic economic ones. Most are artificial. Although a whole chorus (including a few agricultural economists) shouts that big farms are big because they implicitly are more efficient than family-sized ones, I defy them to show conclusive research evidence.

Taxes are the progenitor of much bigness in farming. Big farms, including the newest entrant, the large, confined hog operations, are highly influenced by tax law.

I believe the 1980s will be the decade of decision for family farming...nor do I have more than the slightest glimmer of hope that the trend will be arrested."

TABLE 1. A TALE OF TWO SQUARE MILES

	Selma	Huron
Farm Facts		
Farm owners	11	1
Resident owners	11	0
Farms	9	1
Resident farmers	9	0
Resident owner-operators	7	0
Resident lessee-operators	2	0
Full-time farmers	7	1
People living on land	31	0
Gross value of farm production	$916,000	$590,000
Property value	$1,092,000	$412,000
Property taxes paid	$25,394	$ 8,627
City Facts		
Population	9,036	2,539
Number of businesses	287	55
Manufacturing and processing plants	22	11
Farm corporation offices	0	18
Value of retail taxable sales	$43,317,000	$7,350,000
Hospitals	1	0
Doctors	6	1
Dentists	6	0
Churches	34	2

Farm production on Selma square mile: 120 A of yams--1080 t at $250 = $270,000; 10 A of beans --40 t at $360 = $14,400; 35 A of peaches--490 t at $120 - $58,800; 80 A of cotton--38 t at $1,200 = $45,600; 60 A of alfalfa--420 t at $70 = $29,400; 3.5 A of berries--45.5 t at $610 = $27,755; 280 A of raisins--560 t at $840 = $470,400. TOTAL = $916,355.

Farm production on Huron square mile: 320 A of cotton--152 t at $1,200 = $182,400; 320 A of tomatoes--7,840 t at $52 = $407,680. TOTAL = $590,080.

He concluded with the following two paragraphs:

"Lest this be too sharply worded, let me add that what really is lacking is a sense of history and of destiny. Almost invariably in history, small-enterprise farming has survived only in early years of a nation's history, or where agriculture is so unproductive it is not coveted, or where national policy protects family farming. Early Greece and Rome had family farms. The Roman estates that came later lapsed into desuetude, making years of conquest necessary to gain food. Where soils are poor, there is not much nonfarm interest in acquiring land. France and Scandinavia now require that farmland be owned only by farmers operating on it.

We still think of ourselves as a new nation, with horizons unlimited. In fact, we are neither young nor new; our horizons have limits; and our institutions do not preserve themselves. We can control our destiny only if we set out to do so. In agriculture, that would require protective measures that the majority of our citizens say they want but that call for a degree of social discipline we clearly are resisting."

The age of cheap petroleum and cheap transportation may be over. If so, greater regional self-sufficiency, which in turn would call for a regional diversification of farms, may well become the order of the day again. This is a scenario that the horse and mule fit. How many of you would have bet on wood stoves 10 years ago?

The claim that 3% or 4% of our population now feeds the nation and provides the agricultural exports that shore up our trade deficit holds up as long as you conveniently overlook a few things. To accept this preposterous claim at face value, you ignore the fact that there are far more people farming the farmers than there are farming the land. Without day-to-day succor from this vast group, modern farming would grind to a sticky halt much as Patton's tanks did once. You also look the other way on the topsoil losses we have experienced and are experiencing with our "soil-for-oil" juggling act.

So, if the public does choose policies that make smaller scale, more diversified farming an attractive proposition, here is what our horses and mules have to offer:

1. A source of power that utilizes farm-grown fuels, thereby reducing cash outlay and reducing demands on our dwindling petroleum supplies.
2. Horses and mules return most of the fertility from their feed to the land in the form of manure, from 9 to 15 t annually per animal. Manure, I'm happy to note, has become respectable again.
3. It is power that reproduces itself, not only providing for its own replacement but with surplus animals for sale.

4. With proper care this power plant appreciates, rather than depreciates, for the first 6 or 7 years, and has a working life of about 15 years--longer in the case of mules.
5. Soil compaction, a growing problem, is simply no problem at all with horses and mules.
6. They are, within limits and with a little help from small mounted engines, adaptable to most modern machinery--as witnessed in the Amish areas in the East.

Nor is it just the future of the draft horse and mule that will be found in these larger public policies; the whole of animal agriculture and our rural communities have huge stakes in the shape and thrust of these national policies.

How you regard the inclusion of the draft horse and mule in these proceedings is probably determined by your age and experience. To some, I'm sure, the heavy horse is and will remain an anachronism, pure and simple. To others, I hope you will commence to give some thought to the big fellows as a partial answer to some of the problems confronting our country, and our agriculture in particular.

In conclusion, I would like to reach back to an address made by Donald Hammerly, the Iowa State Supervisor of the U.S. Farm Security Administration at Ames, Iowa, on February 22, 1941. He closed his address with these comments concerning the thousands of farm families who had been enabled to get a new start with a Farm Security loan and program. He said:

"I can't close without saying one thing about the use of horses on our family-sized Iowa farms. The man who operates his farm with horse power primarily maintains the interest that he already has in livestock. It is something that we can't put on paper--something that a man feels--something that a man knows but can't explain. The man who turns to tractor operations in so many, many cases on these small farms finds it necessary to give up all, or practically all, of his horses, and I have seen many of these farmers, who a few years before I had classed as good livestock men, gradually lose interest in their livestock. I have seen them become more and more cash-grain farmers, piece workers if you please, and I have seen their income and their security on the land decrease with that change in attitude and interest. Show me a farmer with enough horses, that are good horses, horses that he is proud of, horses that he sees win glances of envy and admiration from his neighbors and you will how me a farmer that takes good care of his other livestock and makes a profit at it. You will have shown me a farmer that is going to be able to pay his bills and pay his loan and be secure on the farm in the years to come."

Those remarks were made, not by a draft-horse editor, not by a draft-horse breeder, not by a draft-horse breed secretary, but by the supervisor of the thousands of Federal

Farm loans in one of our great agricultural states. His vested interest was not in horses and mules but in the repayment of farm loans, in farm prosperity, if you please. And he was not dealing in great aggregates of farm exports. He was speaking from experience in dealing with farmers on a one-to-one basis. Any industry that prospers at the expense of great numbers of the people engaged in it is a failure. There have been far, far too many successful operations where the farm and farmer died in agriculture in the 40-year period since Mr. Hammerly made those observations. That may be a spectacular way to practice medicine but we have paid a high social cost for some of the razzle dazzle.

As livestock specialists, of one sort or another, I think it is incumbent upon us to view our work and our actions in terms of its impact upon the men and women who make their livings from the land and their livestock--and the communities in which they live. As Ernest Hemingway said, "If you want to know about war, ask the infantry and the dead." Perhaps agriculture should be less concerned with the big picture and consult more with its infantry and its dead. And while you're there, check with the artillery horses, most of them were half-Percheron.

Part 2

GENERAL CONCEPTS AFFECTING AGRICULTURE AND THE INDUSTRY

4
FOUNDATION OF CIVILIZATION: FOOD

Allen D. Tillman

> "And he gave it for his opinion that whoever could make two ears of corn or two blades of grass grow on a spot of ground where only one grew before would deserve better of mankind and do more service to his country than the whole race of politicians put together."
>
> --Johnathan Swift
> The Voyage to Brobdingnag
> in Gulliver's Travels

I chose this quotation because it is apparent to me that man now has the knowledge and power to make two ears of corn or two blades of grass grow on a spot of ground that formerly would grow one or less.

Civilization is defined as "an advanced state of human society in which there is a high level of culture, science, industry, and government." The high level of civilization that we enjoy today has resulted from many technological developments in agriculture that increased the amount of food produced and the efficiency of human labor in producing it. Each innovation freed more people for the development of human society and of the culture, science, industry, and government found in it.

In discussing some of the developments, this paper is divided into major sections as follows: (1) a brief history of agricultural development worldwide; (2) the close relationship of agricultural and industrial developments in modern societies; (3) some characteristics of a successful agriculture at national levels, and (4) reasons for optimism about the world food problem.

HISTORY OF AGRICULTURAL DEVELOPMENTS IN THE WORLD

> "History celebrates the battlefield whereon we meet our death, but scorns to speak of the plowed fields whereby we thrive; it knows the names of the king's bastards, but cannot tell us the

origin of wheat. This is the way of human folly."

--Jean Henri Fabre

For convenience, I have divided this section into three parts, as follows: the gathering and hunting stage (2,000,000 - 7000 B.C.), the low-technology stage (7000 B.C. - 1750 A.D.); and the scientific stage (1750 A.D. - the present).

The Gathering/Hunting Stage

> "Cultural man has been on earth for some 2 million years and for 99% of this period he has lived as a hunter/gatherer. Only in the last 10,000 years has man begun to domesticate plants and animals, to use metals, and to harness energy sources other than the human body. Of the estimated 90 billion people who have lived out a life span on earth, over 90% have lived as hunters/gatherers, about 6% by agriculture, and 4% have lived in industrial societies. To date, the hunting/gathering way of life has been the most successful and persistent adaptation man has ever achieved."
>
> --Lee and Devore (1968)

The first ancestor of man, *Australopithecus*, appeared on earth about two million years ago. His main invention was the knife, which was made by putting an edge on a pebble, an invention that permitted him to kill animals, skin, and cut meat, thereby changing him from an *herbivora* to an *omnivora*. This change was dramatic, because the addition of dietary meat reduced the bulkiness of his diet by about two-thirds, permitting him to leave the trees and to become more mobile to better utilize the rapidly developing savannas. Also, meat required less time to gather, thus he had more time for social activities - improvement in communication skills and in his tools. And so man for the first time released the brake that environment imposes on his fellow creatures. It is significant that this basic tool was not changed very much for the next million years, attesting to the strength of the invention.

Homo erectus came onto the scene about one million years ago. His ability to walk upright freed his arms, which improved his ability to hunt, and his greater ability to adapt to many ecosystems permitted him to spread out from his place of origin, Africa. In fact, the classical discovery of *Homo erectus* was the Peking man, who lived in China about 400,000 years ago. He was the discoverer of fire, which was used for warmth and cooking. The Neanderthal man, who was discovered in Europe, appears to have led directly to us, *Homo sapiens*.

The test of the ability of *Homo erectus* to adapt came about 500,000 years ago when the Pleistocene Ice Age covered much of the earth. Clans of 40 or more moved to caves for protection and work, a move that required a new organization - the young and stronger men (usually 10 per clan - Willham, 1980) were the hunters, while the remainder were assigned duties in keeping with their abilities. Some of the dwellers even had time to paint on the cave walls. Bronowski (1973) felt that these are saying to all - "This is my mark, this is man." Man is now saying that he has the ability to shape the world and is not a mere creature to be shaped by the environment.

The great glaciers began to retreat about 30,000 years ago. Left in their wakes were great savannas that were soon filled with grasses of all kinds and with cloven-hoofed ruminants to consume these. In response, the clans came out of the caves and spread out over the plains following the animals, going north in the summer and returning south in the winter. This was the beginning of a transhumant way of life, which later became dominant in many areas of Eurasia. In fact, there are cases of this way of life even today--East Africa, North Africa, Finland.

The dog was domesticated in about 20,000 B.C., and this greatly increased man's ability to hunt. Also , there developed a symbiotic relationship between man and animals: animals furnished meat, skins, and other products for man, while man furnished some protection and salt to the animals. Urine of meat-eating man contains salt, a valuable commodity to ruminants in many salt-deficient areas. Man developed oars in about 20,000 B.C. and the bow and arrow in about 15,000 B.C.; both inventions increased his efficiency as a hunter. Man domesticated the gregarious sheep and goats during the latter part of this stage, and benefited greatly by the increased food supply from these animals.

At the end of this stage, about 10,000 to 7,000 B.C., and for a long time after man had already established village agriculture, animals continued to be an important source of food for man. The live animal represents, until it is sacrificed, a reserve food supply. This fact is often forgotten or overlooked by planning economists, who plan for the aid programs that are given to the developing countries. Early man recognized the importance of animals: so much so that the root word used for money in many languages reflects the importance of animals to man (Leeds and Vayda, 1965).

Low Technology Agriculture (7000 B.C. - 1750 A.D.)

> "The greatest single step in the ascent of man is the change from nomad to village agriculture."
>
> --Bronowski (1973)

With the receding of the Pleistocene ice, there came great environmental changes in the Old World. The hot and dry winds off the Eurasian Steppes eliminated all but the hardier grasses in much of North Africa and some of Asia, thereby turning some of the lands into semideserts or deserts. As a result, wild animals migrated to the river valleys of the Euphrates, Tigris, Indus, Nile, and Yellow rivers. Agriculture began when the nomads decided to stay put to exploit plants. Whether the nomads planned to develop agriculture or were the benefactors of two genetic accidents is not clear. What is clear is the fact that the early ones came to hunt and to gather wild wheat. By a genetic accident (Harlan, 1975), wild wheat containing 14 chromosomes, crossed with wild oat grass, also containing 14 chromosomes, to produce a fertile hybrid, called Emmer. It contains 28 chromosomes, thus the grain was much larger than wild wheat. Therefore, man began to cultivate it. Emmer's grain is so tightly bound to the husk and chaff that it is easily dispersed by the wind. As a result, it spread over wide areas. It appears that by another genetic acident, bread wheat came to the settled agriculturalists: Emmer crossed with another wild oat grass to produce still another fertile hybrid, bread wheat, containing 42 chromosomes, which has a large grain. In contrast to Emmer, when bread wheat is broken, the chaff flies off, leaving the grain in place, thus it is not spread by the wind. With the advent of bread wheat, which man has to plant and cultivate, man and wheat developed a symbiotic relationship that remains up to this day (Heiser, 1978).

Farming and husbandry in a settled agriculture creates an atmosphere from which technology and science take off (Bronowski, 1973). The first tools used by village agriculturalists were the digging stick, which was invented about 7000 B.C. This later evolved to a crude plow, the footplow that used human labor, which appeared in about 6000 B.C. The cow was domesticated in about 6000 B.C., and when man yoked the ox to the plow he for the first time began to utilize a power source greater than the human muscle. This was undoubtedly the most powerful invention of this early period, making it possible for man to wrest a great surplus of agricultural products from nature. The surplus food released more men to create, invent, innovate, and to build great civilizations--something for which the nomads had never had time. Civilizations, with their specific cultures, developed on the flood plains on the Nile, the Tigris-Euphrates, the Indus, and the Yellow Rivers. Many of their activities, such as irrigation, required cooperation by many men; therefore there developed administrative systems that led to the building of empires. Law and government developed. The great cities of that period thrived on a cereal-based agriculture. Trade between cities developed in order for them to acquire necessities - salt, spice, metals, etc. (Thomas, 1979).

The animals found in the settled villages, up to about 3000 B.C., were goats, sheep, the oxen and the onager, a kind of wild ass. As long as the animals were the servants of agriculture, all went well. But some time after 4000

B.C., the horse was domesticated and the nomads learned to ride in about 2000 B.C. Thus, the nomad was transformed from a poor wanderer to a threat to the settled villages. Warfare of that period was intensified by the discovery of how to ride the horse, and warfare became a nomad activity. The nomads battered on doors of the settled villages from about 2000 B.C. until the early part of the 14th century A.D. Sometimes the nomads were successful and took over villages, but in all instances, the nomads were absorbed into the villages. Historians have made all of us aware of the great wars waged by the nomads, recording the names of the famous nomads--the Huns, the Phrygians, and the Mongols. The Mongols were defeated in about 1300 A.D., thereby ending the threat of their making nomad life supreme throughout sections of the Old World.

When the horse collar was discovered, the horse became an important draft animal, especially in northern Europe were great teams of horses turned the heavy sod. Without these teams of horses, which permitted the Vikings to produce great surpluses of grains, they could not have been such a military power and threat to much of northern Europe.

The low-technology stage continues right on through the European Renaissance (Thomas, 1979), during which time there were many contributions to agricultural innovations, each with its consequent improvements in food production and the efficiency by which man produced it. As there were many such improvements over time, for sake of time and space, let us summarize the advances:

- Man developed methods for the systematic exploitation of plants.
- Man developed methods for the cultivation of plants for the production of grain--wheat, barley, millet, and rice (Heiser, 1978).
- Man domesticated animals--dog, cow, sheep, goat, and the horse.
- Man developed systems of irrigation.
- Man developed some degrees of mechanization--the digging stick, the plow, the ox-drawn plow, the wheel, and others.

The developments in limited technology were not continuous but came in ebbs and flows throughout the period up to about 1750. There appeared to be a de facto technological ceiling upon agricultural production throughout the entire stage. At the core was the simple Malthusian element--population expansion ultimately pressed against the land, thereby producing malnutrition, famine, disease and, finally, a decrease in population. In China, Hung Liang Chi (Rostow, 1978) the predecessor of Malthus, wrote "during a long reign of peace--the government could not prevent the people from multiplying themselves." Rostow said it well--"During this period of limited technology, if war did not get you, peace did." And so ended the second era.

<u>The Scientific State (1750 - present)</u>

The scientific state began in about 1750 and continues until the present. In this period, the western nations for the first time broke the ceilings on agricultural and industrial technology so that invention and innovation came at a regular flow. The key to these, I feel, was the advent of the "scientific revolution which brought with it experimental science." Experimental science, for the first time, permitted and motivated the formulation of scientific laws to describe general and natural scientific phenomena. This led scientists to design experiments that would lead to the manipulation of nature to man's advantage. This exciting time saw the advent of scientific agricultural societies in which agriculturalists met together for the discussion of scientific discoveries and how these could be put to use by farmers. Innovations came at a faster pace, and as in the past, each invention or innovation increased the level of food production and its efficiency of production. The rates increased rapidly and we now have the development of high-technology agriculture. Some of the basic advancements that are characteristic of this stage are as follows:

- Classification of soils, along with estimates of their fertilities.
- Improved plants by gene manipulations (genetic engineering).
- Improved animals by gene manipulations.
- Scientific utilization of fertilizers.
- Proper use of irrigation.
- Proper use of fermentation and other means of food preservation (Tannahill, 1973).
- Use of insecticides, fungicides, herbicides, vaccines, etc.
- Use of modern techniques in farm mangement.

Many names stand out during this third era; however, it is significant to mention that King George III (better known to Americans for other reasons) gave much support to the newly developing agricultural research in England. Some feel that the innovations resulting from this agricultural research greatly increased agricultural production in Great Britain. In fact, some suggest that it was the English agricultural revolution that permitted that nation to defeat Napoleon at a time when agricultural imports were essentially cut off by France.

Agricultural research is so important that every modern nation has now developed a national agricultural production plan in which agricultural research is a powerful component. Those countries that are lagging in agricultural production are the ones that have been slow in developing good agricultural research, teaching, and extension programs.

How about the United States? Our country had its beginning in 1776, or sixteen years after the third stage began. When our Declaration of Independence was signed, the Revolutionary War was fought, and our people set out to build a nation, fully 90% of our people were directly engaged in agricultural production. If time and space permitted, it would be useful to point out the significant inventions, such as the first cotton gin, the first wheat thresher, the first steam or gasoline-propelled tractor and many others, all American inventions or innovations, and to note the effect of each upon food production and efficiency. However, I will end this section by saying that during the 200 years since Independence, our farm population has decreased from 90% of the total population to only 5%. However, these fewer people are producing food of improved variety and quality, providing nourishment for a vigorous population. Our national agricultural program has been one of the modern success stories.

THE CLOSE RELATIONSHIP OF AGRICULTURAL AND INDUSTRIAL DEVELOPMENT IN MODERN SOCIETIES

In reviewing agricultural developments worldwide, Rostow (1978) noted that the modernization of agriculture and industry have gone hand-in-hand in successful national development programs in the past. In fact, one finds that the modernization of agriculture must precede industrialization in most countries. There are many reasons for the close interrelationship; successful agriculture (Foster, 1978) provides:

- An ever-expanding supply of high quality foods to nourish its people, increasing their vigor.
- An adequate supply of high quality food, available at a reasonable price, to combat inflation which, if left uncontrolled, hampers national development.
- Capital for the expansion of the nonagricultural sector of the economy. This is very critical in the early stages of industrial development because at least 90% of the population in every country studied were farmers at this critical period.
- More food for the nonagricultural population. Therefore, all through the modernization process, more and more people are freed for work in the nonfarm sector.
- Educated and motivated people for work in the nonagricultural sector in the developed countries.
- Land for the nonfarm sector--for highways, railroad, airports, shopping areas, etc.
- A market for the nonfarm sector--tools, machines, medicines, insecticides, clothing, gasoline, etc. Again, this market is most critical in the early stages of industrialization.

SOME CHARACTERISTICS OF A MODERN AND PRODUCTIVE AGRICULTURE AT THE NATIONAL LEVEL

In general, a successful national agriculture results from a good agricultural plan or programs that provide farmers with relevant production information and assures an adequate infrastructure that is needed for both production and marketing of agricultural products. In addition, farmers must receive a fair return from their investment of land, labor, and capital. Otherwise, production will be sporadic rather than continuous. Some specific requirements are:

- Adequate government financial support for research, teaching, and extension (public service) for agriculture.
- Infrastructure: The national government also has to provide certain components of the infrastructure needed by farmers--such as roads, harbors, railroads (in some cases). (Many of the infrastructures in the U.S. are now furnished by industry.)

Some inputs needed by modern farmers are as follows:

- Farm machinery and spare parts
- Farm tools and spare parts
- Power source--gasoline, electricity, diesel fuel
- Fertilizers
- Insecticides, fungicides, herbicides, vaccines, etc.
- Irrigation tools--pumps, pipe, valves, etc.
- Credit
- Others

Those who are familiar with the ready availability of inputs on the American scene might well question why private industry hasn't made these available in many developing countries and in developing agricultural industries. In many cases, private industry will not take the risk of production and distribution of many necessary inputs unless the volume and price justifies the risk. In such situations, the government has to subsidize these inputs until the industry is well along toward development. The writer has spent over 10 years in four developing countries and found that the unavailability of certain inputs, such as vaccines, insecticides, and fertilizers, has served as a severe constraint on production. In the same vein, lack of marketing services results in severe losses of the harvested produce.

Marketing services needed to assist production increases are:

- Farm produce collection and storage
- Processing and subsequent storage of the processed produce
- Wholesale distribution of produce
- Retail distribution of produce
- Marketing--farm markets, small stores, supermarkets, and others.

Even with all of the above components in place, farmers will not produce continuously unless the price for agricultural products pays for the costs of inputs and allows them a fair return on their investments.

REASONS FOR OPTIMISM ABOUT THE WORLD FOOD PROBLEM

The world food problem has two components--the demand side (population) and the supply side (production). Many reports today emphasize that the demand side is growing faster than the supply side. Let us analyze the population problem first.

Population

The demographic facts at first glance are frightening! If we plot population against time, we see that the population growth from about 8000 B.C. until about 1650 A.D. was almost a straight line function, a very slow growth rate. In 1650 A.D., there were about 0.5 billion people, and the rate of increase was only 0.3% per year. At that rate, the population would require 250 years to double. Instead we have had an exploding population since about 1700 A.D., and now some authorities estimate that we will have 6 to 7 billion people in the year 2000.

It is my belief that the population estimates have all been wrong, from Malthus down to the United Nations' recent estimates. For example, the U.N. estimated in 1976 that the world population by the year 2000 would be 7 billion. However, in 1979 that estimate was reduced to 6 billion. Therefore, in 3 years, we "lost" 1 billion people. There are many other examples of inaccuracies. Why is this so? Population projections require assumptions about the choices people have and make in regard to the size of their family. Therefore, it is easy to see why all past assumptions have been wrong.

Simon (1981) maintains that there is a built-in, self-reinforcing logic that forces the rate of population growth to respond to resource conditions. It does so by reducing population growth and size when food resources are limited, and expands it when resources are plentiful. Others have studied population changes in Europe for 1400 A.D. until 1800 A.D., and found that the population did not grow at a constant rate, and that it did not always grow. Instead, there were advances and reverses, provoked by many forces--famine and disease, however, were not the major forces. We know that increased incomes, associated with economic developments, reduce birthrate and population growth. For example, the populations in Singapore, Hongkong, Japan, and other places have tended to stabilize within the last 20 years.

In summary, these facts lead me to suggest that population size tends to adjust to the production conditions.

Following a technological advance in agriculture, there is an increase in production followed by an increase in population size. However, the rate of population increase levels off as the new technology is "used up".

Food Production

If I am optimistic about the demand side of population problem, I have even more reasons to be optimistic about the supply side--food production. As pointed out by Schultz (1964), the world food problem exists because of low productivity in the poor countries. It is estimated that the amount of crops produced per ha of land in the poor countries was only about one-fourth of that in the industrial countries, and the production of animal products/animal unit was even lower. Therefore, the potential for greatly increasing production lies in the poor countries where it is needed.

Starting with the success of the Green Revolution in India during the 1965-70 period, there has been an awareness worldwide that a nation can increase its food supply if it has the will to do so.

Since the World Food Congress in Rome in the early 1970s, the FAO, the World Bank, the bilateral aid programs and other organizations have given priority in their aid programs to agricultural development in the poor countries. Up to 1975, the proportion of foreign aid dedicated to increasing food production was less than 10% of the total program of aid and the level has now more than doubled.

Some of the reasons for my optimism about future world food production are:

- The nature of the food/population problem as a whole is now becoming understood.
- The complexity of technology transfer from the industrial countries to the developing countries is also becoming understood; the international research and development centers have made excellent progress in these endeavors.
- The potential for increasing yields per land unit in the poor countries, most of which are located in the tropics, is enormous. We call this "payoff on research." I have made some estimates of the annual rates of return on monetary investments in agricultural research, both by commodities and by countries. If we consider rice research in the tropics, the figures run from 46% to 71%; however, if only Asia is considered the figures vary from 74% to 102%. the International Rice Research Institute, in the Philippines, has done an excellent job.
- Fertilizers and the knowledge of how to use them are now available to farmers in the poor countries. Many national governments are now subsidizing the price of this input.

- New and better adapted plant varieties are now available to most poor countries; many of these plants are resistant to certain diseases.
- Many national governments in the poor countries are now using aid and loans from the FAO, World Bank, bilateral sources and philanthropic organizations to support agricultural production by improving some or all of the following: (a) The infrastructure (such as roads, communication, and harbors), (b) The teaching, research, and extension structure facilities and activities. The research is now directed at solving their own production problems. (c) The credit structure. (d) Water resource utilization (e) Active intervention programs to increase the production of certain critical commodities. For example, in 1976-75, Indonesia initiated an active intervention program in rice production using subsidization of inputs, price stability, increased research, increased extension inputs, etc.

In 1969-71, the average production of padi was 2346 kg/ha in Indonesia. By 1978, this figure increased to 2921 kg/ha, an increase of 27.5% in 8 years, or about an increase of 3.5% per year. Preliminary figures for 1980 show a dramatic increase and give hope that Indonesia is about to gain self-sufficiency in rice production.

An interesting aspect of Indonesia's successful efforts on a single commodity, rice, is the fact that production improvements in other commodities have not improved, and some have decreased. To me, this is a god sign--meaning that with increased effort, Indonesia could increase production in any agricultural commodity chosen. Their leaders now know this and are putting forth successful efforts to increase production of other selected plants and animals.

REFERENCES

Bronowski, J. 1973. The Ascent of Man. Little, Brown and Co., Boston, Mass.

Burke, J. 1978. Connections. Little, Brown and Co., Boston, Mass.

Foster, P. 1978. Food as foundations of civilization. In Food and Social Policy. G.H. Koerseiman and K.E. Dole (Ed.) Iowa State University Press, Ames, Iowa.

Harlan, J.R. 1975. Crops and Man. Crop Science Society of America, Madison, Wisconsin.

Heiser, C.B. Jr. 1978. Seed to Civilization. W.H. Freeman and Co., San Francisco, CA.

Lee, R.B., and I. DeVore. 1968. Man, the Hunter, Aldine, Chicago, Ill.

Leeds, A., and A.P. Vayda. 1965. Man, Culture and Animals. AAAS, Washington, D.C.

Rostow, W.W. 1978. Food as foundation of civilization. In: Food and Social Policy. G.H. Koerseiman and K.E. Dole (Ed.). Iowa State University Press, Ames, Iowa.

Simon, J.L. 1981. World population growth. The Atlantic Monthly 248 (2):70-76.

Tannahill, Rey. 1973. Food in History. Harper and Row, New York, N.Y.

Thomas, Hugh. 1979. A History of the World. Harper and Row, New York, N.Y.

Willham, R.L. 1980. Historic development of use of animal products in human nutrition. Mimeo. Rpt., Iowa State University, Ames, Iowa.

5

WORLD LIVESTOCK FEED RELATIONSHIPS: THEIR MEANING TO U.S. AGRICULTURE

Richard O. Wheeler,
Kenneth B. Young

INTRODUCTION

Livestock producers both in the U.S. and worldwide have an important stake in the operation of the world grain economy. The continued availability of relatively low-cost grain in the world economy would tend to foster further livestock development in grain-deficit countries and provide competitive advantages in all countries for livestock and livestock production practices more dependent on intensive grain feeding. On the other hand, if world grain supplies become more restricted, grain-deficit countries would be more likely to import livestock products rather than grain, and livestock production practices less dependent on grain feeding would gain a competitive advantage.

Prior to the early 1970s, the world trend was toward increased grain supply and continued buildup of large stocks in the industrialized exporting countries, primarily the U.S. and Canada. These large stock levels helped to maintain relatively low prices and assured a stable supply for importing countries. For example, U.S. wheat export prices deviated very little from $170 per ton from the mid-1950s to the early 1970s. This long period of stable grain supply, sold at very attractive prices, encouraged the widespread use of grain feeding to increase livestock production. In addition, some grain exporters offered other inducements, including liberal credit arrangements and Public Law 480 assistance for countries unable to compete in the international grain market.

During this era, many developing countries adopted intensive grain-feeding practices for poultry and swine production, and cattle feedlots became a prominent feature of the U.S. agricultural system. On a worldwide basis, use of all cereals for animal feed increased from 37% in 1961-65 to 41% in 1975-77 (Harrison, 1981). Eastern Europe and the Soviet Union registered the largest increase in grain feeding of all countries--a change from 48% in 1961-65 to 69% in 1975-77. Use in Latin America increased from 32% to 41%.

At the present time, the world grain market has recovered from the 1972/73 shortfall and world stocks have been restored to the level of the 1960s. Nevertheless, the shortfall did mark a major shift in the supply-demand balance of the world grain market, a situation that had not occurred previously.

The structure of the world grain market has changed dramatically since the 1950s. World trade increased over 200% from 1960 to 1980. Currently, over 100 countries are dependent on grain imports from a few exporters. The U.S. dominates the world grain trade, accounting for about 60% of global coarse-grain exports and 44% of world wheat exports. About 40% of the total U.S. grain production is now exported and the annual rate of increase in exports reached 7% per year during the 1970s. Projections of future U.S. crop exports available from the Economic Research Service of USDA (1981) indicate that the growth in foreign demand will continue through the 1980s, although not as rapidly as in the 1970s (table 1). Annual average export demand for corn and rice is projected to increase about 4½% compared with 2% for wheat and soybeans.

TABLE 1. PROJECTED INDICES FOR U.S. CROP EXPORTS, 1981-1989

Commodity	1981	1982	1983	1984	1985	1986	1987	1988	1989
					(1981=100)				
Corn	100	106	111	115	123	127	131	135	139
Wheat	100	96	99	101	103	105	107	110	115
Rice	100	109	113	117	120	124	127	131	135
Cotton	100	107	103	103	103	104	106	106	107
Soybean	100	100	101	104	107	111	113	116	119
Peanuts	100	123	140	147	150	153	157	160	163

Source: These projections have been calculated from Problems and Prospects for U.S. Agriculture, ERS-USDA (1981). They are not official USDA projections.

There is some question now about the U.S. ability to keep up the recent pace of expanding exports. Most of the available cultivated land is currently in production and we are losing about a million acres of cropland per year to nonagricultural uses. The rate of soil erosion has increased substantially with more intensive cultivation and use of marginal cropland formerly not used for crop production. The same problem is occurring in other countries and average crop yields are leveling off over much of the world.

PROJECTIONS ON WORLD GRAIN SUPPLY

A Winrock International study was completed in 1981 on world use of grain and other feedstuffs (Winrock International, 1981). The study was designed to evaluate the interaction between the world livestock system and the feed- and food-grain system.

Estimates of current world use indicate that poultry consume 27% of all grain fed; swine--32%; draft animals--4%; sheep and goats--2%; and cattle and buffalo, including dairy animals--35% (table 2). Feed use in table 2 is expressed in terms of megacalories of metabolizable energy

TABLE 2. ESTIMATED ANNUAL WORLD FEED USE FOR DIFFERENT TYPES OF LIVESTOCK, 1977

Livestock category	Livestock Output		Feed use				
	Meat	Other	Grain	Protein meal	By-products	Forage & other	Total feed
	(million metric tons)		(billion mcal ME)				
Poultry	22.8	23.3[1]	387.9	91.1	73.4	51.7	604.1
Sheep & goats	7.3	--	23.6	5.3	35.8	993.9	1,058.6
Cattle & buffalo	46.8	415.0[2]	507.3	42.8	204.2	4,101.0	4,855.5
Swine	41.0	--	460.9	56.0	213.1	157.2	887.2
Draft animals	13.4	--	57.9	5.9	23.5	1,214.9	1,302.2
All livestock	131.3	--	1,437.6	201.1	550.0	6,518.7	8,707.4

[1] Eggs.
[2] Milk.
Source: Winrock International (1981).

rather than metric tons due to variation in the quality and variety of feed used in different countries. For example, grain-feed use in the Soviet Union is reported on a "bunker weight basis" generally containing excess moisture and extraneous matter. The percentage of grain use in poultry rations is estimated to be similar for developed, centrally planned, and developing countries since the technology of modern poultry production has been readily adopted all over the world. Developing countries feed less grain and more forage and by-products to swine. The ruminants in developing countries subsist almost entirely on forages. However, nearly half of the world grain feeding occurs in developing and centrally planned countries dependent on grain imports.

Grain feeding was projected to continue increasing according to recent trends evaluated in the Winrock study. There will be occasional setbacks for countries with severe foreign exchange problems and domestic recession. Most of the centrally planned countries have set target levels of increased livestock production requiring additional grain feeding. The Winrock study projected that total world feed use of wheat and coarse grains would surpass direct human and industrial consumption by 1985. Recent trends also indicate that total world grain use is increasing at a faster rate than world production. World grain demand has been increasing steadily due to continued growth in world population and rising per capita consumption of livestock products dependent on grain feeding while the growth in supply is slowing due to limitations on development of new cropland and reduced productivity gains on existing cropland. This increased tightening in the world grain market implies that the grain export price should increase substantially by 1985.

GRAIN SUPPLY OUTLOOK FOR U.S.

Winrock International is currently initiating a study of both the potential for and implications of additional crop production in the U.S. The 1977 Natural Resource Inventory compiled by Soil Conservation Service of USDA shows a total of roughly 460 million acres of cropland available in 1977 containing about 70% prime land in the Class 1 and 2 categories. Heady and Short (1981) of Iowa State University have projected that the 1977 cropland base will dwindle to 353 million acres by the year 2000, but that there are 37.6 million acres of high-potential land and 90.1 million acres of moderate-potential land that could be converted to cropland. This land area for potential development is located primarily in the South Atlantic, South Central, Great Plains, and North Central regions of the U.S. However, other economists in the U.S. have serious doubts whether it would be feasible to convert this much additional land to crop production. Some limitations to development of additional cropland and current use of this land are shown in table 3; the data indicate that much of this potential cropland is currently used for pasture and timber production and that there are definite erosion hazards and probable high conversion costs to develop this land area for crop production. Such limitations imply that there will be a major increase in production cost to bring these new lands into crop production after we reach full capacity on existing cropland.

TABLE 3. ESTIMATED POTENTIAL CROPLAND AND LIMITATIONS TO DEVELOPMENT IN THE CONTINENTAL UNITED STATES

Type of limitation	Percent of potential cropland	Present use	Percent of potential new cropland
Erosion	59	Pasture-range	79
Drainage	23	Forest	17
Soil	7	Other rural	4
Climate	4		
No limitation	7		
Total	100	Total	100

Source: 1977 Natural Resource Inventory, Soil Conservation Service, USDA.

IMPLICATIONS FOR U.S. LIVESTOCK PRODUCTION

World population is projected to increase 50% between 1975 and 2000. There is increasing emphasis on livestock production to improve the quality of human diets, particularly in centrally planned and developing countries, and increasing pressure on cropland worldwide. In the short term, there may be temporary swings between shortages and surpluses in the world grain market. This is expected due to greater year-to-year variation in world crop production as a result of expansion of cultivation on marginal lands with increased drouth stress and other climatic variability. Stability of supply may also be reduced due to mounting pressure on exporting countries to reduce the carryover of grain stocks from year to year. Grain prices are projected to increase with gradual tightening of world grain supplies. Increased export volume will require eventual conversion of at least some pasture and timber land in most of the key exporting countries, with an associated rise in grain-production cost.

Some implications for U.S. livestock producers include increased grain-feeding costs eventually rising above the general inflation rate and the loss of some pasture and rangeland area converted to cropland as indicated in table 3. Higher grain prices will be translated into somewhat higher meat prices, particularly for poultry, swine, and fed cattle because these enterprises are highly dependent on grain feeding. However, it will become more profitable to utilize additional crop residues and by-products in livestock feeding, particularly for cow maintenance, to replace present grain use. The biggest deterrent to using these low-quality feeds is cost--primarily for labor, equipment, and interest. To date, the availability of a stable supply price for grain is analogous to the situation we had 10 years ago for oil and natural gas. It has not been cost effective to utilize many alternative sources of feed energy, although there is an abundant physical supply available in the U.S. The amount of corn crop residue physically available was estimated to be 231 million tons in 1977 (Ensminger and Olentine, 1978). This would support 117 million cows for a 4-month grazing period on a purely physical supply basis. The nutritive value of crop residues can be enhanced with special processing techniques, and some very promising research work has been done on ammonia treatment of straw.

Cattle, sheep, and goat producers could potentially utilize these alternative sources of feed to substitute, at least partially, for grain or to enlarge breeding herds even on a drylot basis if it became more economical to do so. If meat prices increase along with grain prices, some livestock producers may regain a competitive advantage over poultry and swine producers who are more vulnerable to rising grain prices. Under the current regime of depressed grain prices, poultry producers, in particular, have been gaining a significant competitive advantage over beef producers. Poultry meat prices have now declined to 30% of average beef prices compared with 80% a few years ago (National Cattlemen's Assocation, 1982).

A reduction in grain feeding of cattle in the U.S. would mean increased competition for use of existing pasture and range lands. With increased grain exports, there would be an associated reduction in the grazing land area, particularly in the Southeast and Great Plains regions of the U.S. Increased dependence on crop by-products and residues implies that more livestock will be produced in traditional cropland areas to utilize these waste products, as was the practice in the U.S. before the feedlot era began. This is the situation now in most developing countries where the bulk of livestock production is found in mixed crop/livestock systems (Winrock International, 1981).

Increased use of crop residues to reduce the amount of grain feeding would have a significant impact on the management system for livestock, especially cattle. Levels of annual offtake would decline as cattle would have to be nearly a year longer to reach market weight on a less intensive feeding program. The cattle operator would be forced to move cattle to crop-production areas and to lease crop residues from crop farm owners as is now done for wheat pastures. Additional use of feed supplements would be necessary as crop residues are generally lacking in total nutrient requirements. The cattle operator would also have to invest in additional fencing and equipment to utilize crop residues. Thus the overall implications are that major adjustments would be required in the U.S. cattle industry, including shifts in the location of production, the composition of herds, and feeding programs.

CONCLUSIONS

The general outlook for the international grain market points toward continued price variability for feedgrains, a gradually rising price level for grains as the world market continues to tighten, and eventual loss of some grazing land in the U.S. when additional cropland is needed to meet expanding export requirements. It is possible that the problem of price variability may be alleviated by additional government intervention such as paid acreage reduction or other methods of supply control on the market, but this does not appear likely in view of the current emphasis on curbing spending for most agricultural support programs.

Expected consequences of the grain-market outlook for cattle producers include continued fluctuations in feeder cattle prices and returns from cattle feeding during the next few years and a general trend toward higher feeding costs. Although price variability is nothing new to livestock producers, the sharpness and range of price movement will likely be increased as long as we continue to be the shock absorber for the world grain market. The U.S. is one of the few nations that exposes domestic producers to price fluctuations of the international market.

Short-term effects of the expected swings in prices will be of more immediate concern to most livestock producers than the longer term upward trend in feeding cost, particularly for those in a weak financial position. To some extent, producers may be able to reduce the financial risk through greater participation in the futures market or by direct contracting. However, their most urgent need to survive in the livestock business will probably be to secure alternative methods of financing to provide more flexibility on repayment of loans. Other possible options for reducing or spreading the risk of price movement include the development of programs for outside investors to assume partial ownership of livestock and other creative financing schemes to shift at least part of the risk from producers to other outside parties. There may also be an opportunity for further revision of the tax laws to encourage more outside investment in the livestock business.

Long-term implications of changes in the world grain market, as well as expected increases in transportation cost, are that the structure of the U.S. cattle production system will change. Projected world food-system trends suggest increasing prices for all livestock products due to rapidly rising consumption in most countries and upward pressure on grain prices. However, there may not be much increase in livestock-product consumption in the U.S. market because per capita consumption rates have stabilized. A continuing problem for beef will be competition from pork and poultry in the U.S. meat market. To recapture its former market share, beef will require more efficient production and marketing throughout the system.

Increasing grain prices may provide some opportunity for beef producers to improve their production-cost relationship relative to pork and poultry by changing to less intensive grain feeding. Additional research and development is needed on the utilization of crop residues and by-products to reduce cost in cattle production.

REFERENCES

Economic Research Service, USDA. (1981). Problems and Prospects for U.S. Agriculture, Washington, D. C.

Harrison, P. 1981. The inequities that curb potential. FAO Review on Agriculture and Development. Food and Agricultural Organization of the United Nations, Rome, Italy.

Heady, E. O. and C. Short. 1981. Interrelationship among export markets, resource conservation, and agricultural productivity," Agr. J. Agr. Econ. 63:840.

National Cattlemen's Association. 1982. The future for beef. Special Advisory Committee Report, Englewood, CO.

Soil Conservation Service, USDA. 1977. 1977 Natural Resource Inventory. Washington, D. C.

Wheeler, R. O., G. L. Cramer, K. B. Young and E. Ospina. 1981. The World Livestock Product, Feedstuff, and Foodgrain System. Winrock International, Morrilton, Ark.

Winrock International. 1981. Report on Livestock Program Priorities and Strategy. Winrock International, Morrilton, Ark.

6
POLITICAL CHALLENGES FOR TODAY'S ANIMAL AGRICULTURE

George Stone

Not all of the challenges to today's animal agriculture are political. Some of the challenges are internal. They relate to the nature of our industry and to the nature of the people in the livestock industry.

We seem to have an allergy to change. And we have a disposition to go our own way as individuals, even if we could improve things for ourselves by working together.

ROLE OF FEDERAL GOVERNMENT

Part of the time, we want the government to leave us alone. Part of the time, we get very impatient when the government is too slow with helping us. This is not the first, nor the last, speech to be made in this nation on the role of the federal government in food and agriculture. We talk about that subject as if we were about to make an original choice. But, as a matter of fact, the choice was made long ago. As early as 1796 and as recently as 1977, and many times in between, the federal government has decided that the family farm should be fostered and encouraged.

Our society has decided that the federal government should take measures to help assure that land remains in the hands of family farmers.

Our society has decided that the federal government should be involved in the conservation and protection of land and water resources.

Our society insists on a federal involvement in environmental protection.

Our society has dictated a federal role in assuring that food supplies are safe and wholesome.

Our society requires federal supervision to see that pesticides and chemicals are safe for farmers and consumers.

Our society provides for federal supervision of the marketing system to try to keep it fair and competitive.

There are more kinds of government intervention in agriculture today than ten, twenty, or thirty years ago. There will probably be more federal involvement in farm and food policy in 1990 or the year 2000 than there is today.

The real question should be in regard to the nature, extent, and purpose of government involvement, and the degree to which farmers have a voice in decisions that will affect them.

Government intervention has often been justified when there was no other effective way to cope with problems. The government's role in agricultural research and education is widely accepted and advocated. Various federal farm credit programs had to be initiated because the private sector could not take the entire risk in financing agriculture. Commodity exchanges and boards of trade had to get federal supervision because they could not be left to police themselves. The federal rural electrification system had to be created because the private sector could not, or would not, do the job.

In many agricultural sectors, government involvement can be good or bad:

- Rules on farmers' handling of pesticides or chemicals can be reasonable or ridiculous, depending on how much farmer input there has been in the process.
- Feedlot pollution abatement rules can be workable and effective, or unrealistic and oppressive.
- Dredge and fill regulations can be a protection or a harassment for farmers.
- OSHA regulations can be a godsend or an aggravation.

Everything depends on how well farmers have involved themselves in the process and how much voice they have had in shaping these laws and regulations. That is not an easy task for us as farmers and livestock producers. The leading codification of laws affecting agriculture now runs to 14 volumes of about 500 pages each. That is 7,000 pages of laws.

MAJOR POLITICAL CHALLENGES

Having said this much in the way of background, let me now turn to what I perceive as some major political challenges for livestock agriculture. One thing that is quite obvious, but not generally appreciated, is that we do not function in a vacuum as livestock people. We depend upon consumer purchasing power and demand for our products, and we cannot expect to be stable and prosperous if there is high unemployment and weak buying power.

Recession and Unemployment

Much of the difficulty that has faced the livestock producer in the past three years has been attributable to recession and unemployment. High interest rates the past three years have diverted about 30 billion dollars a year of consumer purchasing power from food and other necessities. High interest rates have an impact on the cattle producer as well. Some calculations done at Oklahoma State University reveal that the interest cost per head of livestock sold would be $133 per head at 9 percent interest; $237 per head at 16 percent; and $297 per head at 20 percent interest. The difference between the two extremes of 9 percent and 20 percent is $164 a head, more than enough to wipe out any potential profit.

When national economic conditions are difficult and there is a pinch on consumer buying power, the tendency is for a reduction in the higher-priced meat purchases and, to some extent, for the consumer to buy other products. In a time of recession and unemployment, meat purchases are the first thing affected.

Specifically, because of the tendency for the American diet to be hurt by recession and unemployment, the Congress in its wisdom developed and implemented the food stamp program. Of course, one can expect that when times are tough, the food stamp program becomes costly. Each additional one percent of unemployment adds one million people to the food stamp rolls, not because it is a bad program, but because it is doing what it is supposed to do--maintain a healthful diet for the lowest-income people in our society.

About 27 percent of the food stamp benefits are spent by recipients to buy meat and meat products. This means that if the food stamp program makes available 12 billion dollars in food subsidies to low-income families, over 3 billion dollars will be used to buy meat. If the food stamp program is cut by 3.7 billion dollars, as it has been in fiscal budgets for 1982 and 1983, that means a 1-billion-dollar reduction in the demand for meat. That is a political decision and we live with it as livestock producers.

Competition Between Humans and Farm Animals

One of the political challenges that will become more serious as time goes on will be the competition for living space between humans and farm animals. Twenty years ago, a livestock economist at a midwestern land grant university was making the prediction that there would be little room in the world for livestock 100 years in the future. Human population was increasing so rapidly that the land would be needed for living space. There would only be room for enough livestock to provide meat flavoring for synthetic meat substitutes, he predicted. Although there are still 80 years to go to see if the good professor was right, we doubt that he was on the mark. Still, his kind of thinking has surfaced in some other forms.

One noted futurist has looked in his crystal ball and concluded that if we fed the grain to humans instead of to livestock, we could perhaps support a population more than three times greater than at present. Some world hunger activists were suggesting a few years back that if each of us would give up one hamburger a week, it would help feed the starving of the world. Still others have suggested that Americans should quit fertilizing their lawns and golf courses and send the fertilizer to the developing nations to help them grow more of their own food supply. These ideas are simplistic and, even if carried out, would have little measurable effect on hunger in the world. Basically, outside of famine caused by natural disasters, there is no actual shortage of food in the world, nor of land or fertilizer. There are more than adequate supplies of land, fertilizer, and food. What is lacking is the purchasing power to pay for the food. What is lacking is effective consumer demand. Wherever the cash is available to pay for the food, the food becomes available.

Politically Prescribed Diets

I rather expect that we will have increasing frustrations ahead with those who wish to try to influence the diets of the American people by political prescription.

For forty years, the Food and Nutrition Board of the National Academy of Sciences has been the widely recognized and respected source of dietary guidance. The Food and Nutrition Board has been the agency that has issued the "Recommended Dietary Allowances" or RDA that have been the basis for most nutritional education aimed at the consuming public. Just two years ago, the Food and Nutrition Board issued a report, entitled "Towards Healthful Diets," in which it advised that the average adult American whose body weight is under reasonable control should feel free to "select a nutritionally adequate diet from the foods available, by consuming each day appropriate servings of dairy products, meats or legumes, vegetables and fruits, and cereal and breads."

In the same report, the Food and Nutrition Board recommended that dietary change or therapy should be undertaken under a physician's guidance. Aware of some of the political headline hunting being done by self-appointed guardians of American diet, the Food and Nutrition Board plainly warned that it is "scientifically unsound to make single, all-inclusive recommendations to the public regarding intakes of energy, protein, fat, cholesterol carbohydrates, fiber, and sodium."

You are aware, of course, of the rash of studies and reports on diet and heart disease, diet and cancer, and other topics that have singled out animal fats and meat as the causes of human health difficulties. There was a surgeon general's report in 1979 and a study by the Senate Select Committee on Nutrition and Human Needs. Last summer, a special panel of the National Academy of Sciences on "Diet, Nutrition, and Cancer," issued a report which differed in important respects from the position of the Food and Nutrition Board and that had not, in fact, been submitted to the Food and Nutrition Board for review and evaluation. Like many of the other political diet studies, the cancer study issued sweeping recommendations that included a 25 percent reduction in the consumption of fats, fatty meats, and dairy products. The report recommended avoidance of smoked sausages and fish, ham, bacon, frankfurters, and bologna.

The cancer and diet panel admitted it did not know what percentage of cancer risks are attributable to diet or how much the risk could be reduced by modifying one's diet. Still, while admitting that there was considerable uncertainty about the scientific basis for its findings, the diet and cancer panel issued its recommendations anyway and, as you might expect, the press treated it in a sensational manner. It was entirely in order for the livestock industry to ask that a review be held to reconcile the contradictory advice that was originating at the same time from the National Academy of Sciences. In our belief, the practice of medicine should neither be carried out by politicians or advertising agencies. The practice of medicine should be left to the medical profession.

Livestock Marketing Revitalization

At a time when all other industry seems to be centralizing, livestock marketing is disintegrating--breaking up into bits and pieces--with no central system for determining price.

As a result, live cattle prices are being based on data reported in the Yellow Sheet or the USDA meat news--sources that may represent as little as 2 percent of the market volume. Farmers have been looking at other options, such as electronic auction markets, to restore some competition into the system. But, it will take time, considerable capital, and major organizational efforts to establish an effective producer-controlled system of that sort. We may need some federal help and encouragement to get the job done.

International Trade in Meat and Meat Products

Another political challenge we may face will be in regard to international trade in meat and meat products. We appreciate the desire of some in the livestock industry to expand foreign markets for U.S. meat and related products. Foreign market development ought to be pushed in any constructive way. There may well be some potential for gains, particularly as some of the developing countries increase their purchasing power and seek to upgrade their diets. However, care must be taken that nothing we do in seeking to expand world trade reacts to undermine our own meat-import control laws. We ought to recognize that if we attack the quotas, the nontariff barriers and protectionist devices of other countries, this will certainly expose our own Meat

Import Act of 1964, as amended by the Meat Import Act of 1979, to attack from abroad. You may recall that when the 1979 Act was adopted, it was criticized by some who look to the U.S. market to dump their oversupplies. While we now export almost one billion dollars' worth of meat and meat products, we import 2.2 billion dollars in meat and meat products, plus another 1.5 billion dollars in animals and animal products.

The expansion of U.S. meat exports will be a gradual, long-term proposition. It will take a long time before exports offset imports, and this will be particularly true if we go to a free market in meat trade. At any time that world meat supplies are excessive in relation to effective demand, the U.S. will tend to be the magnet for oversupplies. That situation will tend to prevail most of the time. So, in a free market situation, U.S. meat imports will tend to expand more rapidly than meat exports.

We have a good law in the 1979 Meat Import Act. It is a responsible measure that helps us retain a domestic livestock and meat industry. The countercyclical factor that determines the allowable level of imports is particularly important. When the U.S. cattle industry is in the liquidation phase and beef production is relatively high, the countercyclical factor will tend to reduce the allowable level of imports. When the cattle cycle is in the rebuilding phase and domestic production is low, the allowable imports will be increased. If we can hold foreign imports to the minimum figure of 1,250 million pounds or near to the figure, the situation will remain in control. But, in a free market situation, it is easy to imagine that without any controls, meat imports, now subject to the law, would quickly advance to new all-time record levels.

Animal Welfare Concerns

Another political challenge livestock producers may have to face would be from the animal welfare lobby. Up to a couple of years ago, few farmers were concerned about the animal welfare lobby--many had not heard of it. Thus far, there has been no serious effort in the Congress to delete the several words from the Animal Welfare Act that would end the exemption of farm animals and birds from that statute. However, there have been bills in the 96th and 97th Congresses that would regulate confinement feeding of animals. Some activist groups have emerged and have gotten some coverage from farm magazines and the media but, so far, it seems they have been open to dialogue with farm and livestock groups. We should be realistic enough to expect that concerns of nonfarmers about this area of farm production will increase but, if we can keep some lines of communication open with responsible citizen groups, perhaps the discussions can be kept on a reasonable basis.

National Farmers Union has had some concerns over the harmful effects of excessive concentration of poultry and animals. To be frank about it, our concerns have been more in terms of environmental, health, and economic effects of such concentration, rather than with humane treatment of livestock and poultry. Up to this point, here in the U.S., we have been able to avoid the confrontations between consumers and farmers that have been common in Western European countries over animal rights. Most of the difficulty arises with people who are almost totally lacking in knowledge about livestock production methods.

By attempting to carry on a dialogue, we may be able to avoid misunderstandings that tend to polarize viewpoints on their side or ours. If we can win some understanding, we may be able to keep the issue out of the political arena.

Sensible and Effective Regulations

In the time that we have had on the program today, it has not been possible to touch upon all the political decisions that will affect our business and livelihood. There is a whole array of potentially serious problems, if the hysteria for deregulation is carried to the extreme. Reducing needless paperwork burden is desirable, of course, but to eliminate needed regulatory measures is something else. The Packers and Stockyards Administration covers a whole range of competitive issues. What we need is sensible and effect regulation, not the elimination of regulation altogether.

Most of my adult life was spent in Oklahoma as a farmer and farm leader. We found at times that there were governmental decisions so bad that we had to fight them with all our might. But, we also found that we could reason with people and that if we got involved early enough and got a voice in shaping the decisions, we could avoid situations that otherwise might have become desperate. The key is to get involved.

Rest assured, the government is involved and will continue to be involved in our agriculture and our society. If we don't take part in the process, the results will be worse. If we curse the government, the government will not go away. We will just be destined to live under scarcely bearable regulations devised by someone who doesn't really understand livestock farming. Fortunately, we do have a choice and a voice, if we want it.

7

REGULATION OF AGRICULTURAL CHEMICALS, GROWTH PROMOTANTS, AND FEED ADDITIVES

O. D. Butler

Agricultural chemicals, from fertilizers to pheromones, help make U.S. agriculture the most productive in the world. Discovery, testing for efficacy and safety, manufacturing, marketing, and proper use all represent the ultimate in biological sciences, in ingenuity, and in exercise of the free enterprise system.

Some say that in this case the enterprise system is not very "free." Thalidomide, DDT, aldrin, dieldrin, arsenic, and many others did not pass safety tests. The thalidomide tragedy may have aroused the most fear in public minds, but the diethylstilbestrol use in the 1950s for sustaining pregnancy in women, which apparently resulted in increased incidence of cancer in their daughters 20 or more years later, would have to be rated a close second in the world, and first in the U.S.

Public demand expressed through members of Congress the last couple of decades caused ever-more-strict federal regulations on development and use of agricultural chemicals. During the past year, however, the Food and Drug Administration, the USDA, and the Environmental Protection Agency (the major responsible agencies) have shown good evidence of more reasonable postures concerning laws, regulations, and interactions with manufacturers and users of agricultural chemicals.

President Reagan's appointment of a cabinet-level committee chaired by Vice-President George Bush with a mission for reducing burdensome regulations, gave an unmistakable signal to the agencies. Now we see Congress considering revision of the Federal Insecticide, Fungicide, Rodenticide Act (FIFRA), and the Food Safety Laws, especially the extremely strict 20-year-old Delaney anti-cancer clause. This clause was made obsolete by almost unbelievable advances in assay procedures that now detect parts per trillion of materials in foods that were considered to have zero residue with the parts-per-million capability of assays in the 1960s. Assays are now as much as a million times more sensitive.

Strict laws that were formerly written to ban toxic substances on the basis of risk alone are being reconsidered. A couple of reasons derive from the issue of essential elements--such as selenium required by the body at a low level, but toxic at higher levels, and nitrite used for centuries in meat curing to give the characteristic color. Derivatives--for example, nitrosamines that may be developed during cooking of bacon--have been shown to cause an increased incidence of cancer in susceptible laboratory animals. More recently, the finger of suspicion has been pointed at nitrite itself, in a highly disputed experiment with laboratory animals. Nitrite produces color, but more importantly, it protects against the deadly botulism bacteria, so use of nitrites has not been banned, but has been strictly limited. Critics of the regulations point out that many natural foods contain nitrites and that human saliva does also. Avoiding cured meats would reduce nitrite consumption by a very small and negligible amount, critics say. But the "scare" stories certainly reduce demand for ham, bacon, and hot dogs.

What are producers' primary concerns about agricultural chemicals? I believe that you should have a general idea of how they are discovered, tested for efficacy and safety, and used in a safe and effective way. You should also know the direct cost of materials, as well as the indirect cost, if consumer concerns affect demand for products marketed.

Good basic biological research done primarily by public institutions, such as the Land Grant Universities, usually provides the foundation for development of an effective product. The need for products to control pests or diseases usually is expressed by producers reinforced by producer organizations, by extension specialists and research workers who interact with producers, and by supplier representatives.

Because of the similarity of all living cells, there must be a good understanding of the biology of both species affected to be able to kill a parasitic living organism without consequent toxic effect on the host. Then, for food producing plants and animals, there must be great concern about residues that might have an effect on consumers.

Animal producers are served well by a group of competing companies seeking profit by manufacturing and marketing drugs, biologicals, pesticides, and related materials. Most of the companies belong to an industry trade association, the Animal Health Institute (AHI), headquartered in the Washington area. It serves the industry the same as the many other trade associations there, trying in every way to protect the opportunity for the industry to produce products that customers will buy and use because of benefits and thereby earn a profit for investors.

Almost inevitably it seems, any position taken or change advocated by the AHI is opposed by one or more organizations that classify themselves as consumer protectionists. Lawmakers and regulators usually have to make decisions between opposing viewpoints without the benefit of absolutely conclusive evidence. In the last decade such controversy has been a major stimulant to the formation of the American Council on Science and Health (ACSH) and Council for Agricultural Science and Technology (CAST), both of which I support.

"The American Council on Science and Health (ACSH) is a national consumer education association directed and advised by a panel of scientists from a variety of disciplines. ACSH is committed to providing consumers with scientifically balanced evaluations of food, chemicals, the environment, and human health." This is quoted from their March 1982 publication, "The U.S. Food Safety Laws: Time for a Change?"

The Council for Agricultural Science and Technology (CAST) is an organization sponsored and managed by twenty-five scientific agricultural societies. Its major purpose is to assemble and report the scientific information on important issues of national scope for the benefit of law-makers, regulators, and the general public. It is not an advocacy organization. Most of its task force reports, now numbering about a hundred, were prepared at the request of members of Congress, some by government agencies, and some because the 47 officers and directors, all representing the scientific societies, decided that there was a need to assemble and print the scientific evidence on an important issue. CAST celebrated its tenth birthday anniversary in July 1982 at a directors' meeting at its headquarters. I have the privilege of serving as president of CAST in 1981, as did Frank Baker, the Director of this International Stockmen's School, in 1979. (I want to especially recommend CAST task force reports mentioned in the references.)

Some of the scientific societies work directly with regulatory agencies. I served as chairman of the Regulatory Agencies Committee of the American Society of Animal Science for about 10 years until 1981. The Institute of Food Technologists, like the American Society of Animal Science, has been very active in identifying and nominating qualified scientists to serve on CAST task forces and has also produced independent papers on various aspects of food safety.

Drug manufacturers have been very critical of the Food and Drug Administration (FDA) for taking so long to consider new animal drug applications (NADAs) before approval. A recent report entitled "The Livestock Animal Drug Lag" by the AHI describes the problem and suggests solutions. U.S. manufacturers have been able to obtain approval to market their products in the United Kingdom and European countries in a fraction of the time required for U.S. approval. An example is albendazole, a broad spectrum anthelmintic effective against gastrointestinal roundworms, lungworms, tapeworms, and liver flukes in cattle. Approval was obtained in 5 months in England in 1978. The same application filed in the U.S. in 1977 is still pending, though strong producer pressure resulted in limited approval in 1979 under a special investigative New Animal Drug authorization in a limited number of states. After the Food and Drug Administration banned hexachloroethane for liver fluke control, cattlemen had no approved drug. Texas and Florida producers, with pastures along streams and low-lying areas that have snails (the intermediate fluke host), just had to have an effective drug. Cattle producer organizations rallied to the cause and helped obtain the limited approval.

The AHI sponsored a Forum on Regulatory Relief in Alexandria, Virginia, in June 1982. Dr. Arthur Hull Hayes, Commissioner of the FDA, announced there that "I've decided that all activities in the Review of Animal Drug Applications, including issues of Human Food Safety, will be consolidated within the Bureau of Veterinary Medicine." That is certain to allow faster decisions. The Bureau of Food review has been blamed for much of the delay in the recent past.

Dr. Hays gave a definition of safe as "a reasonable certainty of no significant risks based on adequate scientific data, under the intended conditions of use of a substance." More and more we are realizing that there is no such thing as absolute safety, or zero risk. His speech gave some reassurance concerning "sensitivity of method" regulations that have been under consideration for several years by FDA. The bureau now seems willing to accept foreign data in support of New Animal Drug Applications under certain restrictions and also to consider cross-species approvals. It is not a good investment for drug companies to spend several million dollars to obtain approval of an anthelmintic for goats, for instance, that is very important in Texas (which has about 95% of U.S. goats) because the market is so limited. Other minor species, even sheep, fall in that same category. I believe it will be necessary for publicly supported institutions like the Texas Agricultural Experiment Station to assist in developing drugs and obtaining approval for use in such minor species.

The FDA is also considering some liberalization of restrictions on feed manufacturers. Dr. Lester Crawford, recently reappointed to the position of Director of the Bureau of Veterinary medicine (BVM) of FDA spoke to the American Feed manufacturers' 74th Annual Convention at Dallas in May 1982. He reported that "The Subcabinet Working Group, chaired by USDA Assistant Secretary Bill McMillan, has proposed the total elimination of FD 1800s, the notorious application required to authorize manufacturing and sale of medicated feeds. Instead, the BVM would have authority to deny registration of feed manufacturers that lacked adequate facilities and controls to assure safety."

Even the Environmental Protection Agency (EPA) is trying to "simplify the regulatory burden on industry and reduce unnecessary costs." So said John A. Todhunter, Assistant Administrator for Pesticides and Toxic Substances, at the 1982 Beltwide Cotton Production Mechanization Conference, January 1982, at Las Vegas. He described a reassuring response to Vice-President Bush's task force, especially that FDA has instituted a plan to improve the quality of scientific assessment, including a peer review system for major scientific studies and reports. There is, therefore, hope for maintaining availability of the herbicide 2, 4, 5T and even reapproval of compound 1080 for predator control. The states also have regulatory authority and enforcement responsibility. We are all aware of Governor Brown's reluctance in California to institute effective control measures for the Mediterranean fruit fly because of the political pressure of environmentalists.

Food Chemical News, a weekly publication, keeps you up-to-date on what is happening in Washington.

For those of you mixing your own feed, and for feed distributors, I recommend the annual Feed Additive Compendium, a guide to use of drugs in medicated animal feeds with monthly, up-to-date supplements.

In conclusion, I want to make a plea to agricultural producers for closer adherence to label requirements and restrictions on use of agricultural chemicals. The Agricultural Extension Service in every state has a responsibility for assisting producers in the proper use of chemicals. More attention is being devoted to that. Very few people deliberately break the laws, but many are not aware of the precautions necessary to prevent cross contamination of products and elimination of residues in feeds and foodstuffs. The USDA state producers' effort to eliminate sulfa drug residues in pork is an example of the kind of cooperation required to maintain availability of chemicals so important to modern food production. Let us resolve to intensify the effort for safe use of agricultural chemicals in order to gain greater public confidence in the safety of our abundant food supply.

REFERENCES

American Council on Science and Health (ACSH). 1982. U.S. food safety laws: Time for a change? 1995 Broadway, New York, N.Y.

Council for Agricultural Science and Technology (CAST). 250 Memorial Union, Ames, Iowa 50011.

CAST. 1977. Hormonally active substances in foods: a safety evaluation. Report No. 66.

CAST. 1981. Antibiotics in animal feeds. Report No. 88.

CAST. 1981. Regulation of potential carcinogens in the food supply: the Delaney clause. Report No. 89.

CAST. 1982. CAST-related excerpts from U.S. House of Representatives hearing on the Federal Insecticide, Fungicide, and Rodenticide Act (FIFRA). Special Pub. No. 9.

CAST. 1982. CAST-related testimony on the food safety amendments of 1981. Special Pub. No. 11.

Feed Additive Compendium. Miller Publishing Co., 2501 Wayzata Boulevard, P.O. Box 67, Minneapolis, Minnesota 55440.

Food Chemical News. 1101 Pennsylvania Ave., S.E., Washington, D.C. 20003.

8
WORLD AGRICULTURE IN HOSTILE AND BENIGN CLIMATIC SETTINGS

Wayne L. Decker

CHARACTERIZATIONS OF CLIMATE

Health, nutrition, and suffering of the human population are determined, in part, by weather and climate. Regional wealth and the levels of economic development are impacted by the natural resources, including the climate resource. But agriculture and the associated food production industries are more directly affected by weather and climate than any other sector of the economy. Climate determines production potential of both grain and livestock producers, identifies strategies available to the producer for resource allocations and marketing, and determines the feasibility of plans for exports and imports of commodities. An improved understanding by agriculturalists of the nature of the climatic resource is essential if the impacts of climatic risks are to be minimized.

Climate is defined by the space and time distribution of weather events: temperature, precipitation, wind, humidity, and sunshine. In spite of the unpredictability of weather events, climate occurs systematically in both the space and time scale. As a result of these consistencies, climatic zones are easily recognized. For example, in the tropics and subtropics some regions are consistently rain-free in summer, others are smaller areas with an even seasonal distribution and abundant rainfall. In the temperate latitudes, the continental regions also demonstrate regional consistencies in climate. The west coasts of continents are mild with abundant winter rainfall, while the continental interiors tend to exhibit summer maximum of precipitation and marked seasonal temperature variations (Mather, 1974).

In most climatic regions, there are periods during the year with hostile climates for agriculture. These climatic hostilities are associated with temperature stresses (both high and low) and with deficiencies of rainfall. In many of these regions, there are periods of the year during which the weather is consistently dry, thus producing a hostility. This climatic hostility can be removed by irrigation, or avoided by adopting an enterprise with a low water need. For the hostile climates produced by temperature stress, shelters may be constructed to protect animals from the critical temperatures, and crop production can be scheduled to avoid the consistent occurrence of high or low temperatures.

Many regions have climates that are consistently favorable for agricultural production. These climates are usually characterized by dependable water supply (rain or irrigation supply). Benign climates are also characterized by moderate temperatures without a high probability of extreme temperatures during critical times for sensitive plants.

CLIMATIC CHANGE

Climatic change and the impact of climatic change on man have become controversial issues in recent years. Articles on the subjects appear regularly in technical and popular magazines, and both paperback and hardback books have been published dealing with climate change. The written opinion concerning climatic change and its impact are as different as day and night. Even scholars of climatology are confused by the diversity of opinion.

The evidence to support the existence of major climatic changes through geologic time is well documented. Long periods of geologic history are characterized by mild climates, i.e., benign climates. These periods were interrupted by relatively short intervals when glaciers extended into the middle latitudes (the ultimate in hostile climates). It is generally accepted that the current climate of the world is more like that of the glacial period than the warmer "climatic optimum." Climatic researchers do not agree about the mechanisms causing these major climatic changes. Current thinking focuses on long-term oscillations in the slope of the terrestrial axis, but continental uplift and the associate volcanic activity appear to be necessary conditions for the glacial climates.

Variations in the climate of the earth also have been documented from historical records. The rise and decline of civilizations during the past 4,000 years appear to be related to changes in climate. Plagues, famine, and migrations have been linked to shifts in climate. In modern history, the period corresponding to the North American settlement and the establishment of the United States was a period of climatic stress, frequently called "the little ice age." Again, meteorologists do not agree on the physical processes that caused the climatic variations in historical time. Volcanic activity, variability in the solar output, and combinations of both these factors are mechanisms receiving prominent attention.

It was not until the late 1800s that a worldwide network of weather observing stations was established. Although records of weather observations can be traced into the 18th century at selected points, networks of observational stations did not generally exist until the early and mid-nineteenth century. In the United States, for example, it is difficult to find documented weather records prior to the establishment of the Weather Bureau in the Department of Agriculture in 1890. For this reason studies of climatic change based on meteorological observations are confined to the most recent 90 years.

Attempts have been made to establish the trends in climate from the meteorological observations. The best documented estimate of the trend in climate is shown in figure 1 from Waite (1968). During the first 40 years of this century, the average air temperature near the earth's surface increased, but about 1940 this trend was reversed. Figure 1 verifies that these trends in temperature apply to regions of different size and are the most pronounced in the polar and subpolar regions of the northern hemisphere.

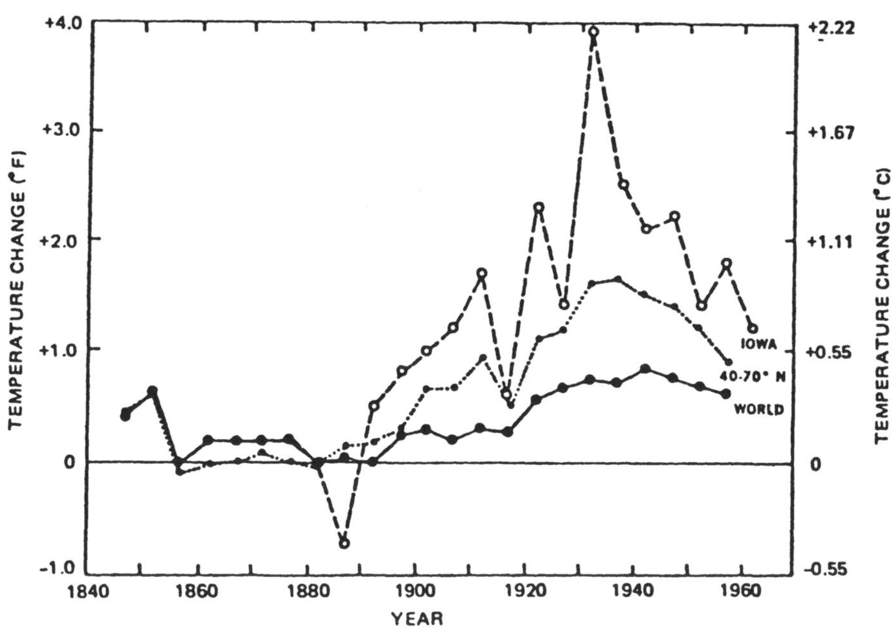

Figure 1. **Worldwide trend in mean annual temperature (Waite, 1968)**

For agriculture it is more instructive to look at the climatic variabilities associated with precipitation. In figure 2 and figure 3 the historical records of summer rainfall are summarized for the prairie provinces of Canada and the winter wheat region of Russia. It is difficult to identify trends from these data, although many interesting pulsations in the annual rainfall statistics, lasting for a decade or so, are apparent.

The atmospheric mechanisms producing the climatic trends and fluctuations that extend for a decade or a few decades have probably all been identified. These mechanisms include ocean-atmosphere interactions, volcanic activity, man's interference (CO_2 and particulate matter), and solar activity. The analytical contribution of each individual mechanism and the interaction between the mechanisms have not been defined; a major objective of the meteorological community to mathematically describe these processes based on known physical relationship efforts.

Efforts to research the physical causes for climate change led Dr. B. J. Mason, Director General of the British Meteorological Office, to observe in a recent article in the New Republic (1977):

> The atmosphere is a robust system with a built-in capacity to counteract any perturbation. This is why the global climate, despite frequent fluctuations, is fairly stable over periods of 10,000 years or so. Sensational warnings of imminent catastrophe, unsupported by firm facts or figures, not only are irresponsible but are likely to prove counterproductive. The atmosphere is want to make fools of those who do not show proper respect for its complexity and resilience.

FLUCTUATIONS IN CLIMATE

Climate variability adds an additional stress to the agriculture system and adds a component to climatic hostility. When the weather of one or more years departs markedly from the expected, a farm management strategy that has been successfully used becomes inappropriate for the agricultural enterprise for a particular year or growing season. Several years of drought (such as the 1930s on the U.S. Great Plains) is hostile to the farm enterprise adopted to nondrought enterprises. On the other hand, periods of years with benign climates often lure farmers into strategies not adapted to the hostile and stressed condition that follows. The type of fluctuation most often used by climatologists deals with the variation of climatic events between years. This variability refers to the variation in annual or seasonal temperatures and precipitation totals. McQuigg (1973), for example, demonstrated the low variability in climate between 1955 and 1970 for the major agricultural production regions of the U.S. McQuigg simulated

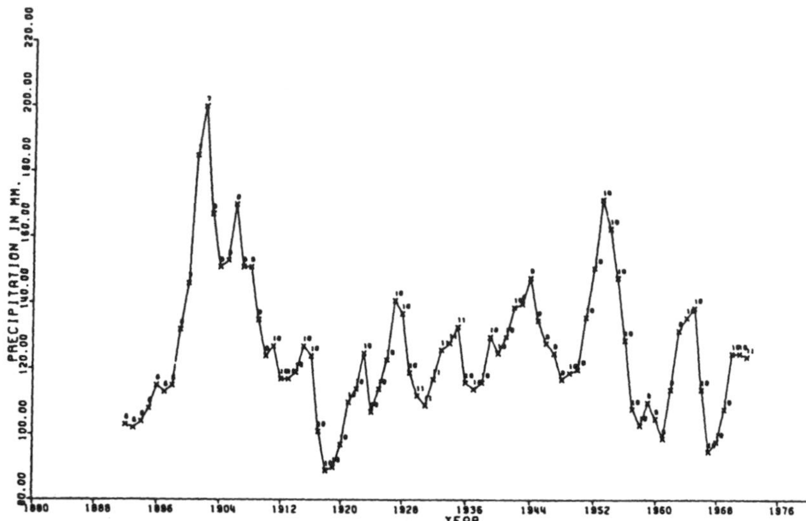

Figure 2. The year to year variability in the total precipitation for May and June (smoothed by a binomial technique) in the spring wheat producing area of Central Canada.

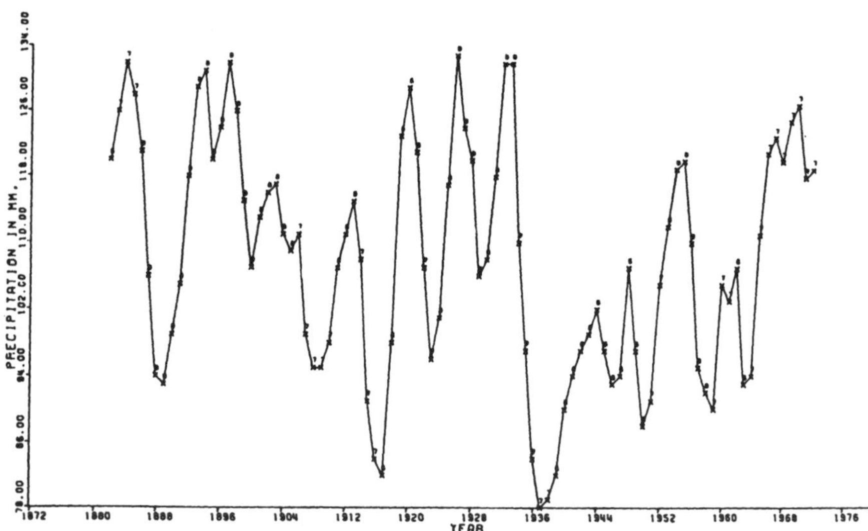

Figure 3. The year to year variability in the total precipitation for May and June (smoothed by a binomial technique) in the winter wheat producing area of the Soviet Union (west of the Volga River).

yields of grain throughout this century from climatic data at a constant technology. He showed (figure 4) that the yields simulated from climate during the period extending from the late 50s through the 60s were remarkably constant and relatively high, i.e., the climate of the U.S. Corn Belt was benign. The year-to-year variability in climate in the U.S. has been greater in the 1970s and the early 1980s than the preceding decade and a half.

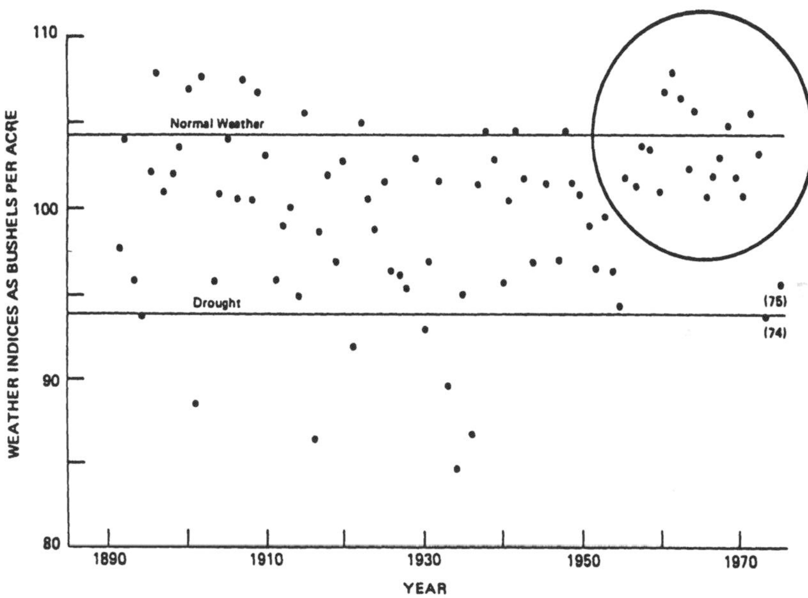

Figure 4. Year-to-year variability in climate expressed in simulated corn yields for the U.S. (NAS, 1976)

To examine how climatic fluctuation varies through time, the variances of climatic elements by decades have been computed for major agricultural production areas in the world. Figure 5 shows the 10 year variances in the May plus June precipitation in the Canadian prairie region, while figure 6 presents the same values for the Soviet winter wheat region. The May and June rainfall totals are vital to wheat production in these two regions. In both cases, the variability in May and June precipitation during the most recent decades were below average; however, the tendency for a lower variance does not appear to depart from that expected from the normal variability. The high year-to-year variability in the May-June rainfall in Canada just after the turn of the century is quite striking. There was also a period of high variability for the May and June precipitation in the Soviet Union between 1925 and 1950.

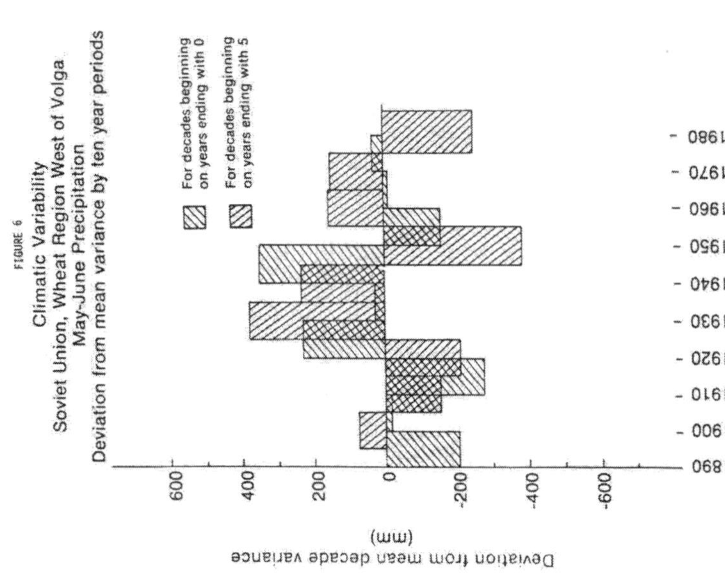

FIGURE 6
Climatic Variability
Soviet Union, Wheat Region West of Volga
May-June Precipitation
Deviation from mean variance by ten year periods

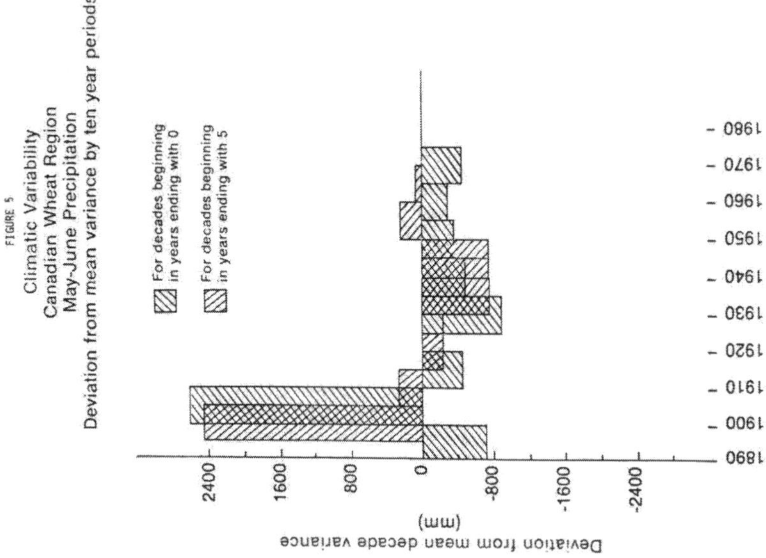

FIGURE 5
Climatic Variability
Canadian Wheat Region
May-June Precipitation
Deviation from mean variance by ten year periods

Figures 5 and 6 show that climate did not experience unusually low variability everywhere in the world during the period 1955 to 1970. A second, and more important, lesson concerns the recognition that periods of 1 or more decades in length with high (and low) year-to-year variability in climate do occur.

A report by the National Research Council (1976) identifies sudden and unexpected climatic fluctuations as the greatest climatic hazard to agricultural production. The managers of agricultural systems are forced to use practices that are not adapted to the regional climate. The farm manager has difficulty in finding the "best" strategy during periods with fluctuating climates. A large year-to-year variability forces the manager to choose management options that are "out of step" with the season's weather.

RESPONSES TO CLIMATE CHANGES AND FLUCTUATIONS

The policy makers and agricultural managers could easily respond to weather and climate fluctuations, if events could be anticipated before they occurred. Although several climatologists, with impressive credentials, regularly make seasonal forecasts, most meteorologists agree that the science of meteorology has not advanced sufficiently for these forecasts to be considered valid. The thirty-day outlook, which projects the general trends in mean temperature and precipitation, is correct only about 55% to 60% of the time. Seasonal forecasts (cold winter, dry summer, etc.) have an even lower accuracy. Complex computer models, which simulate the atmospheric circulation, offer the best promise for the development of rational forecasting schemes. As these atmospheric models are improved, weather and climate forecasts will have improved accuracy and be extended for longer periods. The improvement in forecasting skills will be slow. In the next two decades there appears little hope for major and sudden breakthroughs in our understanding and interpretation of atmospheric circulation.

Weather modification provides an additional strategy for removing weather risks. Modification of the surface energy budget through changes in the surface color, drainage of the land, or shaping of the soil surface through tillage, offers many important options for improved technologies for agriculture. Increased rain through cloud seeding is more often considered as a weather modification option. There are several difficulties that reduce the potential for cloud seeding to respond to fluctuations in climate:

- Proof of small increases in amounts of rain are almost impossible to obtain because variations in area and time affect the amount of rain falling over a region.
- Rain-making only augments the natural rainfall, so it is not a "drought stopper."

- The possibility exists that the manipulation in the clouds will reduce the rainfall from some clouds within a given weather system.

Cloud seeding appears to be most favorable for use in the mountainous regions--to increase the winter snow pack. Increased snow in the mountains improves the water supply for the agriculture of the adjacent semiarid and arid regions.

Disaster insurance spreads the risks of "bad" weather. An international program for grain and food storage should stabilize supply between the "lean" and "bountiful" years. A marketing cooperative, or even the individual farmer, may establish an "ever-normal granary" by withholding grain from the market. For the individual farmer, the best opportunity for spreading the risk is through disaster insurance. A national food policy must include an insurance program using both governmental and private agencies as underwriters. Insurance programs may stabilize farm incomes, but, in the long run, will not provide increased productivity of food for the expected increase in world population.

Technologies developed through private and public research provide an additional response to provide for a stable food supply and farm income under a fluctuating climate. Meteorological science cannot be expected to deliver completely reliable warnings of pending shifts and fluctuations in climate. The meteorologists will not save us from the adversity of a variable and often hostile climate. Agricultural strategies and technologies must be developed to respond to the expected climatic variabilities. These developments will emerge from integrated, interdisciplinary agricultural research.

REFERENCES

Mason, B. J. 1973. Bumper crops or droughts. Mimeo. NOAA, U.S. Dept. Commerce, Washington.

Mather, J. R. 1974. Climatology, Fundamentals and Applications. McGraw Hill, pp 112-131.

National Research Council. 1976. Climate and Food. Report on Climate and Weather Fluctuations and Agricultural Production. National Academy of Science, Washington.

Waite, P. J. 1968. Our weather is cooling off. Iowa Farm Science 23:13.

9
THE IMPACTS OF CLIMATIC VARIABILITIES ON LIVESTOCK PRODUCTION

Wayne L. Decker

CLIMATE AND LIVESTOCK

Regional climates contain factors that need consideration in determining the kind of profitable livestock production for a region. The variability of weather and climate provides a component in determining the profitability of the livestock enterprise. Climate imposes both direct and indirect effects on commercial animal agriculture. The direct effects include weather events producing physical injury (lightning, wind, flood, temperature extremes), occurrences associated with physiological stress (such as heat and humidity), and weather events promoting insect or disease episodes. Indirect climate events are those that impact on availability of forages and the supply of feed grains. These indirect impacts of the weather and climate are generally imposed by chronic deficit in water and occasional droughts.

LOSSES IN ANIMAL PRODUCTION DUE TO CLIMATE EVENTS

Animals, grown commercially on farms and ranches, are normally subjected to ambient environmental conditions. On many occasions, the atmospheric conditions are less than ideal and the animal is subjected to stress. This stress, which is usually related to the heat and energy balance of the animal, reduces the production. The stress causes declines in egg or milk production and reduced weight gains for swine, beef, or broilers.

Over the years there have been repeated attempts to mathematically define the impact of stress imposed by atmospheric conditions (temperature, humidity, wind, etc.) on animal production. The experimental basis of these efforts comes from two sources: (1) barns or chambers with controlled environmental conditions (Brody, 1948) and (2) field experiments measuring animal performance as related to observed ambient environmental conditions. From the observations obtained through the experimentation, functional relationships between the weather event (or events) are derived. The resulting mathematical formulas are usually obtained through standard statistical procedures. Strickly speaking, the expressions are only applicable to the experimental conditions from which the relationship is derived, so each relationship must be tested against independently collected data.

Literature has many examples of relationships between the performance of domestic animals and weather and climate events. Two mathematical expressions for relating cattle performance to environmental conditions are discussed below.

Milk Production

Using data obtained from controlled experiments, Berry et al. (1964) related the decline in milk production to an index involving both atmospheric temperature and humidity. This index, which is called the temperature-humidity-index (THI), is shown in equation (1).

$$THI = T + .36\ T_d + 41.2 \qquad (1)$$

where

T is the temperature in °C,

T_d is the dew point temperature in °C.

The relationship between THI and the decline in milk production (MD) is shown in equation (2).

$$MD = -2.37 - 1.74\ NL + .0247\ (NL)(THI) \qquad (2)$$

where

NL is the normal production of a cow under thermoneutral conditions.

THI is the temperature-humidity-index.

Since negative values for MD do not make sense under the definition in this equation, decline in milk production is assigned the value of zero for all negative values in equation (2). This means that a zero production decline is expected until a critical value of the THI is reached. For higher values of THI, the decline in production decreases linearly. This critical value of THI is between 70 and 74 for normal production levels of between 25 and 30 pounds of milk per day.

Meat Production in Cattle

Bolling and Hahn (1981) and Bolling (1982) report the results of a regression analysis relating climatic variables to rates of gain for beef animals. The functional relationship, which best explains the reduction in weight gains, contained terms related to cold stress, heat stress (THI), precipitation, and wind. The results of the regression analysis are shown in figure 1, as taken from the work of Bolling (1982). The author concluded:

> "Analyses consistently suggested that stress resulting from the direct effects of cold, combined with the effects of precipitation, have a greater impact on feedlot cattle in Nebraska than does any

other type of atmospheric stress studied. Interestingly, heat stress rarely appeared to be a significant factor affecting cattle performance."

Of course, this conclusion concerning heat stress is counter to the one for milk production. This difference may be due to the difference in climate of the regions where the milk production and rate of gain experiments were conducted.

Hahn et al. (1974) noted that beef cattle were able to overcome heat stress. In the Missouri Climatic Laboratory, beef cattle were stressed by being subjected to 5 weeks of temperatures of 30°C. A marked decrease in rate of gain as compared to a control group resulted; when the animals were returned to optimal conditions, the stressed animals outgained the control group. This result, which Hahn calls "compensatory growth," is demonstrated in figure 2. No such compensation occurred after animals were subjected to a greater stress (35°C). Hahn (1976) indicates that similar compensatory growth occurs with swine and broiler chickens. It is significant that (1) the research indicates that compensatory growth does not occur after exposure to a high degree of heat stress, (2) the laboratory experiments on compensatory growth were done at constant temperature and may not apply to temperatures experiencing a diurnal range, and (3) no experiments have been made to discover whether "compensatory growth" occurs after cold stress.

MORTALITY OF ANIMALS DUE TO CLIMATIC STRESS

Hostile climates do cause mortality to domestic livestock. This hostility occurs as a result of both heat and cold stress. In summer, high temperature and humidities (THI values of 80 or higher) produce stress that can lead to death. This condition is, of course, aggravated by other stress factors associated with handling and/or shipping. In winter, the cold stress can cause tissue to freeze and the animal to die. For U.S. cattlemen of the open ranges in the High Plains and eastern slopes of the Rocky Mountains, the cold stress may be further aggravated by high winds with snow. These blizzards cover the winter food supply, make access to the herds difficult (if not impossible), and bury herds under the drifting snow.

Bolling (1982) presents analyses that document the weather impacts on cattle mortality under feedlot confinement. Strong winds and cold stress were the "best" predictors of mortality under Nebraska conditions. The author was apparently unable to document the mortality due to stress imposed by hot and humid weather. Of course, one would not want to use these results to estimate mortality under range or pasture exposures.

The National Weather Service has established policies for issuing weather advisories to stockmen. These advisories are issued by Weather Service Forecast Offices located

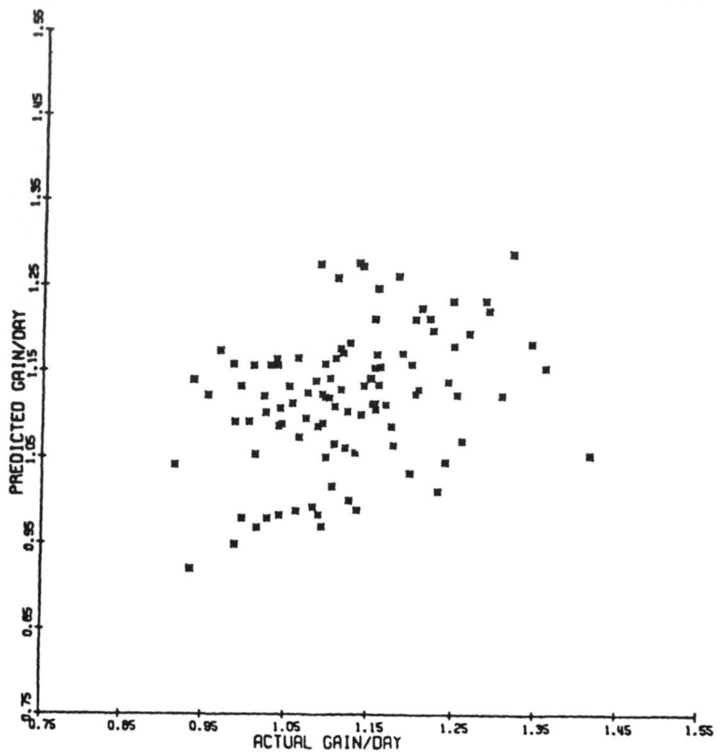

Figure 1. Relationship between measured gain of beef cattle and that predicted from weather data using a statistical relationship (Bolling, 1982)

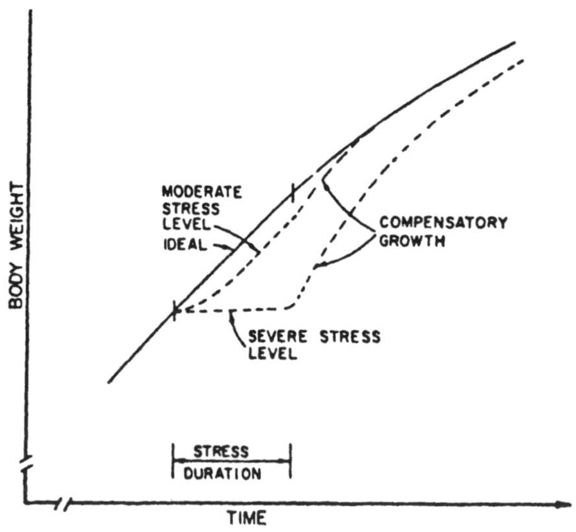

Fig. 2. Schematic to Illustrate the Principle of Equifinality in Animal Growth (Hahn, 1982)

in each state. These advisories are:

- Heat stress advisories are issued whenever high temperatures combine with high humidities to present danger to livestock. The THI is used as the index for danger to livestock. Two categories are recognized: (1) livestock danger (THI 79 to 83) and (2) livestock emergency (THI higher than 83). When air temperatures are below the body temperature, wind provides cooling. Because of this, wind is mentioned in the stockman's advisories when velocities higher than 20 mph are expected.
- Winter storm watches are issued when there are strong indications of a blizzard, heavy snow, freezing rain, or sleet. A blizzard is defined as a condition with high winds (in excess of 35 mph), with falling or blowing snow (visibilities less than 3 miles). A winter storm warning or blizzard warning is issued when the storm's development is a virtual certainty.
- Chill indices are released routinely. This index combines wind and termperature in an effort to approximate the equivalent temperature under light wind (4 mph) conditions. At Columbia, Missouri, in the winter of 1981-82, there were 398 hours (11% of the hours) with chill factors below 0°F, 181 hours (5%) with chill factors below -10°F and 72 hours (2%) with chill factors below -20°F. The number of days with chill factors below 0°, -10° and -20° sometime during the day was 30, 18, and 7 days, respectively.

REFERENCES

Berry, I. L., M. D. Shanklin and H. D. Johnson. 1964. Dairy shelter design based on milk production decline as affected by temperature and humidity. Trans. Amer. Soc. Agr. Engr. 7:329.

Bolling, R. C. 1982. Weight gain and mortality in feedlot cattle as influenced by weather conditions: Refinement and verification of statistical models. Progress Report 82-1. Center for Agricultural Meteorology and Climatology, Univ. of Nebraska, Lincoln.

Bolling, R. C. and G. L. Hahn. 1981. Climate effects on feedlot cattle: Growth and death losses. Proc. 15th Conference on Agr. and Forest Meteorology: 86-89. Amer. Meteorological Soc., Boston.

Brody, S. 1948. Environmental physiology with special reference to domestic animals, I: Physiological background. Res. Bull. 423. Mo. Agr. Exp. Sta., Columbia.

Hahn, G. L. 1982. Compensatory performance in livestock: Influences on environmental criteria. Livestock Environment, Proc. 2nd Internatl. Livestock Environment Symposium. Amer. Soc. Agr. Engr., St. Joseph, Michigan. (In Press.)

Hahn, G. L. 1976. Rational environmental planning for efficient livestock production. Proc. 7th Intern. Biometeorological Cong., College Park, MD. pp 106-114.

Hahn, G. L., N. F. Meador, G. B. Thompson and M. D. Shanklin. 1974. Compensatory growth of beef cattle in hot weather and its role in management decisions. Livestock Environment, Proc. Internatl. Livestock Environment Symposium, pp 288-295. Amer. Soc. Agr. Engr., St. Joseph, Michigan.

Part 3

GENETICS AND SELECTION

10
GENETIC IMPROVEMENTS IN HORSES

Joe B. Armstrong

Few subjects enjoy more discussion, based on fewer facts, than the genetics of the horse.

Mankind has long been aware that differences exist among individuals and that selection and mating systems produce genetic changes in domesticated animals. Animal breeding is actually one of the oldest sciences of which we have written record. According to the Bible, Jacob was effectively selecting for color in sheep and goats as early as about 1746 B.C. Genesis 30:32-43 implies that Jacob practiced selection among the males of Laban's flocks so as to increase the number of animals of specific colors that were to become his by virtue of an agreement with his father-in-law, Laban.

The writings of Varro (116-27 B.C.) indicate that farmers of that era were practicing selection among the animals of their flocks and herds. On the subject of selection in swine, Varro is quoted as follows:

> "A man, then, who wishes to keep his herd in good condition should select first, animals of the proper age, secondly of good conformation (that is, with heavy members, except in the case of feet and head), of uniform colour rather than spotted. You should see that the boars have not only these same qualities, but especially that the shoulders are well-developed."

More familiar to many than the exploits of Laban and Varro are those of Robert Bakewell who lived from 1725 to 1795. Bakewell of Dishley, Leicestershire, England, is perhaps the most noteworthy of early animal breeders and is often referred to as the founder, or father, of modern animal breeding. Bakewell's breeding work was with Longhorn cattle, Leicester sheep, and Shire horses. He achieved considerable success in improving his animals by combining selections with inbreeding. Bakewell's basic principles, which are still widely quoted today, were: "Like produces like or the likeness of some ancestor; inbreeding produces prepotency and refinement; breed the best to the best."

Genetic improvement in horses involves the same basic principles as genetic improvement in beef cattle, sheep, hogs, and dogs. Unfortunately, less factual genetic information is known about the horse. As horse breeders, we tend to use phrases such as "genetic predisposition toward or against" as a scapegoat for our lack of concrete genetic knowledge.

There are many reasons for this lack of specific genetic knowledge of the horse; some relate to the complexities of the horse, and others to man and his inability to plan and direct the horse's future.

DETERRENTS TO GENETIC IMPROVEMENT

Among the major deterrents to genetic improvement in the horse are the following:

Reproductive Efficiency

- The horse is generally recognized as having the poorest reproductive rate of most domesticated animals. A national average for broodmares foaling live foals is generally considered to range from 50% to 80%. The percentage of foals weaned would be smaller.
- The difference in reproductive efficiency between mares and other domesticated females can partially be attributed to the unusual reproductive patterns of the mare. These patterns include 1) the long estrous period within which ovulation can occur at any time, 2) a seasonal ovulatory pattern with a prolonged estrus prior to the first ovulation, and 3) sporadic estrous behavior. The reason(s) for these unorthodox patterns is not readily apparent--genetics? environment? management?

Lack of Selection for Reproductive Efficiency

- The major selection practiced in horses has been based on their ability to walk or run. Economics ensures that other farm animals (cattle, sheep, hogs, etc.) are selected heavily on the basis of reproductive efficiency.
- In addition to requiring two mares to produce one foal annually, lack of attention to reproductive efficiency results in an increased generation interval. Genetic improvement is made only when generations are turned. Breed the same mares to the same stallion for fifteen years and the genetic base will remain the same. The only changes will be in management (environment), except that all the mares and the stallion will be fifteen years older. To maxi-

mize genetic improvement, the superior female offspring from the above matings should be bred to another superior stallion as soon as they reach breedable age.

Failure to Establish and Stay with Clear, Obtainable Goals

- True "breeders" develop realistic goals for their horse breeding operations and earnestly pursue them. Unfortunately, there are few true breeders of horses. Most are "multipliers" who have no planned breeding program. If they do have a plan, it is abandoned in favor of something new every time popular opinion shifts--and popular opinion blows with the winds. Aim at nothing and you will surely hit it! It takes planned perseverance to be a breeder.

Economics

- As compared to research on other farm animals little research has been conducted on horses. The horse, unlike cattle, sheep, and hogs, has had no firm market value in past years because he was not a meat animal. It also costs more to maintain a herd of horses. These factors, when multiplied by poor reproductive efficiency, have made the horse a poor experimental research vehicle.
- However, the increase in interest in horses and number of horses the past 40 years has made the horse an entity of great economic importance; increasing research has been generated over the last 20 years. Nutrition trials and studies can be conducted over relatively short periods, but meaningful genetic studies require generations! Therefore, time and money are major factors related to our lack of knowledge in horse genetics.

Inadequate Use of Information

In spite of the deficiencies in our bank of genetic knowledge, most horsemen are not adequately tapping the store of knowledge that exists today.

Genetic improvement in horses is attained through the use of mating systems and the selection of superior individuals in a given generation to be the progenitors of the succeeding generation. Selection is simply determining which individuals will be the parents of the next generation. For selection to be effective in changing a population of horses, variation--differences between individuals--must exist in the population. Variation is the net result of differences in heredity, environment, and the joint effects of heredity and environment and is often referred to as "the raw material with which the breeder works." That raw material is presently existent in horses and has been in existence through the ages! It is the ingenuity and perseverance of man that is required to shape this material into new and improved horses.

BASIC GENETICS

The gene is the basic unit of inheritance. Countless numbers of genes occur in pairs on threadlike structures called chromosomes in the nuclei of cells. The horse has 32 pairs (64 total) of chromosomes in each cell nucleus. One chromosome from each pair is inherited from each parent at the time of fertilization. While every cell nucleus has 64 chromosomes, each ovum produced by the mare and every spermatozoon produced by the stallion contain 32 chromosomes. The chromosome number is reduced from 64 to 32 during the formation of ovum and spermatozoon by the process of meiosis.

The location of a gene on a chromosome is known as a locus. Each member of a gene series is an allele and there is only one allele at a given locus on the chromosome. There may, in fact, be several alleles for each gene. Our knowledge of coat-color inheritance demonstrates this fact very visibly. During meiosis (the reduction of chromosome number from 64 to 32 in the germ cells) a chromosome pair becomes entwined and exchanges chromosome (genetic) material from one chromosome member of the pair to the other chromosome of the pair. The alleles are mixed randomly in the pair of chromosomes, and as the chromosome pair divides, each germ cell receives one member of the pair of chromosomes. This random mixing of alleles prevents all the genes in a chromosome inherited from a parent from being passed in toto to future generations. This mixing enables a foal to ceive genes from all of its grandparents.

Species Crosses

While the horse has 32 pair or 64 total chromosomes, the ass has only 31 pair or 62 total chromosomes. The resultant cross between the horse and the ass produces superior, rugged work animals (mules and hinnies), but they are infertile. Mules and hinnies have 63 chromosomes, having received 32 from their horse parent and 31 from their ass parent. The influence of chromosome pairs in the sex cells results in sterility to hybrids of these species.

Sex Determination

Sex of the individual is determined by one pair of chromosomes. In the male, the sex chromosomes are designated as XY and are of unequal size or are unmatched.

In the female, the chromosomes are of equal (matched) size and are designated XX. During meiosis, the mare can only contribute the X chromosome to her ova while the stallion contributes approximately 50% X and 50% Y chromosomes to his spermatozoa. It is therefore the male that determines the sex of the offspring. Because the Y chromosome is smaller than the X chromosome, fillies receive more genetic traits from their sires than do colts.

Qualitative and Quantitative Inheritance

As the term indicates, qualitative refers to that kind of inheritance that is controlled by one or a few pairs of genes. That which we can see, the phenotype, may be a direct indicator of the genotype of the horse. Sex of the horse, coat color, curly hair, and hemophilia are examples of qualitative traits. Breeding systems to change and(or) improve qualitative traits are reasonably uncomplicated and predictive and will not be considered in this discussion.

It is considerably more difficult to design breeding systems that will accurately and quickly change those traits that are quantitatively controlled. Quantitative traits are difficult to recognize because the phenotype is controlled by many pairs of genes. Many quantitative traits are difficult to measure or define. Examples of traits that are obviously controlled by many genes and are difficult to measure are "cow sense," soundness, and jumping ability. Measurement of these traits is highly subjective and varies according to the person making the judgment. Quantitative traits that are more easily measured are racing speed, wither height, and weight at a given age. The stop watch and weights and measures provide definite, objective measurements.

Heritability

Most, if not all, traits are genetically controlled at least to some degree. The degree to which traits are genetically controlled is the heritability estimate for that trait. The heritability value indicates the amount of the average differences between the individuals selected to be the parents of the next generation and the population from which they were selected that we expect to see in their offspring. With all selection there tends to be an incomplete selection or a regression back toward the breed average rather than an averaging of the superiority of the selected parents. The degree of this incompleteness of selection is directly related to the estimated heritability value of the trait. The heritability estimate is the measure of the differences measured or observed between horses that is controlled by genetic influences.

Observed differences are made up of both genetic and environmental factors: Phenotype = genotype + environment. Because the phenotype is what we see or observe, it is important to know a horse's environmental influence if we are to be able to make accurate genetic evaluations. The more a trait is influenced by environment, the less genetic change the breeder can make in a given period of years.

Change resulting from selection for quantitative traits may be formulated as follows (change can be either in a positive or negative direction): Change = heritability x selection differential. Selection differential is defined as the difference between the average merit of the stallions and mares selected to be the parents of the next generation and the average merit of the entire horse population from which they were derived. The stallion and the mare each contribute equally to the genetic makeup of their offspring. For an individual foal, each parent is equal in genetic importance. Extra emphasis is generally placed on the genetic merit of the stallion because he will normally sire 10 to 200 offspring annually compared to the mare's production of, at best, one foal annually.

$$\text{Genetic change} = \text{heritability} \times \frac{(\text{Sire superiority} + \text{mare Superiority})}{2}$$

A simplified example of selection for a quantitatively controlled trait would be selection for wither height. Assume the average wither height for a breed of horses is 14 hands and you wish to increase wither height as rapidly as possible. Wither height is generally regarded as being about 40% heritable. From a population that averages 14 hands (56 in.) you select a 16 hand (64 in.) stallion and mate him to mares that average 15 hands (60 in.). The stallion is +2 hands (8 in.) and the mares are +1 hands (4 in.) superior to the average of the horse from which they were selected. Plugging their values into the formula, we would expect the offspring from these matings to have a wither height of:

$$\text{Change} = .40 \times \frac{(8\text{ in.} + 4\text{ in.})}{2} = 2.4\text{ in. above the breed average}$$

Wither height of the selected offspring should average 2.4 in. higher than the average height of 56 in. of the original population of horses. The offspring resulting from this selection would measure 56 in.+ 2.4 in.= 58.4 in. (or 14 2 1/2 hands) at the withers.

The genetic change demonstrated here is permanent and will be carried from one generation to another and may be added to or subtracted from depending upon whether or not future electrons are positive or negative for wither height. It should be readily apparent that the higher the heritability value of a trait the more rapidly a breeder can make progress in that particular trait.

Maximum progress is made when heritability is high and selection is for a single trait. Therefore, horse breeders should be discreet in their breeding programs and attempt to select for as few traits as possible in a given generation. Thoroughbred breeders have been most successful in selecting for speed at a given distance. Speed and(or) money earned are the primary selection criteria. Money earned or average-earnings index values take into account both speed and soundness and enable breeders to make considerable genetic progress. Attention to smaller "fancy" traits, such as length and shape of ear, etc., serve only to slow down the genetic improvement in the thoroughbred breeders' quest for faster, sounder horses. The following chart lists heritability estimates for some traits.

TABLE 1. HERITABILITY ESTIMATES

Trait	Heritability
Wither height	0.25 - 0.60
Body weight	0.25 - 0.30
Body length	0.25
Cannon bone circumference	0.19 - 0.30
Trotting speed	0.40 - 0.50
Racing speed	0.25 - 0.67
Reproductive traits	Low?
Cow sense	Medium to high?

It should be noted that the above values are referred to as "estimates." Because many traits that horse breeders desire to change and improve are difficult to measure or define, it should be apparent that the accuracy of the heritability estimate fluctuates in direct proportion to the breeder's ability to measure and define the trait. This is the reason that few estimates of cow sense, soundness, and reproductive traits exist in the literature today.

Heritability estimates are broadly classified as low, medium and high. Low heritability estimates range from .10 to .25, medium from .30 to .50 and high from 0.50 to .80. Highly heritable traits are generally easy to measure and positive change is made by utilizing the individual's own record of performance. Conversely, traits with low heritability are difficult to select for because environmental factors (training, feeding, etc.) are responsible for most of the observed differences between individuals for the trait.

Cutting ability ("cow sense") is a good example of a lowly heritable trait. Cutting-horse breeders must utilize pedigree or family records in addition to the individual's own performance record when selecting for cutting ability, but they must also rely upon the training and horsemanship abilities of their riders. The training and management are environmental factors! Few good genetic studies have been made of cutting ability, but those that have been made indicate that the heritability for cutting ability is quite low.

Most cutting and reining horse breeders depend heavily on pedigree selection, and pay particular attention to the maternal grandsires as well as the sires.

The increase in number of large futurities for the different performance events when all horses are exhibited at a standard age, coupled with the use of high speed computers, should yield valuable genetic information on these traits in the next decade. As this more accurate genetic information becomes available, rapid and more predictable improvements will be in reach of the astute horse breeder. It should be noted, however, that computers are no better than the data put into them, i.e., "garbage in, garbage out!"

The net effect of selection is to change the average merit of the population for a particular trait. Most biological measurements in livestock populations are considered to be normally distributed. A theoretical normal distribution curve in which the values are clustered at the midpoint, thinning out symmetrically toward both extremes, is illustrated below.

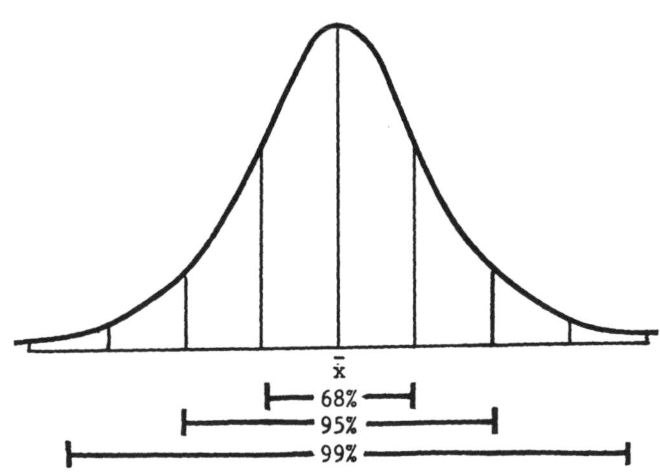

Figure 1. Theoretical normal distribution curve

The bracketed areas indicate the percentage of the total population under a theoretical normal curve as they move around the average or mean ($\bar{x}$) of the population.

If selection for a trait is effective, the mean value or merit of the population will be shifted to the right. For selection to be the most effective, the breeder should choose his stallions from at least the upper 10% of all stallions and the mares from the upper 60% of all mares. The following normal curve illustrates this type of selection.

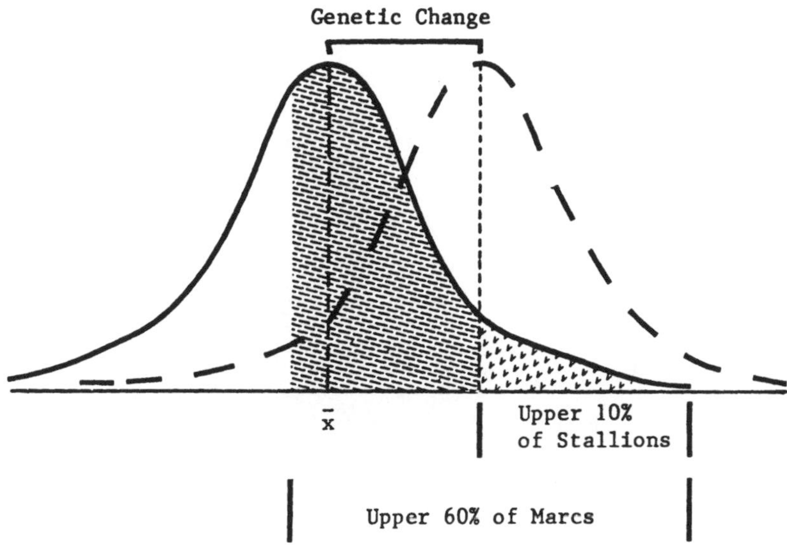

Figure 2.

The population resulting from this selection intensity should result in a normal curve whose mean is increased as illustrated by the curve formed by the dotted line.

The genetic change that results is permanent and is passed from generation to generation. It is equivalent to building a wall with bricks and strong mortar.

Number of Traits Considered

If the breeder is only concerned with improvement in one trait, all of his selection efforts can be made for the one trait. Everytime he considers another trait in his selection program, the selection pressure is lowered for both traits.

SUMMARY

Genetic improvement in horse breeding is slow at best and the responsibility for breed improvement rests upon the shoulders of the few tried-and-true breeders that have bred for specific goals over a period of generations. Too many so-called "breeders" are only flashes in the pan, have no breeding program, change with every whim of the industry, and generally disperse their herd every 7 years. Dedication to the pursuit of breeding outstanding horses is a must. The dedicated breeder has a well-defined, obtainable goal, and does everything within his means to reach that goal in the shortest period of years. He must select and use stallions from the upper 10% of his breed, based on accurate and intelligent measures of merit. He must consider the stallion's own record, those of his parents and, if possible, the records of his offspring (most breed associations have these records readily available to their breeders). The dedicated breeder must be extremely discriminating in the selection of his broodmares. Poor producers should be culled from the herd and replaced with young fillies that are the result of mating superior stallions to superior mares. It is best if the replacement mares have proven records of performance and production. The more knowledge one has about his breeding stock, the more efficient and rapid will be his genetic improvement.

Breeding superior horses may well be the greatest challenge in the livestock industry. There are no shortcuts. Variation, "the raw material with which the breeder has to work," is abundant. The source is virtually untapped!

Plan, establish goals, work hard, stick to the plan and success will follow. But, aim at nothing and you'll surely hit it.

11

GENETICS OF THE PERFORMANCE HORSE

Joe B. Armstrong

Pick up most any textbook on the horse and you automatically turn to a brief chapter in the back, if you're looking for the section on genetics. And, if you're looking for genetic information on performance, you're lucky if you find one paragraph with any specific information. There are valid reasons for this dearth of information.

With the exception of a horse's speed, few performance events have objective measurements. Most performance events are judged by subjective standards depending upon what each judge visualizes as the ideal for each particular event. The fact that certain breeds require performance classes to be judged on conformation, quality, substance, and appointments, and not performance alone, further clouds the lack of objectivity in judging performance claims.

Selection for conformation is a must if performance horses are to lead long, sound, useful lives. Conformation should be defined as "form to function."

The inheritance, or heritability, of speed is known to be medium to high. Therefore, breeders have been successful in increasing the speed of racing stock by breeding the fastest mares to the fastest stallions of their era.

Joe Thomas, vice president of Windfields Farm, presented some extremely interesting thoughts on selecting thoroughbred yearlings in the July 3, 1982, issue of The Blood Horse. "Fashion in type, however, is a passing fad and reflects only what the most popular and successful horses of the day look like." Two of the most knowledgeable writers in the history of the turf, Joe Estes and Joe Palmer, wrote in a pamphlet published in 1942 that: "One of the oldest axioms of the race track is, "They run in all shapes and sizes." This is true to the extent of creating no end of confusion among those who would specify the best shapes and sizes. There is a tremendous range in size among good horses, roughly from about 14 hands, 2 in. to 17 hands; and from 700 lb to about twice that weight. As to shape, good horses can be found to argue against the importance of even the most universally accepted "faults," such as ewe necks, flat withers, straight shoulders, calf knees, strait postures, club feet, long backs, goose rumps, sickle hocks, and so on. The great majority of good thoroughbreds are from 15.1 to 16.2 hands in height (a range of 5 in.) and weigh (under training conditions) from 900 lb to 1,150 lb. The great majority of the good ones will be found comparatively free of those characteristics that are most confidently set down as "faults." But the occurrence among the good ones of the undersized, the oversized, and the faulty is frequent enough to dissuade most horsemen from being positive about what makes a good horse.

What was pertinent 40 years ago is still pertinent today, so when judging or valuing a horse, probabilities must be dealt with, not rules and exceptions. Thus a sound knowledge of conformation makes the probabilities more favorable in estimating the chances of a horse standing intensive training without breaking down and in estimating its mechanical efficiency."

This is exactly what we do when we estimate heritabilities and use them in making selections of stallions and mares that should produce the fastest offspring. In the science of genetics, we are always dealing in the realm of probabilities and estimating the fraction of the sire's and dam's "good" genes that will be passed on to their offspring.

Those stallions that consistently sire winners are said to be "prepotent," indicating that, more often than the average, their offspring receive a higher percentage of genes favorable for outstanding speed or other performance events.

Heritability value estimates are simply indicators of what we can expect. Because our genetic knowledge is based on, and revolves around, average values, it is logical to expect some sires and dams to perform well above average while a like number will perform below average.

It is only after the actual proving of a stallion or a mare by the performance of their offspring that we can call them proven--and they may be proven to be superior, average, or inferior! After progeny test information of this nature is known, our selections are much easier and more effective. The disadvantage is that it takes years to accumulate progeny information. It takes even additional years to prove a sire's merit as a sire of outstanding broodmares.

This progeny information is available today from all the major herd associations and is generally shown on the pedigrees of racing individuals. Breeders must also study and research those families or strains within their breeds that are known to be prone to certain conformation faults and other weaknesses. Breeders must select against these individuals with faults.

Few, if any, selection indices have been formulated and adhered to in selecting horses. Joe Thomas of Windfields Farm has formulated his own scale for selecting yearling thoroughbreds for conformation, as conformation relates to performance in the race horse. His scale represents five areas of conformation and functions that must influence selections made by yearling buyers. Each part was given an arbitrary weight based on the assumed market criteria.

Thomas' five values are:

Overall appearance	25
Top line	10
Forelegs	35
Hindlegs	15
Gait	15
TOTAL	100 points

The foreleg was given the most weight because "serious faults here completely compromise the racing machine."

Pedigree will always be important because, when coupled with performance records of two or three generations, it increases the breeders' probability of success.

Racing speed is reasonably easy to select for because it is medium to high heritability.

Cow sense, cutting, reining, and roping ability are generally considered to be of low heritability. Few good studies have been made of these traits. One study from two years of National Cutting Horse Association Futurity scores estimated the heritability of cutting to be only 4%. This would indicate that factors (environment) other than genetics are vitally important in the cutting-horse world. These factors are known as environmental factors and are mainly training and the horsemanship of the riders.

If, in fact, the heritability values of these performance traits are very low, it proves that cutting, reining, and roping horse breeders have been proper in their heavy reliance on pedigree in selecting their breeding stocks.

Fundamental rules of selection are:

- For low heritability traits the breeder should rely heavily on family or pedigree information.
- Medium heritability traits should combine individual and performance information.
- High heritability traits may be effectively selected for and improved by using the individual's own performance records.

Cow sense and reproductive ability are examples of low heritability traits; conformation is an example of a trait with medium heritability, and racing speed is a highly heritable trait.

This helps explain the dependence of breeders of cutting, roping, and reining horses on pedigree information. When one studies the pedigrees of modern cutting, reining, and roping contenders, it is clear that certain families and relationships exist.

Within the quarter horse breed, most of the top contenders in these events will trace to the sire families of King, Old Sorrel, Leo, Three Bars (TB), Oklahoma Star, and Hollywood Gold. Some of these families excel on the distaff or maternal side of the pedigree. The offspring and grandoffspring of these sires are in greater demand today than when these sires were living. Their progeny have proven these sires to be genetically superior for these traits. This holds true for sires for these same events in all breeds.

There are other families within each breed that excel as the sires of halter horses. Unfortunately, too many halter-horse families do not prove to have the genetic and, consequently, athletic ability to become great performance horses. Some sires of leading halter horses apparently do not transmit the mental capacity to their offspring to make performance horses even when they do possess sufficient athletic ability. Breeders should study the yearly summaries of the leading halter horses and see how many of their offspring do or do not surface in the lists of leading performance horses.

Most horse breeders appear to be selecting two different kinds of horses--halter horses or performance horses; but few can excel in both. This is a constant source of frustration. Halter-horse trends tend to swing with the pendulum of time, and the breeder who attempts to always have popular halter horses is changing his breeding program every 5 to 10 years.

On the other hand, the performance-horse breeder has a more defined and controlled objective in his breeding program. The same sires that were popular in the 1950s are even more popular today. Not so with the halter horse.

In truth, cow sense and working ability probably do not have heritability as low as the few studies in the literature indicate. Ask any breeder who raises and trains cutting or reining horses. They will let you know in no uncertain terms that this ability is transmitted from one generation to the next.

The nature of our definition of heritability and the population from which we estimate these values are at least partially responsible for the low estimates, and they are simply that--estimates. Heritability is defined as that fraction (genetic) of the observed differences (genetic + environmental) between individuals that is passed to the next generation. If one was to study an entire breed of horses, the variation (observed differences) between horses in their ability to cut cattle or rein would be great. Some would have little or no ability while others would have great ability and a whole host of horses would be intermediate in ability. Our horses are identified and selected for their ability to work cattle or rein so that breeders do not introduce families that are recognized as having little or no cattle ability. Therefore, when cutting or reining futurity horses are measured and studied, we are dealing with a very narrow segment of the total horse population or breed. Because of the selectiveness of the group being studied, most of the variation observed is because environment, training, riders, etc, and the resulting estimate of heritability is low. In studying pedigrees, one should be cautious of going too far back in an attempt to introduce the blood of a certain horse into his breeding program. Each horse receives a sample half of his genetic make up from each of his parents. The following diagram illustrates the percentage of genes received from each parent, grandparent, etc.

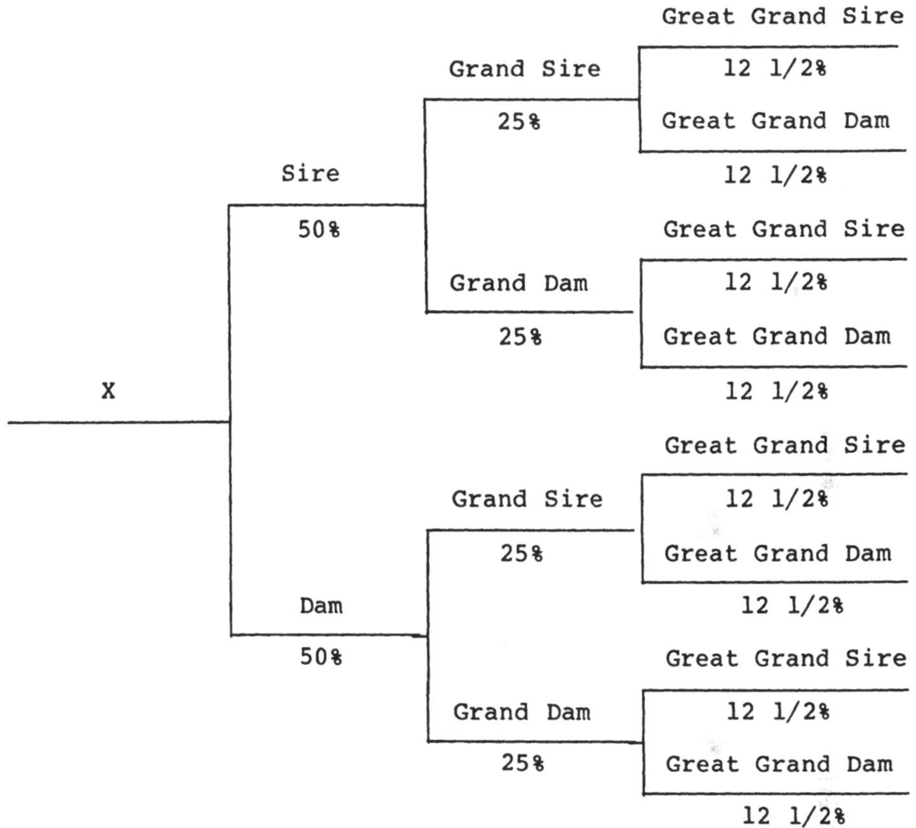

When one refers to an ancestor in the fourth generation, you are speaking of 6 1/4% genetic influence on the offspring in question. This is the reason most breeders look at the first two questions and only occasionally at the third generation. Past the third generation there is little genetic influence.

Linebreeding or mild inbreeding is an exception to the above stand. Linebreeding is mating animals that are more closely related than the average of the breed. Inbreeding is the mating of closely related animals.

Both linebreeding and inbreeding, when practiced with careful culling, have the effect of increasing the desirable genes for which the breeder is selecting. While the desirable genes are being concentrated so are the undesirable genes--if they exist in the family. This is the reason for the necessity of a rigid culling program when linebreeding or inbreeding.

The beneficial effect of linebreeding is that the relationship between individuals is increasing above that shown in the previous diagram and enables the offspring to receive a higher percentage of the gene of a desired ancestor. The King Ranch was highly successful in linebreeding to Old Sorrel. Linebreeding tends to fix a type within a breed. Many of our most superior performance horses are linebred to some other.

Robert Bakewell's principle of breeding the best to the best is just as much in effect today as it was when he made it in the late 1700s. It is up to the breeder of performance horses to establish the objective standards that will enable him to identify the best, and then to design breeding programs that will enable him to concentrate the greatest number of favorable genes for performance while purging those genes that are detrimental.

There is an old adage that states "More great horses have made trainers great than great trainers have made great horses." Truth or untruth?

12

CONFORMATION AND STRUCTURE: A BASIS FOR SELECTION IN HORSES

James C. Heird

Conformation is the physical appearance of the horse related to the make-up of muscle, bone, and other body tissue. Almost every horseman has attended a lecture on conformation where function is related to form, lines are drawn, and correctness is emphasized. The purpose of this presentation is not to tell you why we want a certain conformational structure but to acquaint you with some basic facts that will help in evaluating two or more animals.

To understand conformation, we must understand the skeleton--the framework of the horse. Horsemen tend to talk about long heads, short backs, long backs, short hips, low knees, and high hocks. However, when we look at the skeleton, we find that all long bones are a standard proportion of total height or length. In other words, tall horses have long legs and short horses have short legs. Of course, there are exceptions, but when we take a group of horses of the same age and divide long bone length by height or length, we find the percentages to be the same (tables 1, 2, 3, and 4). We have to remember that every horse has the same number of bones and each bone is attached to a bone just like that of any other horse.

Obviously, there are some skeptics to this statement who say they know that they have seen short-neck, long-backed horses, etc. I agree that they have seen such animals. The reason for this is due to the movable angles that can be found in the skeleton. What I want is for you to be able to evaluate a horse based on the framework found in the skeleton. To begin with, examine the two skeletons found in figure 1. Which skeleton is the taller? Obviously, the one on the left. The amazing thing is that each skeleton has exactly the same length of bones. In one, we have straightened the shoulder and other movable angles. A straight shoulder and other angles have several effects on over-all appearance. First of all, the animal is taller. Secondly, the top to bottom line ratio of the neck changes. A horse with a long sloping shoulder will have a neck that is considerably longer on top than on the bottom. The straight-

TABLE 1. LEAST SQUARES MEANS FOR THE EFFECTS OF SEX ON BODY PROPORTIONS (PERCENTAGE OF BODY LENGTH)

Measurement	CV%	Stallions	Mares	Geldings
Length of head	3.55	38.05[a]	37.51[a]	38.03[a]
Width of head	3.93	14.44[a]	14.43[a]	14.28[a]
Length of neck	6.58	48.18[a]	47.94[a]	47.99[a]
Length of elbow to ground	5.07	60.97[a]	58.92[b]	59.13[b]
Length of knee to ground	4.27	31.08[a]	30.07[b]	30.12[b]
Length of hock to ground	4.02	40.71[a]	38.69[b]	39.29[b]
Height at withers	3.09	99.32[a]	98.42[a]	98.16[a]

TABLE 2. LEAST SQUARES MEANS FOR THE EFFECTS OF SEX ON BODY PROPORTIONS (PERCENTAGE OF HEIGHT AT WITHERS)

Measurement	CV%	Stallions	Mares	Geldings
Length of head	3.30	38.31[ab]	38.13[b]	38.75[a]
Width of head	3.94	14.55[a]	14.67[a]	14.55[a]
Length of neck	6.33	48.57[a]	48.73[a]	48.89[a]
Length of elbow to ground	3.70	61.38[a]	59.87[b]	60.25[ab]
Length of knee to ground	3.42	31.29[a]	30.55[b]	30.69[b]
Length of hock to ground	2.80	40.99[a]	40.32[b]	40.03[b]

TABLE 3. LEAST SQUARES MEANS FOR THE EFFECTS OF AGE ON BODY PROPORTIONS (PERCENTAGE OF HEIGHT AT WITHERS)

	Age Months			
Measurement	18-24	24-36	36-48	48 & Above
Length of head	38.05[a]	39.14[a]	38.47[a]	38.27[a]
Width of head	14.50[a]	14.53[a]	14.70[a]	14.73[a]
Length of neck	47.02[a]	48.52[a]	48.67[ab]	50.17[b]
Length of elbow to ground	59.75[a]	61.02[b]	59.63[a]	59.92[a]
Length of knee to ground	31.83[a]	30.77[b]	30.65[b]	30.28[b]
Length of hock to ground	40.99[a]	40.65[a]	40.07[b]	39.88[b]
Length of body				

ab = Means on the same line with different superscripts are different ($P<.05$).

TABLE 4. LEAST SQUARES MEANS FOR THE EFFECTS OF AGE ON BODY PROPORTIONS (PERCENTAGE OF BODY LENGTH)

Measurement	Age Months			
	18-24	24-36	36-48	48 & Above
Length of head	38.66^{a}	37.91^{a}	37.83^{a}	37.01^{b}
Width of head	14.78^{a}	14.34^{a}	14.46^{ab}	14.24^{b}
Length of neck	47.74^{a}	47.88^{a}	47.85^{a}	48.56^{a}
Length of elbow to ground	60.17^{a}	60.24^{a}	58.65^{ab}	57.95^{b}
Length of knee to ground	60.17^{a}	60.24^{a}	58.65^{ab}	57.95^{b}
Length of hock to ground	60.17^{a}	60.24^{a}	58.65^{ab}	57.95^{b}
Height at withers	101.70^{a}	98.71^{b}	98.35^{bc}	96.75^{c}

abc = Means on the same line with different superscripts are different (P<.05).

Figure 1.

shouldered animal's neck will have more of a 1:1 top to bottom line ratio of the neck. The straight-shouldered horse will have a longer back due to an upright shoulder. In addition, if we also straighten the angle of the hip, we further lengthen the back and add a short hip. If the pastern or the hock, or both, are made more upright, the knees and hocks appear higher.

Now, with this basic approach to conformation as viewed from the standpoint of the skeleton, look at specific conformation traits. The easiest place to start is at the head. The length of head is just as proportional as any other long bone. Tall horses have longer heads. You cannot logically compare the head of a 15.2 hand horse to the head of a 14.3 hand horse. It is true that young horses will grow to their heads. Because a horse has obtained approximately 80% of his length of head by 1 year of age, he will increase little in length as he gets older. Remembering that at the same age all long bones are the same proportion of height, this would also indicate that the little, short-headed yearling is probably not going to be very tall. I am not saying that the pig-eyed, roman-nosed, flop-eared yearling is going to have a beautiful head as an adult. I am only saying that length of head is an indicator of size or height. The head will tell us more about the horse than just height. We know that the eye is an indicator of disposition. The larger and quieter the eye, the calmer the horse. The sullen, unwilling horse is almost always accompanied by a small, unattractive eye. We know that the more white that is visible in the scalaria (unless it is a breed characteristic), the more nervous that horse is. The depth of mouth is a good indicator of potential lightness or softness of that horse's mouth. The more shallow the mouth, the more responsive the horse.

Before we continue, it is necessary to talk about proportion of muscling. All muscles are of the same proportion of total muscle mass in one horse as in another. This simply means that a horse with a large forearm will have a large gaskin and wide stifle in comparison to another horse of the same age with a smaller forearm. The exception to this would be those muscles developed by exercise. However, it is difficult to increase the muscle mass of the horse through exercise because to increase muscle mass, you must have muscle tissue break-down. To have muscle break-down, an animal must work beyond fatigue. The horse, unlike the human athlete, stops when he gets tired. It is possible to tone muscle in horses and give the fit horse the appearance of a tighter, fitter muscle. It is incorrect to call a horse a long-muscled horse or a short-muscled horse. All muscles are attached at exactly the same place on the bone. Short bunchy-muscled horses are nothing more than a short horse that is heavy muscled. A long-muscled horse is just a taller, lighter-muscled horse. The perfect example is the old-time quarter horse as compared to the thoroughbred. One animal was short and heavy muscled and the other was tall and light muscled. However, the muscles attached at the same place. This information should be nothing new, only a new way to look at the situation.

The neck is one of the most talked-about anatomical points on the horse. Certainly we would all like horses with nice trim necks; however, there are at least two areas that are more important than just overall trimness. One is the throatlatch. If a horse is going to bridle and be flexible, he must be trim and neat in the throatlatch. Too often, however, we look only at the bottom of the throatlatch and not the entire groove from the bottom of the ear to the bottom of the neck. The second important place to evaluate the neck is at the neck-shoulder junction. As previously mentioned, the shoulder is most important to overall conformation. Certainly those of us who ride know the form-to-function aspect of a sloping shoulder vs a straight shoulder. One of the most common mistakes made in horse judging is to look at a short-legged, heavy-muscled horse and compare him to a long-legged, lighter-muscled horse, and then conclude that his neck is too short and thick compared to the other horse. Muscle and length of neck is proportionate with the rest of the body.

The ideal horse is sharp at the wither and short in his top line. The secret to a short top line lies in the sloping shoulder already discussed.

The ideal horse is said to stand perfectly straight when viewed from behind. However, we find that most horses, unless artificially straightened, tend to be slightly toed-out behind. This allows the horse to have a longer stride because he will clear his rib cage more easily than the horse that is perfectly straight behind. Perhaps much of the short stridedness we often see in so-called halter horses is because they are "too perfect" in their conformation.

Regardless of breed, we do want to see a horse that is full at the quarter when viewed from the side and wide in the stifle when viewed from the rear. Fullness of quarter is initially due to the angle of the stifle. The horses that have the hind-quarter appearance of the nickel ice cream cone certainly are predisposed to stifle unsoundnesses. When a horse is viewed from the rear, the widest point should be the stifle. If a horse is wider at the top of his hip than he is at the stifle, it indicates one or a combination of three things: (1) he is too fat (horses tend to deposit fat from the top down), (2) he stands too close behind, and (3) he is light-muscled. Certainly any combination of these factors is possible.

Now, let's review the muscle types. In the quarter horse industry, we tend to work under the assumption that "more is better": more muscle, more height, more total volume. Have you ever seen a horse that was too light muscled? You automatically say you have. Let's consider the two basic types of muscle fibers in human athletes: fast twitch and slow twitch. Those with slow-twitch muscles tend to be lighter-muscled individuals with much more endurance. The fast-twitch muscled individuals tend to be heavier muscled and have more power in short bursts. The same is true in horses. We also know that muscle mass decreases mobility. A weight lifter is strong but generally lacks maneuverability.

We have spent little time talking about straightness of travel. Everyone knows that the ideal horse is straight and stands on a straight column of bone. We know that the horse is an athlete and must stand straight to travel straight. The age-old argument is that we've all known of a crooked horse that stayed sound under the most strenuous of conditions. What I would like to challenge you to do is evaluate the horses that have broken down and then tell me how many straight ones you found. Most horses that break down are the result of some deviation away from ideal feet and leg structure.

In summary, this presentation has not been an effort to review the age-old straight-legs, and true-travel talk. What I hope to provide is a way to (1) think about the way you evaluate a horse, (2) think about why you evaluate it the way you do, and (3) see if your criteria for evaluation are based on sound facts.

Part 4

ANATOMY, PHYSIOLOGY, AND REPRODUCTION

13

SIMILAR ANATOMICAL AND DEVELOPMENTAL CHARACTERISTICS OF SHEEP, CATTLE, AND HORSES

Robert A. Long

Each species of animal with which we concern ourselves in animal science is constructed according to a fixed plan. Therefore, we consult a textbook on the anatomy of the bovine and have confidence that it applies to all cattle. There are variations, of course, due to differences in genetics, nutrition, age, sex, and even disease. However, the two basic tissues of bone and muscle are always present in the same general pattern. This constancy of structure or organization of tissues is of great value to us in the evaluation of cattle both alive and dead.

We should also realize that there is great similarity in this overall plan of structure among different species. All mammals have essentially the same skeleton and the same muscles attached to the skeleton. The similarity between cattle, sheep, and horses is very pronounced and, even though small differences exist, the factors that affect the growth and development of the three species are identical. Let us examine these common traits and discuss how they might help us in evaluating livestock.

THE SKELETON

All cattle, sheep, and horses are made according to the same general plan or design. Their skeletons are composed of the same number of bones, and the general shape of each bone is the same in all three species. Also, the percentage of total weight or linear size that each bone represents of the whole skeleton is constant. Butterfield (1964), Kauffman (1973), Ramsey (1976), Heird (1971), and Mukhoty (1982) are in agreement with this statement. This simply means that, for all practical purposes, if we can measure one bone in the skeleton, we can determine the dimensions of the whole skeleton. In fact, in the case of the long bones of the limbs, the measurement of one bone is much more accurate than an overall measurement of height. Most people currently measure frame size (or think they do) by measuring height at the withers and/or hips in cattle and sheep or hands at the withers in horses. Figure 1 illustrates how

Figure 1. Identical skeletons showing the effect on height at the withers and hips of changing the angle of movable joints of the long bones of the legs.

misleading this can be. Note that these are identical measurements of these two skeletons, which is exactly the case. If one bone is longer, every other bone in the skeleton will be longer and proportionately so. Remember, this is only valid if the cattle are of the same age and sex. You must compare bulls with bulls, steers with steers (castrated at the same age), and heifers with heifers, because at puberty the level of sex hormone production changes greatly; this results in closure or calcification of the epiphyseal groove, and the length of the leg bones stops increasing. Thus, steers and wethers generally are taller and shorter bodied than their identical twins that have been left sexually intact. We do not notice this as often in horses since they are usually castrated after sexual maturity. Of course, nutrition has an effect on the age at which animals reach puberty; thus, for comparisons to be legitimate, they must be made under uniform conditions of nutrition (or for that matter under uniform total environment).

This is of less concern in castrates such as steers and wethers. Skeletal growth or bone formation in growing animals takes priority for nutrients over that of muscle growth and fat deposition. Therefore, regardless of plane of nutrition, if we compare animals of the same age, their frame size has probably increased according to genetic potential and is a good measure of what their mature size will be. When compared at the same age, the larger the frame, the larger it will be at maturity and the longer it will take to reach that point. Also, we know that as an animal approaches maturity, he begins to deposit fat in the muscle, which is the marbling that gives his carcass a quality factor. This is the very basis of the USDA feeder cattle grades that are discussed by the author in another presentation at this school.

About ten years ago, a national field day was jointly sponsored by a different breed association and the University of Wisconsin in each of three consecutive years. Each breed selected different "types" of steers which were placed on feed and were slaughtered when ready for market. A field day was built around the data collected. Some good things came out of these sessions but, unfortunately, the most attention was received by the profile drawings in figure 2, which is entitled "Body Types." Note that "body type #1" is shortbodied and lowest and shows heavy development in the dewlap, brisket, and belly, and great proportional depth of body. Also, observe that "body type #5" is tall and long and is trim-fronted and tight-middled. The implication here is that all small-framed cattle are wasty and fat and that all large-framed cattle are trim and desirable. Nothing could be farther from the truth.

I do not believe that such a thing as a body type exists. I believe, and will offer evidence to prove, that every frame size of beef animal can and does occur with every possible combination of fat and muscling. Some small-framed cattle are highly desirable in composition--some are not. Some large-framed cattle are desirable in composition--some are not. The same can be said for any frame size (Long, 1982).

I want you to look at the data from three steers in table 1. Their weight is very different but their skeletons are practically identical in size, which is, of course, their frame size. Now examine the dissection data in table 2. Not only were their skeletons identical in linear measurements, but their skeletons weighed the same. However, here the similarity stops. Note the tremendous difference in muscle both in total weight and as a percentage of the carcass of the #1 steer. This gives a muscle:bone ratio of just twice as much for the heavily muscled steer as for the thinly muscled one. Fat varies only a little in this case, but keep in mind that it would be easy to put together a large group of steers with identical skeletons that vary widely in fat and muscle composition. Table 3 lists the conventional carcass measurements. This table makes two major points.

1. The Yield Grade formula ranked these three steers essentially the same, which is obviously in error. This is because the formula was constructed with conventional British breeds that did not offer the range in muscling that we have here. It underevaluates the heavily muscled #1 steer, overevaluates the thinly muscled #3 steer, and does a good job on #2.
2. The frame size or skeletal size of these steers had nothing to do with the desirability of their carcasses.

I would hope that your conclusion would be something like mine that simply stated is: Why anyone would use frame size in the evaluation of cattle for slaughter is beyond me. Yet, that is exactly what takes place in the majority of steer shows in this country--they put the tall ones up. Think what this means. The cattle are shown by weight and

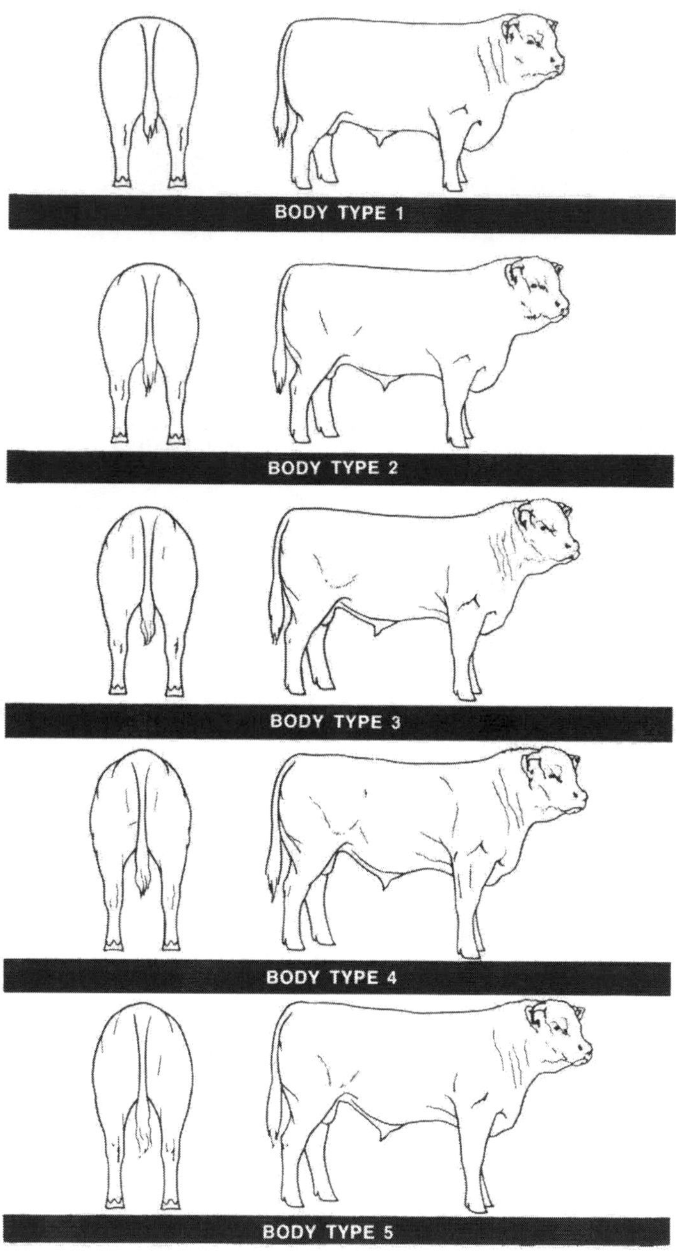

Figure 2. Body types.

TABLE 1. MUSCLE:BONE RELATIONSHIPS AMONG SLAUGHTER STEERS LIVE MEASUREMENTS

Steer #	1	2	3
Live wt (lb)	1450	1300	1005
Length of body (in.)	60.23	60.23	59.84
Rump length (in.)	20.07	20.07	20.47
Ht. withers (in.)	51.96	51.57	52.36
Ht. hips (in.)	53.54	53.14	53.93

TABLE 2. MUSCLE:BONE RELATIONSHIPS AMONG SLAUGHTER STEERS DISSECTION DATA

Steer #	1	2	3
Lb of bone	64	68	67
% bone	13.1%	16%	23%
Lb of muscle	320	262	168
% muscle	66%	63%	59%
Lb of fat	104	81	53
% fat	21%	19%	18%
Muscle:Bone	5.01	3.88	2.52
Mucle:Bone IM fat included	5.16	3.94	2.61

TABLE 3. MUSCLE:BONE REALTIONSHIPS AMONG SLAUGHTER STEERS CARCASS MEASUREMENTS

Steer #	1	2	3
Carcass wt	976	820	570
Dress %	67%	64%	57%
Maturity	A^{75}	A^{50}	A^{75}
Marbling	$Small^{30}$	$Slight^{80}$	$Slight^{60}$
Quality grade	Ch^{-}	Gd^{+}	Gd°
Fat thickness (in.)	.3	.3	.12
Rib eye area (sq in.)	18.1	14.3	9.9
% KHP	3.0%	2.5%	2.5%
Yield grade	1.8	2.3	2.3

most of them have been fed and managed in such a way that they are not excessively fat. Therefore, placing the tall, big-framed steers up in class and the small-framed ones down means that selection is against muscle or meat, which makes no sense at all in the beef-production business. The placing of the tall ones of the same weight on top of the class further complicates the situation. Large-framed cattle mature later, which fact decreases the chances of the large-framed steer making the choice grade.

Unfortunately, it is commonly believed that size of the skeleton is of great value in evaluating other classes of livestock as well. As already stated, it tells us something about expected time of maturity but nothing about composition. Sheepmen are particularly prone to make the statement, "He has a longer hindsaddle with more weight in the high-priced cuts." Sheep are like cattle. They are all in the same proportion. Therefore, if one has a longer hindsaddle the rest of his skeleton is also longer. However, this does not mean that there is more meat or muscle on it (or more or less fat, for that matter.)

THE DETERMINATION OF MUSCLING

Now let us devote our attention to differences in muscling. First, dispose of the old, often-used phrase, "More weight (or more meat) in the high-priced cuts." This statement originated years ago when some cattlemen decided that more muscle in the rib, loin, and round--and less in the rest of the carcass--would be a great thing. It would be a good thing but, unfortunately, it isn't possible. The research data of Butterfield of Australia, Berg of Canada, and several people in this country show that different breeds of cattle (British, European, Brahman, dairy breeds, wild cattle, and water buffalo) have essentially the same relationship between the various muscles. This does not mean that we cannot increase their muscle, but it does mean that we cannot increase one muscle or group of muscles in a steer without increasing all of them. This should not be discouraging; indeed, it is most fortunate. If we can measure the amount of muscle in one part of the animal, we can depend on proportional development in all other parts. Therefore, we can appraise the muscling of a steer by looking at the forearm, at the muscle working in his top as it moves, and at the thickness through the lower quarter or stifle in proportion to that through the rump. Wide stance both in front and from the rear tells us the steer has a lot of muscle.

We know that the approximate shape of the forearm bone is that of a cylinder or piece of pipe which is approximately uniform in circumference and width. Since this is true, any change in shape or increase in width of the forearm region comes from muscle, because observation and experience have shown us that little fat is deposited here. Likewise, practically no fat is deposited over the outside, lower round; so, as we stand behind a steer, a horizontal plane that would pass through the stifle joint should be the thickest place in the steer. If it is not, he lacks muscling, or has heavy deposits of fat over the rump and on top of the round, or a combination of both.

We also know by looking at the bovine skeleton that the foreleg is attached to the rest of the skeleton only by muscle. The amount of muscular development determines the space between the leg bones and the rib cage and, therefore, how far the front legs are held apart. Likewise, the muscular development between the hind legs determines how far apart the steer stands as viewed from the rear. Therefore, the heavily muscled animal will stand wide when viewed both from the front and from the rear.

We hear a great deal about the "kind" of muscle on cattle, sheep, and horses, and the favorite terms are "the right kind of muscle" or "that good, long, smooth muscle." Fortunately, there is only one "kind" of muscle. It is composed of muscle fibers bundled together by connective tissue and attached by connective tissue and tendons to other muscles and to the skeleton. The "length" of the muscles is determined by the size of skeleton since each muscle is attached to the skeleton at the identical spot in each specie. Therefore, animals of equal frame size have the same length of muscle. "Smooth Muscle" is a term used to describe animals that have a layer of subcutaneous fat, or are thinly muscled, or both.

Just as the skeleton is in the same proportion, each muscle in its anatomic entirety represents a constant percentage of the total muscle mass. This is well-established by both Butterfield (1964) and Kaufman (1976) and is the basis for estimating total muscle by examining an animal for degree of muscling over the forearm or through the stifle. Let's face it, a steer or wether cannot produce an excellent carcass without being well muscled. This, of course, adds to his weight and when finish is constant the heavily muscled beast far outweighs the smooth-muscled one of the same frame size. Therefore, a large-framed animal will be considerably heavier than the packer wants if his composition is correct.

You will recall that we have pointed out that both the skeleton and musculature occur in essentially the same proportion in each specie. This results in a near constant percentage of carcass weight in each of the wholesale cuts. For example, a heavily muscled Limousin steer has the same percentage of hindquarter as the thinnest-muscled Jersey steer. Cattle just don't possess "more weight in the high priced cuts." The difference is in the percentage of meat, fat, and bone in each cut. The data that illustrate the constant proportionality of skeleton and muscle have often been misinterpreted to mean that all cattle are the same--and, if you measure them, the longest or largest are the best. This is in complete error. You must know muscle:bone ratio and degree of fatness to know composition. This is also true in sheep and horses.

REFERENCES

Butterfield, R. M. 1964. Relative growth of the musculature of the ox. In: D. E. Tribe (Ed.) Carcass Composition and Appraisal of Meat Animals. The Commonwealth Scientific and Industrial Research Organization, Melbourne, Australia.

Heird, J. C. 1971. Growth parameters in the quarter horse. M.S. Thesis. Univ. of Tennessee.

Kaufman, R. G., R. H. Grummer, R. E. Smith, R. A. Long and G. Shook. 1973. Does live animal and carcass shape influence gross composition? J. Anim. Sci. 37:1142.

Kaufman, R. G., M. D. Van Ess and R. A. Long. 1976. Bovine compositional interelationships. J. Anim. Sci. 43:102.

Long, R. A., and C. B. Ramsey. 1982. Visual scores and linear measurements of slaughter cattle as predictors of carcass characteristics. Proceedings, Western Section, American Society of Animal Science. Vol. 33.

Mukhoty, H. and H. F. Peters. 1982. Influence of breed and sex on muscle weight distribution of sheep. Proceedings, Western Section, American Society of Animal Science. Vol. 33.

Ramsey, C. B., R. C. Albin, R. A. Long and M. L. Stabel. 1976. Linear relationships of the bovine skeletons. J. Anim. Sci. 42:221 (Abstr.).

14

HORMONAL REGULATION OF THE ESTROUS CYCLE

Roy L. Ax

INTRODUCTION

All reproductive events are regulated by hormones. In simple terms, if the organs of reproduction correspond to the "plumbing," the endocrine system can be called the "wiring." A delicate balance exists between the nervous system and the endocrine system. We are entering an era in which artificial control of the estrous cycle with hormones promises to become more commonplace. Thus, hormones can be considered valuable management tools. If producers are to use the tools effectively, we must develop a better understanding of the complex hormonal interrelationships between the hypothalamus, pituitary, and ovary.

HORMONES DEFINED

A classical definition for a hormone is that it is a substance produced in one tissue that is transported to another tissue to exert a specific effect. (Some of the confusion about hormone actions should be clarified in the next section). Hormones have many chemical classifications; some of the most common reproductive hormones are briefly described here. Gonadotropin-releasing hormone (GnRH) is composed of amino acids and is thus a polypeptide in nature. The follicle-stimulating hormone (FSH) and luteinizing hormone (LH) are glycoproteins. This means they are composed mostly of protein, with some carbohydrate attached to the protein. Estrogen and progesterone are steroids that are synthesized from cholesterol. Prostaglandins are produced from a fatty acid--arachidonic acid. The diversity of the composition of hormones leads to the variation in their biological functions. Most hormone concentrations are in billionths or trillionths of a gram per milliliter of blood.

HOW DO HORMONES WORK?

The fact that a hormone is produced by a tissue does not necessarily imply that it will exert a physiological effect somewhere else. The ability of one tissue to respond to a particular hormone rests in whether that tissue possesses a receptor to the particular hormone. A receptor functions as the lock, and the hormone functions as the key that fits the lock. Therefore, as an example, if a tissue is going to respond to estrogen, its cells must possess estrogen receptors. After a hormone is bound to its receptor, a cellular response is initiated in the target tissue. A target tissue may possess receptors for several different hormones, and exposure to the various hormones can modulate the final response.

THE HYPOTHALAMUS

The hypothalamus is located at the base of the brain. It contains nerve endings to integrate sensory information and sorts out hormonal signals as well. The major reproductive hormone of the hypothalamus is gonadotropin-releasing hormone (GnRH) that is sometimes called luteinizing-hormone-releasing hormone (LHRH). For purposes of this discussion we will use GnRH nomenclature. GnRH is transported in blood vessels to the pituitary gland to regulate secretion of FSH and LH from the pituitary.

THE PITUITARY

The pituitary is positioned underneath the hypothalamus directly above the roof of the mouth. The major reproductive hormones produced in the anterior lobe of the pituitary are called gonadotropins, which means to stimulate the gonads. Follicle-stimulating hormone (FSH) and luteinizing hormone (LH) are the two gonadotropins that regulate the ovary. They are secreted by the pituitary and transported in the circulation to the ovary where they interact with their respective receptors to affect ovarian functions. The main action of FSH is to initiate growth of follicles on the ovary. Continued follicle growth depends on the presence of both FSH and LH. The major effect of LH is to promote ovulation, but there is increasing evidence that FSH can exert a major influence to facilitate ovulation.

THE OVARY

The ovary has two biological functions: (1) to provide the eggs (ova) for the female genetic contribution to the next generation, and (2) to produce hormones to coordinate behavioral changes with ovulation and prepare the reproductive tract for pregnancy.

Estrogen is the hormone produced by follicles as they develop on the ovary. As the predominant follicle or follicles approach ovulatory size, the increased amounts of estrogen are transported to the hypothalamus to cause behavioral heat. The pituitary also responds to the elevated estrogen by releasing a surge of LH which leads to ovulation. Thus, estrogen coordinates behavioral acceptance of a male when the egg will be released into the female tract. This is Mother Nature's attempt to ensure that the probability of fertilization occurring is maximized.

After ovulation has occurred, the tissue that a moment ago was a follicle starts a dramatic change into becoming a corpus luteum (yellow body). The corpus luteum produces progesterone to prepare the female tract for a possible ensuing pregnancy. The corpus luteum forms, regardless of whether or not mating occurs in farm animals. If the corpus luteum remains functional due to pregnancy occurring, the sustained production of progesterone prevents cyclicity. Until the corpus luteum regresses, the typical pattern of cyclic hormonal changes is absent.

THE FOLLICLE

In a simple sense, the follicle is the dwelling of the ovum. Ovulation of a follicle is the exception rather than the rule because over 99% of all potential oocytes are never shed from the ovary. This loss is called atresia. Atresia can occur at any time during follicle growth. When animals are injected with gonadotropins to induce superovulation, some follicles that would have undergone atresia are rescued. This supports the hypothesis that continued follicle growth is dependent upon continued exposure to gonadotropins and the presence of gonadotropin receptors in the follicle.

OVULATION

Once the LH surge has been elicited from the pituitary, the follicle starts to undergo a series of changes to prepare for impending ovulation. The cells lining the inside of the follicle begin to luteinize and secrete progesterone as the major steroid rather than estrogen. The oocyte commences its maturational steps to get it in the proper meiotic configuration for chromosome pairing with the meiotic contribution from a sperm.

Enzymes are activated to degrade the follicle wall and permit the egg to pass into the oviduct. Biochemical studies have pointed to FSH being responsible for stimulating production of those enzymes. FSH also promotes the spreading apart of cells that are tightly surrounding the egg, which then leads to some of the subsequent maturational changes in the egg. Prostaglandins are required for normal ovulation. Substances known to inhibit prostaglandin formation prevent ovulation. Due to the enzyme, steroid, and prostaglandin effects, a hypothesis was formulated comparing the ovulatory process to an inflammatory reaction.

THE CORPUS LUTEUM

The scar tissue remaining after ovulation becomes the corpus luteum; this endocrine tissue has been studied extensively. Low amounts of LH from the pituitary are essential for establishment and continued function of the corpus luteum. In all livestock species, the corpus luteum functions to maintain early pregnancy by secreting progesterone. Progesterone prevents cyclicity. The placenta of the developing fetus eventually sustains the pregnancy by producing progesterone in the bovine, equine, and ovine. In the porcine, the corpus luteum is required to support the entire gestational period.

If cyclicity is to resume, the corpus luteum must regress and cease progesterone production. In pregnant animals, an embryonic signal leads to maintenance of the corpus luteum, (discussed in the next section). In nonpregnant animals, the uterus recognizes the absence of an embryo and secretes prostaglandins. Those prostaglandins are transported to the corpus luteum and cause it to regress. Thus, the inhibitory effects of progesterone are removed, new follicles start to develop, and heat occurs in a few days.

PREGNANCY RECOGNITION

The corpus luteum continues to function to provide a signal from the embryo to the dam. These signals are hormones identified in the human as chorionic gonadotropin (hCG) and in the mare as pregnant mare serum gonadotropin (PMSG). These two gonadotropins can be used to regulate the estrous cycle of animals. PMSG is biologically similar to FSH, and hCG exerts an action similar to LH. Thus, PMSG can be used to induce follicle growth and hCG will promote ovulation of these follicles. However, both PMSG and hCG are recognized as foreign proteins by livestock, and the animals build up antibody resistance to them, if they are used too frequently.

An ideal pregnancy test would be the identification of the embryonic signal in the bovine, ovine, and porcine. Experiments have shown that embryonic extracts will maintain a corpus luteum in an animal that has not been bred. However, even with sophisticated biochemical tests, specific signals from embryos in the dam's circulation have not been detected. The livestock industry could benefit significantly from pregnancy tests of these types if they are ever developed successfully. Since an embryonic signal would have to be apparent to prevent a subsequent heat in the dam, a pregnancy test would also pinpoint which animals would be returning to heat within a few days.

HORMONAL REGULATION OF THE CYCLE

The preceding sections indicate that follicle growth, ovulation, and corpus luteum formation are a dynamic process of sequential steps in an intricate balance. Administration of a hormone to mimic the effect of that hormone in the animal can be used to regulate the cycle. If hormone administration is to produce the desired result, it must be given at a time that is physiologically compatible with the cycle. The common hormones that have been used experimentally or commercially are progesterone-like drugs (progestins), GnRH, and prostaglandins.

Progestins

These compounds were the first to be experimentally employed to regulate the cycle. Progestins have been injected, fed, implanted, or administered via vaginal sponges. Regardless of what stage of the cycle an animal is in when the progestin commences, cyclic fluctuations in other hormones are arrested. As long as progestin is administered, cyclicity ceases. Removal of the source of the progestin results in renewed follicle growth, and estrus, within a few days. Field trial data suggest that fertility at the first estrus after progestin withdrawal is lowered. Thus, it is usually recommended that breeding be done at the second estrus, since the cycles of the animals will still be in close synchrony.

Prostaglandins

These compounds have largely replaced the use of progestins because (1) only one or two injections are required, and (2) fertility is not affected by use of prostaglandins. Prostaglandins are only effective if the animal possesses a functional corpus luteum. Contrary to some opinions, prostaglandins are not a heat-inducing drug. Rather, they cause a corpus luteum to regress, and the animal secretes her own gonadotropins to regulate the ovary and cause a physiological heat. Success has occurred regularly by breeding animals at a predetermined time after prostaglandin injection. Greater success in conception rates can occur if animals are watched for estrual behavior after receiving prostaglandins and are bred in relation to standing heat. Care must be exercised with prostaglandins, because injections into an animal with a functional corpus luteum sustaining a pregnancy could induce an abortion.

GnRH

GnRH is composed of 10 amino acids. It can now be chemically synthesized in a laboratory, and this has permitted chemists to develop some powerful analogs. There are no noticeable ill effects from administering GnRH; its action is to promote a release of gonadotropins from the pituitary. Maximum gonadotropin output occurs approximately 2 to 4 hours after GnRH injection.

A common use for GnRH is to initiate cyclicity in animals with anestrous. GnRH is the most widely used therapy for treating cystic ovarian degeneration. Cystic ovaries usually result from inadequate gonadotropin production. Thus, GnRH triggers release of gonadotropins to restore ovarian function.

A new use for GnRH is for injection after prostaglandin administration; the interval to the gonadotropin surge, and hence, ovulation, can be coordinated more closely. This reduces the variation in time between prostaglandin injection and standing heat that is ordinarily seen among animals.

We have an ongoing study at the University of Wisconsin to evaluate the efficacy of GnRH injections at the time of insemination in dairy cattle. The heifers receiving GnRH have shown no advantages over heifers receiving the saline control. In lactating cows, administration of GnRH 14 days postpartum or at the first artificial insemination has improved first-service conception rates by 15% to 19%. In cows presented for third service (and thus classified as "repeat" breeders in commercial herds) conception rates were about 30% higher for cows that received the GnRH. The physiological effect elicited by GnRH has yet to be experimentally established. We have postulated that gonadotropins produced in response to GnRH cause a corpus luteum to form that may have otherwise been deficient and led to early embryonic death. GnRH could also promote what would have been a delayed ovulation to occur sooner or have a direct effect on the ovary. The lactational stress imposed on a dairy cow may make her unique to respond to GnRh in this manner. Experiments with other farm animals are needed to determine if similar effects result.

SUMMARY

The reproductive cycle is regulated by fluctuations in different hormones. The cycle can be regulated by administering hormones to mimic the effect that would occur in the animal. Therefore, producers have endocrine tools to assist them in managing their animals. For maximum success the producers must understand how the hormones work biologically and realize that they are powerful drugs. We will see an increasing frequency of producers regulating the reproductive cycle to maximize reproductive efficiency.

REFERENCES

Britt, J. H. 1979. Prospects for controlling reproductive processes in cattle, sheep, and swine from recent findings in reproduction. J. Dairy Sci. 62:651-665.

Britt, J. H., N. M. Cox and J. S. Stevenson. 1981. Advances in reproduction in dairy cattle. J. Dairy Sci. 64:1378-1402.

Foote, R. H. 1978. General principles and basic techniques involved in synchronization of estrus in cattle. Proc. 7th Tech. Conf. on Artif. Insem. and Reprod., Nat'l Assoc. Anim. Breeders, pp 74-86.

Hansel, W. and S. E. Echternkamp. 1972. Control of ovarian function in domestic animals. Amer. Zool. 12:225-243.

Jones, R. E. (Ed.) 1978. The Vertebrate Ovary. Comparative Biology and Evolution. Plenum Press, New York.

Lee, C. N., R. L. Ax, J. A. Pennington, W. F. Hoffman and M. D. Brown. 1981. Reproductive parameters of cows and heifers injected with GnRH. 76th Ann. Mtng. of the Amer. Dairy Sci. Assoc. Abstract P228.

Maurice, E., R. L. Ax and M. D. Brown. 1982 Gonadotropin releasing hormone leads to improved fertility in "repeat breeder" cows. 77th Ann. Mtng. of the Amer. Dairy Sci. Assoc. Abstract P233.

Nalbandov, A. V. 1976. Reproductive Physiology of Mammals and Birds. W. H. Freeman and Co., San Francisco.

15

PROSTAGLANDIN F2 ALPHA AS AN AID IN BREEDING MARES

James W. Lauderdale

This paper presents information concerning the use of Prostin F2 alpha® as a tool in breeding management programs.

USE OF PROSTIN F2 ALPHA TO MANAGE ESTRUS

Even good mare-management programs have problems. One problem is failure to detect estrus in mares even though the mares are having ovarian cycles. In other words, these mares are forming corpora lutea in the ovaries but do not express behavioral signs of estrus. If the "good management" is decreased, the percentage of mares that are not detected in estrus increases. Some mares form a corpus luteum that persists for extended intervals. It is important to determine whether or not the mare is pregnant if she has a retained corpus luteum. Pregnancy is the best way to maintain the corpus luteum, but if Prostin F2 alpha is administered to the pregnant mare, she probably will abort.

Another management problem is associated with estrus detection and efficient use of the stallion. On any given day, if there is a large band of mares, the efficiency of estrus detection is much more accurate in the morning when the stallion is fresh than by late in the day when the stallion is tired. After teasing many mares the stallion may lose interest, thus we do not know if the mare was not in estrus because the stallion was not aggressive or if she really was not in estrus. Also, Prostin F2 alpha can be used to regress the corpus luteum of estrous cycling mares to predict, or preprogram mares to be in estrus. This program is most effective if the mares are injected between days 6 and 8 after ovulation.

Prostin F2 alpha can be a valuable asset to breeding management program. By producing luteolysis (corpus luteum regression) in mares, it aids in the detection of estrus and influences the onset of estrus in estrous cycling mares and in difficult-to-breed mares that have a functional corpus luteum (CL).

The estrous cycle must be understood to determine when Prostin will be effective. During the first 4 or 5 days after ovulation, when the corpus luteum is developing, Prostin F2 alpha is ineffective--it will not alter the life span of the CL or alter the length of the estrous cycle. However, once the CL is formed, after about 5 days after ovulation, Prostin F2 alpha will regress the corpus luteum just as though the signal had been turned on in the uterus, which normally will regress the CL and the mare will return to estrus about 3 days after injection. Prostin F2 alpha approaches 100% effectiveness when administered on day 6 or later after ovulation, but we know that Prostin F2 alpha will be effective on days 4 or 5 after ovulation in some mares. With accurate measurement of ovulation, administration of Prostin F2 alpha on days 5 to 6 after ovulation is almost 100% effective. In the absence of palpation and daily teasing, Prostin F2 alpha should be administered 8 or 9 days after the last day of estrus to be sure to include those mares who will normally ovulate after they go out of heat. Thus, Prostin F2 alpha is effective only in mares with ovaries that have a CL.

Prostin F2 alpha has been investigated extensively for treatment of clinically anestrous, or difficult to breed, mares. We have classified arbitrarily various types of difficult breeding mares into four general categories: 1) barren or maiden, 2) lactation anestrus, 3) bred but neither pregnant nor returned to estrus (pseudopregnant), and 4) anestrus following abortion or resorption. Generally, we are talking about clinically anestrous mares, i.e., mares that had not been detected in estrus for various extended intervals of time during the breeding season.

In one study, 73 mares had not been detected in estrous-type activity for an average of 73 days. Estrous activity was being tested approximately three times a week by teasing with the stallion and routine rectal palpation by the veterinarian on the stud farm. Of the 73 treated mares, 73% were detected in estrus as defined by standing or expressing signs to the stallion. Remember that these mares had not been detected in estrus for over 2 months prior to treatment. The average duration of estrus was 3.3 days, and mares were detected in estrus between 2 and 20 (average = 4.4) days after injection. The percentage of mares ovulating was 86, or about 13 percentage points higher than the percentage of mares detected in estrus. Ovaries of these mares were being palpated per rectum, usually at every-other-day intervals. Thus, a portion of the mares were determined to have had a developing follicle that reached ovulatory size and ovulated even though the mare was not detected in estrus. Of the 73 mares originally treated, pregnancy data was available on 54. The difference between 54 and 73 was that some of the mares were lost to follow-up because they were moved from one farm to another, some of the mares were not inseminated or bred because they had two follicles as opposed to one follicle, and some of the mares were not inseminated for other various reasons. Of the 54 mares, 56% were pregnant to breeding at the estrus that occurred following administration of Prostin F2 alpha.

A second study was designed to investigate further the effects on the reproductive cycle of clinically anestrous mares administered Prostin F2 alpha. A total of 157 Standardbred, Thoroughbred, and Arabian mares under supervision of investigators in seven locations were injected intramuscularly once with Prostin F2 alpha. Mares had not been detected in estrus prior to treatment for an average of 58 days during the breeding season and were classified as clinically anestrous. Follow-up data were available on 134 of the 157 mares. Average intervals after injection were: 3.6 days to first day of estrus; 6.8 days to ovulation; and 5.6 days to first breeding. Average duration of estrus was 4.6 days.

One hundred-eight (108) of the 134 mares (81%) were detected in estrus within 10 days after treatment and were bred in association with that estrus. Fifty-seven (57) percent (62 of 108) were pregnant to that breeding. This percentage (57%) reflects fertility of mares treated with Prostin F2 alpha. Pregnancy rate was 46% based on all mares treated for breeding in association only with estrus detected within 10 days after PGF_2alpha treatment (62 pregnant of 34 treated). This percentage (46%) reflects the effectiveness for establishing pregnancy in clinically anestrous mares within a few days after treatment with Prostin F2 alpha. Average number of breedings was 1.3 during estrus detected within 10 days after treatment.

Consistency of results among the various studies and level of response as measured by return to estrus and pregnancy rates support the conclusion that Prostin F2 alpha is effective. To date, no data are available to indicate that Prostin F2 alpha will grow follicles in the ovaries of mares that are anestrous because of failure of ovarian function. If a mare does have a functional CL, Prostin F2 alpha will be effective no matter what month of the year treatment is given.

Based on data derived from toxicology studies and from biological studies, Prostin F2 alpha has at least a 10- to 20-fold safety margin in the mare. The active component of Prostin F2 alpha, prostaglandin $F2\alpha$, is a natural compound that is found normally in the body; the mare removes the injected Prostin F2 alpha rapidly (within an hour or so) from her body.

CONCLUSIONS

Data presented demonstrated that Prostin F2 alpha is luteolytic in the mare and also that Prostin F2 alpha has applicability to assist in management of reproduction in the mare. One broad application is in the population of mares that are under good teasing programs, are not detected in estrus, but do have a corpus luteum in one ovary--those mares that we call clinically anestrous. Such corpora lutea may be retained, or they may have been produced in a normal 21-day cycle, but the mares have not been detected in behavioral estrus. Prostin F2 alpha administered to these mares will regress the corpus luteum and a high percentage of the mares (70% to 80%) will be detected in estrus. If the mares are bred, acceptable pregnancy rates can be achieved by breeding at estrus detected within 10 days after injection of Prostin F2 alpha. During the breeding season, good success can be anticipated in terms of return to estrus, percent ovulating, and pregnancy rate. However, we expect that percentage of treated mares pregnant after Prostin F2 alpha treatment will probably be lower than that reported if mares are treated during seasons of either anestrus (September to December) or transition from anestrus to estrus (December to March)--because of a lower percentage of clinically anestrous mares would be expected to have a corpus luteum. Prostin F2 alpha can also be used to schedule estrus of mares to more efficiently manage a breeding program.

Prostin F2 alpha is a natural compound and is safe for use in mares at luteolytic levels (5 mg to 10 mg). There is a safety margin of at least 10- to 20-fold. The side effects observed have been transient in all cases and have not been detrimental to the animal.

16

ANALYSIS OF REPRODUCTIVE PATTERNS IN MARES BY RADIO TELEMETRY AND ESTROPROBE

Joe B. Armstrong

Mares typically show the poorest reproductive efficiency of most domesticated animals. This can be attributed to many physiological factors peculiar to the mare and also to an inability of breeding-farm managers to deal effectively with these problems. A national average for brood mares having live foals is about 50% according to Evans (1977b). However, Ginther (1979) states that a more accurate average would be approximately 80%. The basis for the difference between these two figures is that Evans considered all mares and breeding operations in the U.S. whereas Ginther only considered those breeding operations that are a business and not those operations that are pleasure oriented. Ginther (1979) attributed the difference in reproductive efficiency between mares and other domesticated females to the unusual reproductive patterns of the mare. These patterns include the long estrous period within which ovulation can occur at any time, a seasonal ovulatory pattern with a prolonged estrous prior to the first ovulation, and sporatic estrous behavior to name a few. The reason for these unorthodox patterns is not readily apparent.

The major selection procedures practiced in horses have related to the horses ability to walk or run (Stanbenfeldt et al., 1975). Selection for high reproductive performance in mares does not and has not received adequate attention by horse breeders.

The study reported here was designed to investigate the feasibility of detecting estrus and ovulation in mares by measuring changes in either body temperature or the resistance of the cervical secretions to an electrical charge. Both of these techniques have been used in the dairy industry with good results.

This study was conducted at the New Mexico State University Horse Center. Eighteen registered mares of either Arabian or Quarter Horse breeding, all in foal for the spring of 1981, were selected for and assigned to one of three treatment groups: rectal palpation (RP), deep body temperature transmitter (radio transmitter, RT) or electrical resistance of the cervical mucus (estroprobe, EP).

Observations were taken once daily, between 7:00 and 10:00 a.m., with an attempt to maintain a consistent order of observations for all mares.

The estroprobe mares were assigned as the first treatment group, the radio-transmitter mares as the second treatment group, and the rectal-palpation mares as the control group. As each mare foaled she was placed in the appropriate group, i.e. the first mare to foal went into group one, the second mare into group two, the third into group three.

The daily schedule for each mare consisted of: 1) being teased by a teasing stallion, 2) being observed for the respective treatment (EP or RT) 3) being evaluated for the position and physiological condition of the cervix. Figure 1 graphically depicts the expected response of an average, "textbook" normal mare to teasing and the physical changes of the cervix during the estrous cycle.

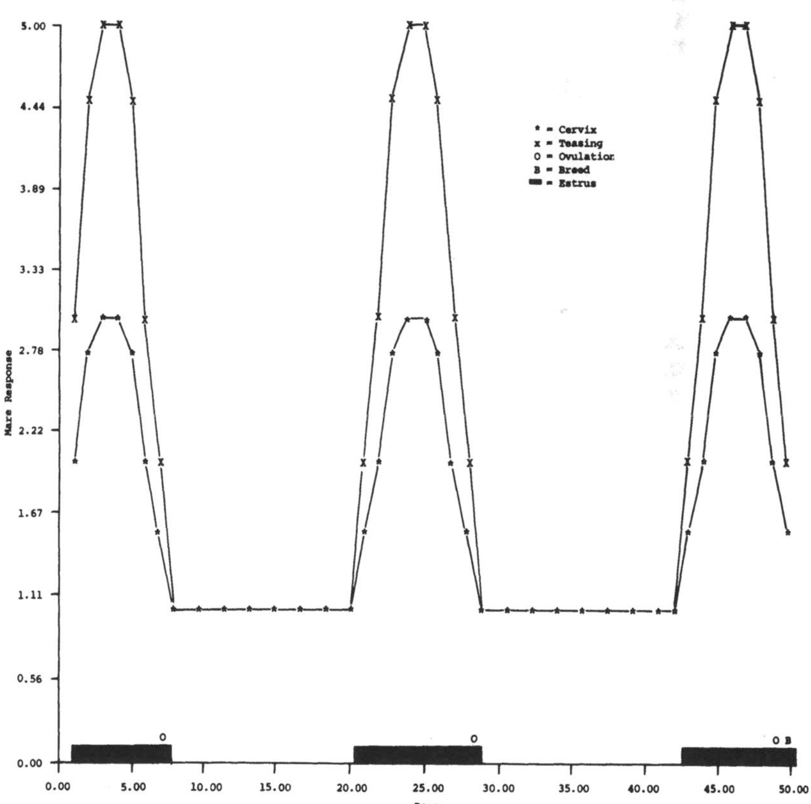

Figure 1. Expected Data for a "Textbook" Normal Mare

PROCEDURES

Teasing

Teasing consisted of leading each of the mares to a stallion to check for estrus (or heat activity). Estrus activity was evaluated on a scale of 1 to 5. The value of 1.0 indicates diestrus (no heat activity) antagonistic; 1.0 to 3.0 = definite interest in the stallion and winking of the vulva; 3.9 to 4.0 = mare will wink heavily and urinates heavily; 4.0 to 5.0 = the mare will urinate, wink heavily, and will present herself to be mounted by the stallion. Mares with teasing scores of 2.0 and higher were considered to be "in heat" and would allow the stallion to mount.

Cervical Observation

After the teasing and treatment scores (RT or EP) had been determined, the cervix of each mare was examined with an equine vaginal speculum to evaluate the muscular condition of the cervix.

The cervical evaluation is a subjective determination of the position of the cervix in relation to the normal diestrous cervix position (table 1). The cervix in the normal diestrous position is pale in color, constricted, and protruding. The diestrous cervix is assigned a value of 1.0. The cervix at the peak of a mare's estrus is vascular, completely dilated, and lying on the floor of the vagina. The peak estrus cervix was assigned a value of 3.0. A position in between these two extremes was assigned a value of 1.5, 2.0, or 2.5, depending on the position of the cervix relative to the extremes. This observation was then recorded for that mare for that day.

Estroprobe

The Estroprobe Model 712 was designed and marketed for use in detecting estrus activity in dairy cows. It was not marketed as a replacement for the current estrus detection methods but as an aid in detecting those cows with a silent estrus. The probe works by measuring differences in resistance to an electrical charge by the cervical mucus. Research done in the U.S., Great Britain, and West Germany has shown that there is an increase in the fluidity of the cervical secretions and a resulting decrease in resistance of these secretions to an electrical charge just prior to ovulation. This decrease in resistance is evidenced in a lowered reading on the digital output of the estroprobe.

The estroprobe consists of two parts: a meter box that contains a rechargeable battery-operated voltmeter with a digital output and a stainless steel probe 22.5 in. in length. The probe has two isolated contact points in the tip that are connected via shielded cable to the meter box.

Table 1 Clinically Detectable Changes in the Cervix, Vagina and Uterus During the Estrous Cycle and the Assigned Scoring.

Method of Examination	Organ	Mid-diestrus	Estrus: Early	Estrus: Middle	Estrus: Late
Palpation per rectum	Uterus	Maximal tone and thickness	DECREASING TONE AND THICKNESS		
	Cervix	Firm and distinct	Beginning to flatten	Flatter, shorter, and wider	Barely discernible, very flat
Palpation per vagina	Vagina	Dry	INCREASING IN WETNESS		
	Cervix	Firm, protruding	Admits 1 finger	Admits 2 fingers	Admits 3 fingers or entire hand
Speculum	Vagina and cervix	Viscous fluids Vaginal walls stick together	INCREASING FLUID WITH DECREASING VISCOSITY		
		Dull, yellow-grey	Pink	Bright pink	Glistening Red
		Decreased vascularity	INCREASING VASCULARITY		
	Cervical orfice	Protruding, centrally located, tight	Beginning to open and drop	Dropped below center	Dropped near floor of vagina
Cervical Score (Visual observation)		1.0 1.5	2.0	2.5	3.0

The mare was placed into an inspection chute where her vulva and anal regions were washed with a disinfectant solution. Special care was taken to remove all fecal material and any discharge from the vulva. Once this was completed the probe was inserted into the vagina in an attempt to establish contact with the cervix. When the probe tip was in contact with the cervix or the anterior end of the vagina, the lowest stable reading (the reading which remained on the display for at least one second) was recorded. This was repeated five times and an average of these five readings was recorded for that day.

Radio Telemetry

Radio transmitters approximately 2 in. long and 1 in. in diameter were surgically placed in the vaginas of the mares when they began foal heat. The transmitters consisted of four components tied together into a single circuit. These components were: 1) a power source, which in this case, was a miniature lithium battery; 2) a thermistor; 3) a resistor; and 4) a transmitter.

As the vaginal temperature of the mare increases, the resistor opens and closes with greater frequency allowing the transmitter to pulse faster. This causes the transmitter to emit more signals per minute. These signals were monitored though an AM radio. The operation of the transmitter can be compared to that of a nerve fiber in the body. The stimulus increases until the threshold level is attained and the nerve fires and sends a wave of depolarization down the fiber. Once the nerve fires, it returns to the resting state until the stimulus reaches the threshold level and fires the nerve again. The transmitter signals that were monitored through an AM radio were recorded on a portable cassette tape recorder for three minutes. These signals or "beeps" were counted and the resulting number was established as the data point of that mare for the day. Readings were taken between 7:00 and 10:00 a.m. each day.

Rectal Palpation

The rectal palpation mares were palpated every other day according to typical veterinary procedures. Each ovary was palpated for follicle corpora lutea, corpus hemorrhagica, and any changes in size or shape from the previous palpation two days before. When a follicle was present, it was evaluated on the basis of how many fingers could be placed across it. This was expressed as a number 1 follicle (one finger, approximately 10 mm), a number 2 follicle (two fingers, approximately 25 mm), through a number 5 follicle (five fingers, approximately 50 mm). A number 1 follicle was considered an immature follicle and a number 3 was considered a mature follicle. When a corpus luteum was palpated, an ovulation was recorded for that mare for the previous day. The location of the follicle (right or left ovary) and the size of the follicle were recorded for that mare for that day. After each ovary was palpated, the veterinarian moved to the body of the uterus and determined its tone and condition.

DETERMINATION OF TREATMENT EFFECTIVENESS

Treatment observations were started on each mare as she began her foal heat period. Teasing, cervical, estroprobe, and/or radio transmitter scores were taken daily. Rectal palpation was done on an every-other-day basis. Each mare in the two treatment groups was carefully observed through the foal heat and the first regular estrus following foal heat. An attempt was made to determine the optimum time to breed the mare on her next cycle on the basis of the relationships among the estrobe or radio transmitter readings and the cervical, teasing, and rectal palpation scores during these two estrous cycles. The optimum time to breed was considered to be one day prior to ovulation. The mares in the estroprobe and radio transmitter groups were to be bred one time only by natural service when the data indicated ovulation was to occur within 24 hours.

A positive pregnancy examination was the criterion of the success for each detection method.

RESULTS AND DISCUSSION

Estroprobe

The success of the estroprobe in detecting ovulation was much greater than that of the radio transmitters. Five of the six mares bred on the basis of data indicated by the estroprobe became pregnant as the result of a single service. The probability of the pregnancies occurring due to simple chance is 0.0000022. This extremely low probability allows us to reject the null and accept the alternate, which states that there is a correlation between the estroprobe ovulation detection method and ovulation.

The six mares in the estroprobe group were determined to be in estrus (by palpation, in the first phase of the study) 13 times with eight ovulations. Three of the eight ovulations occurred when the mares were not in estrus and were classified as silent heats (ovulation occurring when the mare was not receptive to the stallion). Of the five remaining ovulations, four (80%) were recorded during the last half of the estrus period and one (20%) was recorded in the first half of the estrus period. Of the eight palpated ovulations that occurred in the first phase of the study, the estroprobe reading was depressed in seven cases.

Radio Telemetry

Of the six mares in which radio transmitters were implanted, none became pregnant as a result of the breeding. The probability of all six mares not becoming pregnant under the conditions set forth by the null hypothesis (detection technique had no correlation to ovulation) is 0.62. Because of this high probability value we cannot reject the null hypothesis that there is no relationship between body temperature and ovulation. This is in agreement with the findings of Evans et al. (1976a) that indicated no correlation between body temperature and ovulation.

During the initial portion of the study in which palpation was conducted to determine ovulation (foal heat to first estrus period), only five ovulations occurred at the same time a temperature rise occurred, and two ovulations occurred simultaneously with a noticeable temperature drop. Within this period there were 12 total ovulations, nine of which occurred during a recognized estrus period (teasing score of 2 or higher). The three remaining ovulations were classified as silent heats. Of the nine ovulations during a recognized estrus three (33%) occurred during the first half of the estrus period. The remaining six ovulations (66.7%) occurred in the second half of the estrus period. This approximates values set forth by Ginther (1979), which indicate that 69% of all ovulations were observed on the last two days of estrus and 14% occurred after the end of estrus. It was found during this study that the radios were useful in determining illness or infection. One mare received a minor puncture injury from a wire and developed an infection and temperature. This was evidenced the day after the injury by an extreme rise in data recorded from the transmitter for that day.

Rectal Palpation

Two of the six mares in the control group (rectal palpation) became pregnant. This group exemplified many of the problems that face the breeding-farm manager in dealing with the eccentricities of each mare's cycle. Some of these eccentricities were: persistent corpus lutea that can prevent a mare from cycling normally, the inconsistency of the length of the estrus periods from mare-to-mare, and a lack of repeatability form one cycle to the next cycle in a single mare.

The mares were bred the morning following the palpation of a number 3 or larger follicle (mature follicle). In the case of two of the mares, both were palpated and had immature number 2 follicles and were not considered for breeding because of the size of their follicles at ovulation in the previous estrus period. When these mares were palpated 2 days later both had ovulated and it was impractical to breed them. A third mare that ovulated off a number 4 follicle in the first phase of the study was bred when she had a follicle of the same size. Two days later she was palpated to check for a corpus luteum and instead had a number 5 follicle. She maintained the number 5 follicle for 2 days and then ovulated. Even though this horse had not ovulated when rechecked after breeding she was not bred a second time and was considered to have completed the study, since the experimental design called for a single breeding on the third estrus period after foaling.

The final result of the rectal palpation portion of the study was that two of the six mares became pregnant. The probability of the two pregnancies occurring due to chance is 0.98. This indicates a very high probability that the two mares became pregnant simply due to chance. This is very likely because when only one breeding is allowed according to palpation data you are simply guessing that the follicle you palpated and consider mature will ovulate the next day. Additionally, the use of rectal palpation has been found to be detrimental to first-service conception rates (Irwin, 1975; Voss et al., 1975a; Voss et al., 1975b).

SUMMARY

The estroprobe was found to be very beneficial in detecting ovulation in mares. An 83% conception rate on a single breeding was obtained when the mares were bred according to estroprobe data. The probability is 0.0000022 that this occurred simply due to chance. It was also found that the results (zero percent conception) of the radio transmitter group agree with the current literature that reports no correlation between body temperature and ovulation in the mare. The results (33 percent conception) of the rectal palpation portion of the study also agree with the current literature that states that frequent palpation may delay ovulation and thereby reduce the first-time conception rates.

In summary, it is believed that monitoring changes in resistance of the cervical mucus to an electrical charge by mechanical means may be beneficial in increasing pregnancy rates and also may result in a savings of time and effort in the equine breeding industry.

REFERENCES

Evans, J. W., C. W. Winget, C. DeRoshia and D. C. Holly. 1976a. Ovulation and equine body temperature and heart rate circadian rhythms. J. Interdiscipl. Cycle Res. 7:25.

Evans, J. W. 1977b. Horses. W. H. Freeman and Co., San Francisco.

Ginther, O. J. 1979. Reproductive biology of the mare. McNaughton and Gunn, Inc., Ann Arbor, Mich.

Irwin, C. F. P. 1975. Early pregnancy testing and its relationship to abortion. J. Reprod. Fertil., Suppl. 23:485.

Stabenfeldt, G. H., J. D. Hughes, J. W. Evans and I. I. Geschwind. 1975. Unique aspects of the reproductive cycle of the mare. J. Reprod. Fertil., Suppl. 23:155.

Voss, J. L. and B. W. Pickett. 1975a. The effect of rectal palpation on the fertility of cycling mares. J. Reprod. Fertil., Suppl. 23:285.

Voss, J. L., B. W. Pickett, O. G. Back and L. D. Burwash. 1975b. Effect of rectal palpation on pregnancy rate of nonlactating, normally cycling mares. J. Anim. Sci. 41:829.

17

INFLUENCE OF BODY CONDITION ON REPRODUCTIVE PERFORMANCE OF MARES

Don R. Henneke, Gary D. Potter, Jack L. Kreider,
B. F. Yeates, Doug Householder

Horse breeders have many more questions than answers about how stored body fat i.e., body condition can affect a mare's reproductive ability. Do thin mares that are gaining weight breed more efficiently than do fat mares? Do fat mares have more problems during foaling? Can fat mares utilize stored energy to breed efficiently and produce adequate milk for a foal even when losing body weight?

It is often stated that a mare's fertility is adversely affected if she becomes too thin or too fat; however, there is little data to support such claims. (Unfortunately, very little conclusive research has been conducted on the subject of nutrition - reproduction interrelationships in the mare.) Some contend that mares should be brought into the breeding season in thin condition (with their ribs showing), and then fed (on an inclining plane of nutrition) to produce weight gains for maximum breeding efficiency. Another often-stated theory is that fat mares have lower fertility than thin mares and that there are more foaling problems among fat mares. To provide more definitive research information on the subject, an in-depth research project was recently completed at Texas A&M University to study the relationships between pre- and postpartum body condition and reproductive performance of mares (Henneke, 1981).

THE CONDITION SCORE SYSTEM

To evaluate the influence of body condition on reproductive performance, a condition score system was needed that would allow on-the-ranch comparison of mares for the amount of stored fat in their bodies. The best measure of stored body energy is fat content, and it can be measured experimentally. The condition score system described herein was developed to reflect relative differences in fat content of mares.

Mature Quarter Horse mares fom the Texas A&M University (TAMU) Horse Center were examined, areas of the horse's body that reflected changes in body fat content were selected, and the condition score system was developed based on visual

SCORE	DESCRIPTION
1	*Poor*. Animal extremely emaciated. Spinous processes, ribs, tailhead, and hooks and pins projecting prominently. Bone structure of withers, shoulders, and neck easily noticeable. No fatty tissues can be felt.
2	*Very Thin*. Animal emaciated. Slight fat covering over base of spinous processes, transverse processes of lumbar vertebrae feel rounded. Spinous processes, ribs, tailhead and hooks and pins prominent. Withers, shoulders neck structures faintly discernible.
3	*Thin*. Fat built up about halfway on spinous processes; transverse processes cannot be felt. Slight fat cover over ribs. Spinous processes and ribs easily discernible. Tailhead prominent, but individual vertebrae cannot be visually identified. Hook bones appear rounded, but easily discernible. Pin bones not distinguishable. Withers, shoulders, and neck accentuated.
4	*Moderately Thin*. Negative crease along back. Faint outline of ribs discernible. Tailhead prominence depends on conformation; fat can be felt around it. Hook bones not discernible. Withers, shoulders, and neck not obviously thin.
5	*Moderate*. Back level. Ribs cannot be visually distinguished but can be easily felt. Fat around tailhead beginning to feel spongy. Withers appear rounded over spinous processes. Shoulders and neck blend smoothly into body.
6	*Moderate to Fleshy*. May have slight crease down back. Fat over ribs feels spongy. Fat round tailhead feels soft. Fat beginning to be deposited along the sides of the withers, behind the shoulders, and along the sides of the neck.
7	*Fleshy*. May have crease down back. Individual ribs can be felt, but noticeable filling between ribs with fat. Fat around tailhead is soft. Fat deposited along withers, behind shoulders, and along the neck.
8	*Fat*. Crease down back. Difficult to feel ribs. Fat around tailhead very soft. Area along withers filled with fat. Area behind shoulder filled in-flush. Noticeable thickening of neck. Fat deposited along inner buttocks.
9	*Extremely Fat*. Obvious crease down back. Patchy fat appearing over ribs. Bulging fat around tailhead, along withers, behind shoulders and along neck. Fat along inner buttocks may rub together. Flank filled in flush.

Figure 1. Description of the condition score system

and palpation appraisal. This system is presented in figure 1, and a diagram of the areas of fat deposition referred to in the scoring system is given in figure 2. After the scoring system was completed it validity was challenged by comparing the condition score of 32 mares against an ultrasonic procedure used to measure body fat (Westervelt et al., 1976). Correlation between the condition score and measured body fat was high and statistically significant, suggesting the condition score system was an accurate indicator of the amount of stored fat in the mare's bodies.

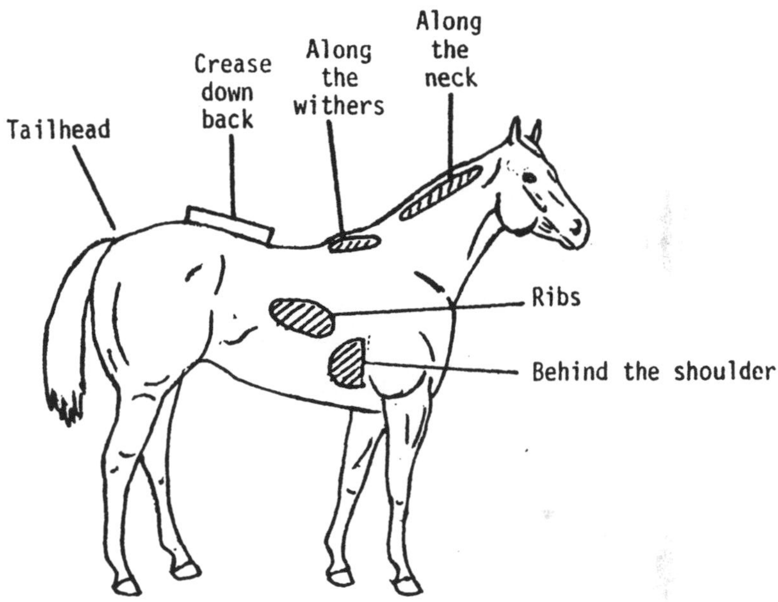

Figure 2. Diagram of areas emphasized in condition score

Although this condition score system is based primarily upon visual appraisal, palpating the areas indicated for fat cover was beneficial in accurately determining the final score. During the winter months the presence of a long, heavy hair coat may interfere with accurate visual appraisal, thus handling techniques were used in those instances.

Conformation differences between horses can make certain criteria within each score difficult to apply to every animal. Some mares had more prominent withers or were flatter across the loin than others. Therfore, more emphasis was placed on fat cover over the ribs, behind the shoulder, and around the tailhead when evaluating these types of mares.

When evaluating pregnant mares, adjustments were required when scoring mares as they approached parturition.

The weight of the fetus and associated products of conception tended to pull the skin and musculature tighter over the back and ribs. Therefore, emphasis was placed upon fat deposited behind the shoulder, around the tailhead, and along the withers in mares near term.

Maiden mares tended to feel fatter over the ribs than older mares that had already produced foals. Therefore, fat cover over the ribs was not a good single indication of body condition in these mares.

In summary, all the conditions for a given condition score were evaluated. When one or more of the conditions for a given score were not present, the scores were moved upward or downward as indicated.

The condition score system described here has been evaluated by several horse producers. All have indicated that it is easily learned and is useful to monitor changes in the amount of stored fat in mares. This system promotes more accurate evaluation of mares on a given farm, and its use could improve communication regarding mare condition in exchanging information between farms and between breeding seasons. It should be emphasized that this condition scoring system refers to differences in body fat content and not to quality of the mares. Scores are given in .5 unit increments to provide increased accuracy in comparing horses.

TEXAS A&M UNIVERSITY STUDY

The level of stored fat in the mare's body may be an important source of energy when energy requirements of the mare exceed that supplied in the feed. Potential beneficial effects of energy reserves are very apparent in the broodmare when she makes the transition from late gestation to early lactation. Previous research at TAMU (Gibbs, 1979) has shown that unlike the cow, maximum milk production in the mare occurs in the first week after foaling. Accordingly, the energy requirement increases two-fold during this time. If feed intake is doubled during this short period, there is an increased risk of colic and founder. Since thin mares lack stored energy reserves, rebreeding efficiency and milk production may suffer during the early lactation period because feed intake cannot be safely increased sufficiently to meet all requirements without causing digestive problems.

An intensive study was conducted at TAMU to study the influence of body condition at foaling on rebreeding efficiency and foal growth and development. Sixteen Quarter Horse mares were fed to foal in fat condition (condition score 8), and 16 mares were fed to be thin (condition score 4) at foaling. At parturition the mares were further divided into 4 postfoaling treatments for the first 90 days of lactation: 8 fat mares were fed to maintain their weight and remain fat; 8 fat mares were fed to lose excess weight; 8 thin mares were fed to meet lactation requirement but gain no weight; and 8 thin mares were fed to gain weight. Treatments were designated as fat-fat (F-F), fat-losing (F-L), thin-thin (T-T) and thin-gaining (T-G) relative to pre- and postpartum periods.

As seen in table 1, pregnancy rates over three cycles were significantly higher (100%) in the F-F, F-L, and T-G groups than in the T-T group (50%). As seen in table 2, maintenance of pregnancy to 90 days was also significantly higher in the F-F and T-G groups (100%) and in the F-L group (88%) than in the T-T group (25%). Foal growth at 90 days of age was similar among all groups, indicating that the energy requirement for lactation was met in all treatments.

TABLE 1. PREGNANCY RATES OVER THREE CYCLES

Treatment	1st Cycle	2nd Cycle	3rd Cycle	Total
	(%)	(%)	(%)	(%)
F-F	62 (5/8)[a]	66 (2/3)[a]	100 (1/1)[a]	100 (8/8)[a]
F-L	57 (4/7)	75 (3/4)	100 (1/1)	100 (8/8)
T-T	37 (3/8)	20 (1/5)*	0 (0/4)	50 (4/8)*
T-G	60 (3/5)	100 (2/2)		100 (5/5)

[a]Number of mares pregnant/number of mares bred.
*Significantly lower (P<.05) than other treatments.

TABLE 2. MAINTENANCE OF PREGNANCY

Treatment	60 Days	90 Days
	(%)	(%)
F-F	100 (8/8)[a]	100 (8/8)
F-L	88 (7/8)	88 (7/8)
T-T	75 (3/4)	25 (1/4)*
T-G	100 (5/5)	100 (5/5)

[a]Number of mares pregnant/number of mares conceived.
*Significantly lower (P<.05) than other treatments.

Results of this study indicate that mares foaling in thin condition had impaired reproductive efficiency even when energy requiements for lactation were met. Increasing the energy fed to the thin mares during lactation improved rebreeding efficiency but required dangerously large amounts of feed. To produce the changes in body condition observed in this study, mares in the T-G group were fed an average of 35 pounds of grain (3 feedings at 8-hr intervals) and 15 pounds of hay per head per day during the 90-day lactation period! Mares in the F-F, F-L, and T-T groups received 20, 10, and 15 pounds of grain, repectively, and 15 pounds of hay per head per day. It is evident that mares foaling in fat condition utilized stored fat for efficient reproduction and milk production, even when energy intake did not meet lactation requirements. Excess fat did not impair rebreeding efficiency and no foaling problems were observed in any mares in the study.

FIELD STUDY

To further investigate the relationship between mare condition and breeding efficiency, a field study was conducted with the cooperation of McDermott Ranch, Madisonville, Texas; Phillips Ranch, Frisco, Texas; Shelton Ranches, Kerrville, Texas; and Wilson Ranch, Pattison, Texas. All types of Quarter Horse mares were represented at these ranches--racing, halter, and arena performance. Therefore, results of this study should be applicable to different types of breeding operations.

A total of 927 mares were evaluated for body condition and monitored for reproductive performance. These included 52 maiden mares, 378 open mares, and 497 foaling mares. Results revealed no significant influence of mare status, age of mare, or farm influence on reproductive performance. Therefore all data presented will include all mares, unless otherwise stated.

Open and maiden mares were scored for body condition on February 1, or when they arrived at the ranch (initial condition score), whereas foaling mares were evaluated at parturition (initial condition score). All mares were scored once during each heat period and a final score was placed on each mare when she was diagnosed pregnant or sent home. Mares were classified as thin (condition score 4.5 or less), moderate (condition scores 5.0 to 6.5), or fat (condition score 7.0 and above) based upon their initial condition score.

Table 3 shows the average change in condition score, pregnancy rates, and cycles bred per conception for the thin, moderate, or fat mares in the field study. Pregnancy rates and cycles/conception are illustrated graphically in figures 3 and 4. Pregnancy rate was signifcantly lower in the thin mares (71%) than in the moderate (93%) or fat (96%) mares. In addition, the number of cycles bred per conception was twice as high in the thin mares when compared to the moderate and fat mares. However, the thin mares were gaining more body condition than the moderate or fat mares; thus, the concept of bringing mares into the breeding season in thin condition and then feeding them to gain weight was not supported in this study. In contrast, this study suggests that mares that entered the breeding season or foaled in fatter condition achieved higher reproductive efficiency.

TABLE 3. AVERAGE CHANGE IN CONDITION SCORE, PREGNANCY RATE AND AVERAGE CYCLES/CONCEPTION FOR THIN, MODERATE, OR FAT MARES IN THE FIELD STUDY

Initial body condition	No. of mares	Change in condition score (%)	Pregnant (%)	Cycles per conception
Thin (≤4.5)	158	49	71*	2.8*
Moderate (5-6.5)	667	12	93	1.4
Fat (≥7)	102	1	96	1.4

*Pregnancy rate and average cycles per conception are significantly different for the thin mares (P<.05).

In 74 maiden and open mares that were present on the ranches on February 1 and maintained under natural light, the average intervals from february 1 to the onset of estrus (heat) and to the transition ovulation are given in table 4, grouped by mares that had initial condition scores 5.0 less than those in higher condition. The transition ovulation is defined as the ovulation preceding the first normal cycle of the breeding season. The onset of estrus was significantly delayed in those mares entering the breeding season at a condition score of less than 5.0 (39 days) when compared to the fatter mares (26 days). In addition, the transition ovulation was significantly later in the breeding season in the thin group of mares (63 days) than in the fatter mares (37 days). Therefore, thin mares did not begin normal cycles until 26 days after the fatter mares.

TABLE 4. MEAN INTERVALS FROM FEBRUARY 1 TO THE ONSET OF ESTRUS AND THE TRANSITION OVULATION FOR MARES ENTERING THE BREEDING SEASON AT LESS THAN 5.0 AND 5.0 ABOVE CONDITION SCORE

Initial condition score	Onset of estrus (days)	Transition ovulation (days)
less than 5.0	39	63
5.0 and above	26*	37*

*Interval is significantly shorter (P<.05).

CONCLUSIONS

The results of this study suggest the following conclusions:

- Mares foaling in thin condition had impaired reproductive performance even when fed to meet energy requirements for lactation.
- Increasing the energy fed to thin mares during lactation improved rebreeding efficiency, but the large amount of feed required to produce weight gains may increase the risk of colic and founder and was very expensive.
- Mares foaling in fat condition utilized stored body energy for efficient reproduction and lactation even when losing weight.
- No increase in foaling problems was observed in mares foaling in fat condition.
- Open and maiden mares entering the breeding season in moderate or higher condition achieved higher reproductive efficiency.
- Foaling mares that foaled in moderate or higher condition achieved higher rebreeding efficiency.
- Excess fat did not reduce breeding efficiency of lactating or nonlactating mares.
- Increasing body condition above moderate levels prior to the breeding season did not impair rebreeding performance in mares but was of no benefit unless mares were losing weight during the breeding season. Therefore, breeding efficiency was apparently maximized by maintaining mares in moderate or higher body condition (condition score 5.0 or above) throughout late gestation and early lactation.
- Results from this study with mares are similar to those previously reported in beef cattle (Donaldson, 1969; Wiltbank et al., 1962; Wiltbank et al., 1964) and dairy cattle (Butler et al., 1981).

After analyzing the results of this study, breeders who cooperated in the field study concluded that they would like open and maiden mares to arrive at their farms in at least moderate to fleshy condition (condition scores 6.0 to 7.0) and foaling mares should foal in fleshy to fat condition (condition scores 7.0 to 8.0). Mares often lose weight when they first arrive at the breeding farms because of the extra stress associated with hauling, breeding - palpation, teasing, different feed, etc. This transient weight loss would probably be handled better by a fatter mare than by a thin one. Therefore, the extra condition may improve the reproductive performance of the mare by reducing the adverse affects of the extra stress as well as by improving reproductive function.

While excess fat did not adversely affect reproductive performance in this experiment, it is not recommended that mares be allowed to reach condition score 9. Obese mares are unusual under the management practices of most producers and are not a major problem. There were no pathologically

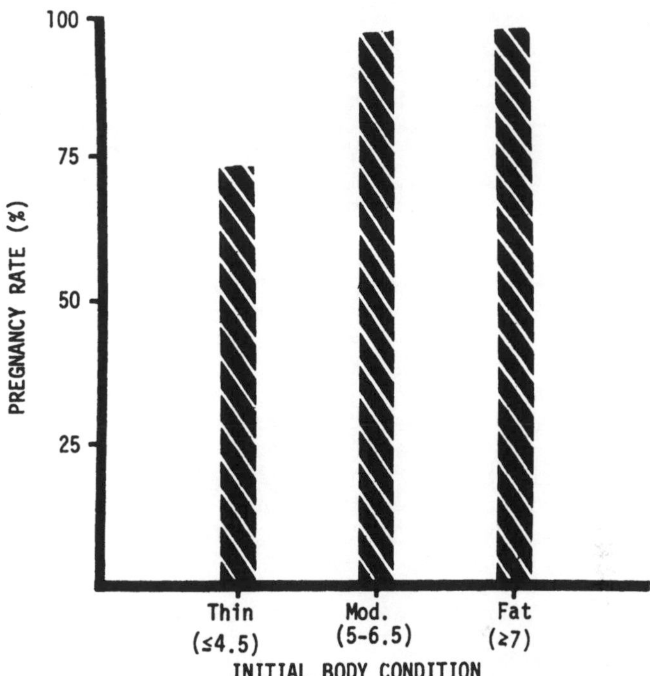

Figure 3. Pregnancy rates of mares entering the breeding season or foaling in thin, moderate, or fat condition

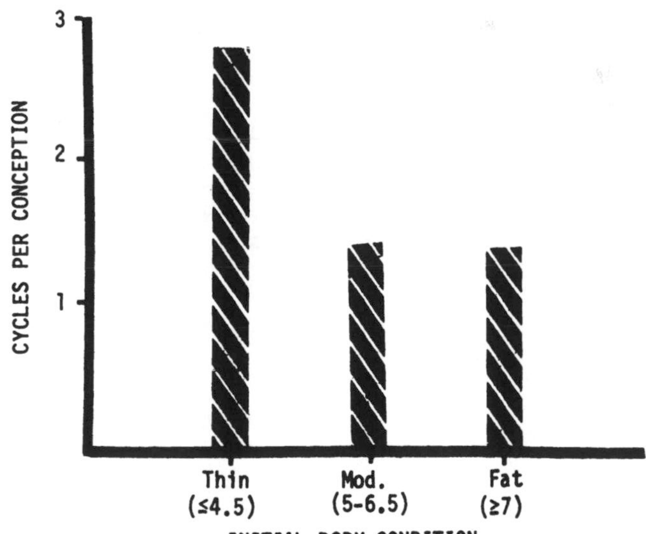

Figure 4. Average number of cycles per conception for mares entering the breeding season or foaling in thin, moderate, or fat condition

obese mares represented in this study, but it is logical that such a mare could have reproductive problems, which could be due to endocrine disorders, since extreme obesity is often associated with hormone imbalances. Thus, the extremely obese mare may have reproductive problems that are not related to body fat content but rather to some other physiological dysfunction.

From an economic standpoint, maintenance of mares in moderate to fat condition may be cheaper than trying to increase the weight of thin mares during the breeding season. Previous research conducted at Texas A&M (Brewer, 1968) showed that adding fat to a thin mare required more energy than maintaining an extra fleshy mare. In addition, feeding costs are usually at their highest point from December through March, and attempting to place mares on an inclining plane of nutrition at this time can be extremely expensive. Also daily amounts of feed required to allow a lactating mare to gain weight are dangerously high. It seems much more logical to allow mares to gain weight during the summer and fall when pastures are plentiful and feed costs are lower, and then to maintain these mares in good condition throughout the winter.

In addition to demonstrating the potential daily savings in feed costs, the data from this study indicate that mares arriving in thin condition stayed at the stud farm significantly longer than did moderate and fat mares. The extra time required to get the thin mares in foal would cost significantly more in total board charges than would be required for a mare arriving in better condition.

The data from this study, enhanced by the cooperation of the four large commercial breeding operations, suggest that mares should enter the foaling/breeding season in moderate to fat condition to achieve maximum reproductive performance. Mares with significant initial stores of fat cycled earlier, had higher pregnancy rates, lower number of cycles per conception, and more sustained pregnancies than did thin mares. Therefore, to achieve maximum reproductive efficiency, mare owners and stud managers should pay careful attention to meeting the energy needs of broodmares in late pregnancy and early lactation. Results of this study offer management alternatives that can be employed to allow the mare to store energy in times of low need and recall that energy in times of high energy requirement.

ACKNOWLEDGEMENTS

The authors wish to thank Tom Gibbs, Tom Gibbs Quarter Horses, College Station, Texas; Mike Klem, former Broodmare Manager of Matlock Rose Quarter Horses, Gainesville, Texas; and Gary Webb, former Stallion Manager of Wilson Ranch, Pattison, Texas for their help in developing the condition score system. In addition, sincere appreciation is extended to Vernon Glass and Bill Weston, McDermott Ranch, Madisonville, Texas; Bruce Hill and Mike Stuart, Phillips Ranch, Frisco, Texas; David Trevino and Rick Colson, Shelton Ranches, Kerrville, Texas; and Steve and Martha Vogelsang, Wilson Ranch, Pattison, Texas for their assistance in collecting the data from the field study.

Very special acknowledgement and appreciation is extended to Dr. G. L. Morrow, Shelton Ranches, and Dr. Jerry Rheudasil, Phillips Ranch, for their professional assistance and input to this study. Their interest and cooperation was invaluable and will lead to further advancement of equine research.

Finally, the authors wish to express thanks to Mr. Joe McDermott, M. B. F. Phillips, Mr. Bobby Shelton, and Mr. Sam Wilson for the opportunity to collect the field data and for the very gracious hospitality extended the authors during this project.

REFERENCES

Breuer, L.H. 1968. Energy nutrition of the light horse. Proc. Equine Nutr. Res. Symp. U. of Ky.

Butler, W.R., R.W. Everett and C.E. Coppock. 1981. The relationship between energy balance, milk production, and ovulation in postpartum Holstein cows. J. Anim. Sci. 53:742.

Donaldson, L.E. 1969. Relationships between body condition, lactation and pregnancy in beef cattle. Aust. Vet. J. 45:577.

Gibbs, P.G. 1979. Yield and composition of milk from mares fed soybean meal or urea as protein supplements. M.S. Thesis. Texas A&M University Library, College Station.

Henneke, D.R. 1981. Body condition and reproductive efficiency of mares. Ph.D. Dissertation. Texas A&M University Library, College Station.

Participants in the TAMU Horse Breeders Schools, fall 1981.

Westervelt, R.G., J.R. Stouffer, H.F. Hintz and H.F. Schryver. 1976. Estimating fatness in horses and ponies. J. Anim. Sci. 43:781.

Wiltbank, J.N., W.W. Rowden, J.E. Ingalls, and D.R. Zimmerman. 1964. Influence of postpartum energy level on reproductive performance of Hereford cows restricted in energy intake prior to calving. J. Anim. Sci. 23: 1049.

Wiltbank, J.N., W.W. Rowden, J.E. Ingalls, K.E. Gregory and R.M. Koch. 1962. Effect of energy level on reproductive phenomena of mature Hereford cows. J. Anim. Sci. 21:219.

18
STALLION MANAGEMENT AND SEMEN HANDLING

Gary W. Webb,
Jack L. Kreider

On a breeding farm, to obtain maximum conception rates, a great deal of time is necessarily spent to ensure that the broodmares are receiving the proper nutrition and care. Often, however, too little attention is given to the management of stallions and the actual processes of semen collection and insemination. This is unfortunate because improper management and/or poor techniques in these areas can have extremely detrimental effects on conception rates.

Good stallion management includes a proper nutrition and exercise program tailored to each stallion's needs. The proper training of the stallion for breeding purposes also is necessary to obtain good ejaculates on a routine basis throughout a long breeding season.

STALLION CARE

In addition to proper nutrition, stallions require the proper amount of exercise. Most stallions do well when kept in an open-housing situation, which is a stall with free access to a paddock. Other stallions may require forced exercise, and some can only be turned out for a few hours each day. A good exercise program will go a long way in eliminating many bad habits in the breeding shed.

HANDLING THE STALLION

Individuality of stallions is an important consideration when determining how a stallion should be handled. Some horses require a great deal of discipline, but over restraint of a young or timid stallion may result in a slow breeder or even a refusal of the animal to mount and ejaculate.

There are two things that aid or deter the training of a young stallion to breed: 1) his fear of abuse from you or the mares and 2) his libido. If a stallion loses his libido, you are out of busines. During the first few trips to the breeding shed it is often better to approximate the desired behavior of the young stallion rather than demand he work like an older, trained stallion. Remember to always reward him to service the mare or artificial vagina.

In training a stallion to jump a breeding dummy, it is helpful to pour urine from a mare in estrus on the dummy. This olfactory stimulation will stimulate the stallion to achieve erection and mount the dummy.

Another helpful procedure is to place a lead rope on the near ring of the stallion's halter, then place a mare in estrus on the off side of the breeding dummy and allow the stallion to tease the mare across the dummy. When the stallion jumps for the mare, the lead rope, rather than the stud chain, can be used to physically pull the stallion onto the dummy. As with training the young stallion to jump a mare, it is sometimes necessary to collect the semen when he comes close to the desired position.

SEMEN COLLECTION

Collection of stallion semen can be accomplished most efficiently by use of an artificial vagina (AV). Three models of AV are available commercially: 1) the Colorado model, 2) the Japanese model, and 3) the Haver Lockhart or Missouri model. When filling any model of AV, water temperature should be 115 to 120° F. Make certain it is no hotter or the stallion may be injured. Since some time will elapse between filling of the AV and actual collection, the water will generally have cooled to 110 to 112° F. However, just in case it has not, check it with a thermometer. The amount of pressure required to stimulate ejaculation will vary between stallions and must be adjusted accordingly.

A good practice is to keep a record of the AV temperature and pressure at each collection, along with the number of mounts per ejaculate so that this information can later be used to determine at what pressure each stallion works best.

The ejaculate of the stallion has three fractions. The first fraction is clear and contains no sperm; the second is called the sperm-rich fraction and contains most of the sperm; the third fraction contains a gelatinous material with few if any spermatozoa. For ease in handling and extending the semen, the third fraction should be removed from the ejaculate. This can be accomplished by placing an in-line pipeline milk filter in the collecting bottle so that the fraction is removed at collection, or the semen can be strained through cheese cloth after collection. The AV should be well lubricated with KY jelly to aid in penetration by the stallion.

Semen collection requires three people: the mare handler, the stallion handler, and a third person to do the actual collection. The procedure should begin by wrapping the tail and putting breeding hobbles on the collection mare. Preparation of the mare is followed by preparation of the artificial vagina, making sure that the water temperature is correct and the AV is adequately lubricated with a sterile, water-soluble lubricant. Next, the stallion handler should bring the stallion into the presence of the mare, approaching from the near side as in natural service. He should be allowed to tease the mare until erection is achieved (if the stallion is to be washed it is done at this time). The mare handler, stallion handler and person doing the collection (collector) should all work on the near side of the mare and stallion. The stallion should be allowed to mount with the collector remaining back out of the way. The collector should then step in and direct the stallion's penis into the AV. The AV should be braced against the mare's quarter in line with the vulva. The stallion should be allowed to serve the AV as he would a mare. However, if he does not achieve good penetration before the glans begins to flower, a collection may not be obtained. In such a case, the stallion should be pulled away from the mare and the procedure started over. Although older well-trained stallions can be collected from either side, in the interest of safety all three people involved in the collection should be in view of each other. Further, since all horses are trained to be handled from the near side, it is logical to suppose that both mares and stallions would be more at ease and respond more naturally if the collection is done from this side. Once collection is obtained, the gel portion of the ejaculate should be removed immediately. If a filter has been used, this can be removed and thrown away. Since spermatozoa are temperature sensitive, it is important to protect semen from cold shock by placing the ejaculate in an incubator or a warm water bath at 37°C until insemination.

SEMEN EVALUATION

Since there is considerable variation in the quality of semen from different stallions and even between ejaculates within the same stallion, it is advisable to evaluate each collection to be used for artificial insemination. Semen evaluation criteria include volume, motility, concentration, and percent abnormal spermatozoa per ejaculate.

Semen volume can be determined immediately after collection since the collection bottle is calibrated. Motility must be evaluated under a microscope and is a measure of the viability of spermatozoa. Motility is estimated by examining a sample of the semen under a microscope and evaluating several fields. Motility may be estimated as the total percentage of cells that are moving with a separate rating for rate of forward movement, or these criteria may be combined into one index of motility referred to as percent progressively motile spermatozoa. Sperm cell concentration can be measured either with a hemacytometer under a microscope or by use of a spectrophotometer. The spectrophotometer must first be calibrated using a hemacytometer or commercially available latex particles. The percentge of abnormal spermatozoa must be determined under a high power microscope. Also, the sample must be stained for this evaluation unless a phase contrast microscope is used. Because of this requirement, the percentage of abnormal spermatozoa is usually not determined except when an extensive fertility exam is being done. The total number of spermatozoa in the ejaculate is obtained by multiplying the concentration by the volume. An outline of the specific procedures for semen evaluation is presented below (Sorenson 1977).

I. Semen Characteristics
   A. Volume -- measured in milliliters (ml), same as cc.
   B. Motility -- expressed as percent moving cells. These cells moving in any direction at any speed.
      1. Visual appraisal for motility
         a. Place a drop of raw (unextended) semen on a microscope slide and place a cover slip over it.
         b. Observe the concentrated sample for swirling motion at low power magnification (100x).
         c. Switch to 200x or 400x and observe for moving cells by differentially counting 10 sperm, distinguishing between moving and still. Count 10 cells in 10 different areas. Number of live cells counted will be percent motility.
   C. Rate of Forward Movement (RFM) - Expressed on a scale of 1-4 with 4 being most desirable. <u>Forward</u> is the key word.
      1 - No forward movement, all dead
      2 - Slow forward movement
      3 - Moderate forward movement
      4 - Fast forward movement

      Movement in any direction other than forward is not considered. Some swim circles, some backwards, and some beat their tails without forward motion. Only <u>forward</u> movement counts.
   D. Morphology - expressed as percent normal cells.
      1. Most easily observed after sperm have been killed and stained or viewed under phase contrast microscopy.
      2. Count 10 cells in 10 different fields as in determining motility. This is then expressed as percent.
      3. Morphology includes:
         a. Normal sperm cells.
         b. Coiled tail--may be an abnormality or may be caused by cold-shock in handling.

c. Looped tail--tail folds back on itself. Movement usually backwards.
d. Bent tail--tail bends at right angle at any level from neck to the posterior tail.
e. Immature--sperm with cytoplasmic droplet still attached to tail at any level. Presence of these indicates overuse of male.
f. Other abnormalities that may be seen.
1) Giant head
2) Small head
3) Headless
4) Tailess
5) Crooked tail
6) Double head
7) Double tail
8) Short tail

4. Staining spermatozoa--Place one drop raw bengal stain on slide. using an additional slide, make a smear as in doing blood smear. Allow to dry and examine under microscope.

E. Concentration--number of cells, usually given as sperm per unit volume.
1. May be estimated visually with some accuracy following extensive practice on known concentrations.
2. Microscopic counting--make with a diluting pipette and a hemacytometer.
a. Diluting pipette--an instrument used to accurately dilute sperm cells for counting. A disposable unopette or a standard blood diluting pipette may be used.
1. Depress plug into the reservoir until it drops inside
2. Remove pipette from shield and submerge capillary tube tip into sample of semen (unextended)
3. Allow semen to fill capillary tube
4. Remove from semen and wipe off excess semen on outside of capillary tube
5. Squeeze reservoir between thumb and forefinger and place the capillary end of the pipette into it; release pressure on reservoir and allow semen to be siphoned into

it; rinse pipette by squeezing reservoir several times, but do not force fluid out the tip
6. Place finger over the upper opening of the pipette and gently mix semen and diluent
7. Remove and invert the pipette
8. Hold reservoir under hot water for 1 min to kill sperm cells

b. Hemacytometer - instrument used to determine concentration of sperm cells.
   1. Locate grid lines on hemacytometer
   2. Notice that cover slip does not rest on grid, but on shoulders on either side allowing a 0.1 mm clearance
   3. Locate under microscope the 25 layer squares with triple lines; the triple lines outline an area 1

c. Counting
   1. After disposing of a few drops, touch pipette to the angle of the coverslip and grid surface and allow the diluted semen to flow under the coverslip until the area is filled; do not overfill
   2. Locate counting area under low power (100x) and then switch to high power (400x) for counting
   3. Count number of cells in 5 of the large squares; these should be counted diagonally from top left to bottom right

d. Calculation of concentration per cc of raw semen
   1. The number of sperm counted multiplied by 10^7 = concentration per cc of raw semen (The usual expression of concentration is sperm/cc or sperm per/ml). The calculation follows:
   2. Dilution rate with the pipette (a. above) = 1:200
   3. Hemacytometer factor (b. above) = 1/50 mm^3
   4. X number of sperm were counted in the 1/50 mm^3 (c. above)
   *5. Dilution rate X hemacytometer factor X sperm counted = sperm/mm^3, 1cc = 1000 mm^3 therefore the calculation continues: *200 X 50 X 1000 = 10,000,000(x) sperm/cc

3. Spectrophotometric counting
   a. Photometer is zeroed with a blank tube of 37% formalin solution.
   b. Semen is added to a tube of 37% Formalin solution in a 1:20 or 1:40 dilution
      1. 1:29 dilution = 7.6 ml Formalin
         .4 ml semen
      2. 1:40 dilution = 7.8 ml Formalin
         .4 ml semen
   c. Tube is toppered and inverted gently several times to mix dilution.
   d. Tube is placed into photometer (lid closed) and % transmittance is read after a few seconds as needle settles.
   e. On precalculated chart, % T is read and gives corresponding concentration of semen in number of sperm/ml.

II. Insemination Dose Calculation
   A. Total Cells = sperm/ml X volume (ml)
   B. Total Motile Cells = total cells X % motile cells
   C. Number of live motile cells needed for insemination. (500 X 10^6 is considered best for optimum conception.)
   D. Insemination Dose = motile $\frac{\text{sperm/ml X } 10^6}{500 \text{ X } 10^6}$

      = ____ ml of semen for each insemination

EXTENDING SEMEN

Semen may be used with or without an extender. The primary beneficial effects of using an extender are to control pathogenic organisms in the semen and to extend the volume of an ejaculate having low volume and high concentration. Other reasons given for using extenders are that they enhance fertility of ejaculates of subfertile stallions, prolong sperm survival, and protect sperm from unfavorable environmental conditions. However, further research is needed to fully substantiate these effects of extenders on semen. Thus an extender only needs to be used in cases where the stallion has become infected with a pathogenic organism, has low semen volume, or when the mare is suspected of having an infection.

If a stallion gives 30 ml of gel-free semen with sufficient concentration and motility to obtain 500 million (500 x 10^6) live motile sperm with 3 ml of raw semen, 10 mares could be bred with this ejaculate. However, the insemination dose would be only 3 ml and a substantial amount of this would be lost in the insemination pipette and syringe. If the semen were extended at a 4:1 ratio, 10 ml of extended semen would be used to inseminate each mare. This procedure would lose a lower percentage of the total cells in the insemination equipment. On the other hand, if there were only two mares ready to be inseminated to this horse, the ejaculate could be split so that the total insemination dose would be 15 ml, in which case there would be no need to add extender to increase insemination dose volume.

If an extender containing antibiotics is to be used, it should be added to the raw semen as soon as possible after collection. This is necessary to ensure an adequate time for the antibiotics to kill any bacteria present in the semen--whether it originates from the stallion or the collection equipment. Research has shown that a period of 30 min at 37°C will provide an adequate kill of bacteria present in the semen. A good management practice is to routinely culture the semen at 0, 15, and 30 minutes after addition of the extender to ensure that the antibiotics are in fact inhibiting bacterial growth.

The following are two formulas commonly used for stallion semen extender. They can be frozen in 20, 50, or 10 ml aliquots depending on which is most commonly used for a particular horse. Before adding the extender to raw semen it should be warmed to 100°F.

Skim Milk Extender

a. 12gm nonfortified nonfat dried skim milk is dissolved in 500 ml 5% dextrose solution and heated to 92°C for 10 min.
b. The solution is cooled to approximately 35°C and 500 thousand units of penicillin and 500 thousand units of streptomyin are added.
c. The extender is frozen in 10 ml, 20 ml, and 50 ml doses until needed.

Knox Gelatin Coffee Cream Extender

a. 1 pint half-and-half and coffee cream - heat to 100°F in a water bath.
b. 1/2 oz Knox unflavored gelatin in 40 cc. water; heat to 145°C in a water bath.
c. Mix gelatin solution and warm coffee cream and add: 1,000,000 I.U. buffered or crystalline penicillin; 3.75 cc gentocin; store frozen in 10 ml, 20 ml, and 50 ml doses and heat to 100°F in warm water bath before use. May be used from 1:1 to 4:1 extender to semen.

REFERENCES

Burns, S.J. 1977. Equine reproduction notes. College of Veterinary Medicine. Texas A&M University, College Station.

Kreider, J.L. 1981. Horse breeders' school manual. Animal Science Department. Texas A&M University, College Station.

Sorensen, A.M. 1977. Repro lab a laboratory manual for animal reproduction. Kendall/Hunt Publishing Co. Dubuque, Iowa.

Part 5

BEHAVIOR AND TRAINING

19

EXPRESSIONS OF THE HORSE

James P. McCall,
L. R. McCall

The language of the horse is the language of expression. Without the ability to communicate through a wide variety of vocal sounds, the horse developed a language capability based upon his evolution-derived sensitivity to movement. The ability to perceive slight movements in his environment led to the survival of the species from predatory attack. This trait that Nature developed for survival became the basis for herd communication.

Throughout the entire spectrum of the animal kingdom, communication takes on many faces. Communication may take place through sounds, scenting, visual displays or the highly complicated patterns used by Homo sapiens. Regardless of the expression of communication, all variations have one thing in common: information is transmitted to another individual, who, in turn, responds. Communication is the giving and sending of information--a two-way street.

One question that scientists have discussed is whether or not intraspecies communication exists. Is it possible to "carry on a conversation" with a member of another species? Oftentimes the parameters used to define language behind the ivy-covered walls of universities seems to cloud the issue. Is it really important to argue whether a species must creatively put together new phrases and ideas to have a language capability?

The function of language is to transmit information. That which is communicated relates to the species' frame of reference to the environment. In the world of the domesticated animals close to man, such as the dog, the cat, and the horse, the important lines of communication deal with whether or not master and pet approve of each other's actions. Did you ever misread the communication of a strange, snarling, snapping dog with his teeth bared? Did you ever fail to understand the message of a cat with outstretched claws, humped up and hissing? To most of us, these forms of communications are well recognized. Are you able to understand these animals because of a close association? How many people unfamiliar with horses would fail to understand the equine expression of ears flat back, head extended, and lips stretched tight against teeth?

We believe that the language of expression is a universal language that transcends species barriers. It is recognized that there are specific differences among species for expression of the same emotion; however, there seem to be common elements that can be read by all. Fear is easily recognized by the facial expression of dogs, cats, horses, and humans. Aggression is often expressed by the baring of teeth, arching of the back and neck, and flattening of the ears. Of course, it is very difficult for man to lay back his ears, but gritting his teeth during an aggressive discussion is a normal response, as is "bowing up" in the heat of an argument.

Since it is well outside the capability of the horse to learn human language, it falls to our sophisticated, adaptive intelligence to return to the common ground of the language of expression to communicate to the horse. Once on his ground, the horse can become amazingly sensitive to the messages being sent, both consciously and subconsciously. As an animal whose comfort and stability within the herd are a direct reflection of his ability to "read" the expression of other horses, it is an easy transition for the horse to seek information from human expressions. In fact, the horse can become so "in-tune" to his trainers that he can know them better than they know themselves because in the world of the horse, there is no rational thought. He does not recognize the ability to cover up true expressions by the process of rationalization. Fear is fear regardless of the mask a man may wear to disguise it. The man-made facade, therefore, is not seen by the horse. He only views the expression of fear and the <u>strange</u> behavior that occurs when man tries to cover up his fear.

Deviations from how the horse expects an emotion to be expressed only confuse and frighten him. When other animals feel fear, they try to flee from it. Only when cornered do they turn and fight. The horse sees man as being able to escape from a fearful situation, yet he remains and tries to cover up his fear behind a coat of cruel and overwrought aggression.

The example of the expression for fear is used to make a point: Horses and other animals in close association with the man "read" the emotional expressions of their masters. Instead of this being a "tale," trainers need to use this language of expression to better communicate their desires to their pupils.

To further describe the language of expression as it relates to the horse, let's divide it into two sections: the feelings (peaceful, spirited, aggressive...) and the actions (walk, run, eat, stop, go away...). In essence we have a language of verbs (the actions) that are modified by feelings. A horse walking peacefully is very different from one walking aggressively. Training a horse to respond to cues for actions is mastering only half of the horse. Communication through the language of expression will give you the ability to modify the action to be performed in a special way.

What are the "modifiers" of the language? Using Maslow's model for the basic human emotions, it was decided to break down the expressions of the horse into their primary components. In order for an expression to be considered a primary emotion, the language component had to be expressed by a young foal. This supposition led to the recognition of 6 basic expressions: peaceful, curious, submissive, spirited, confident, and desirous (wishful). These are the raw expressions from which the horse develops the complicity of his language.

As with human language development, it takes time and maturation for the horse to develop more sophisticated expressions such as trust and arrogance. Don't be misled into thinking that a language that is based on six expressions is a relatively simple one. We begin equine communication on a foal because of the simplicity of his dialogue at this age. We are not dealing with the recognition of only six simple emotional states because expressions have intensities.

Just as the Eskimos may have 50 words in their langauge to describe every conceivable kind of snow, the horse has a wide range of intensities that can be expressed for each emotion. We humans who rely upon our sentence structure to define the specificity of an emotion by the addition of phrases, adverbs, and other qualifiers, find it difficult to use multiple single words to characterize different intensities of an expression such as peaceful.

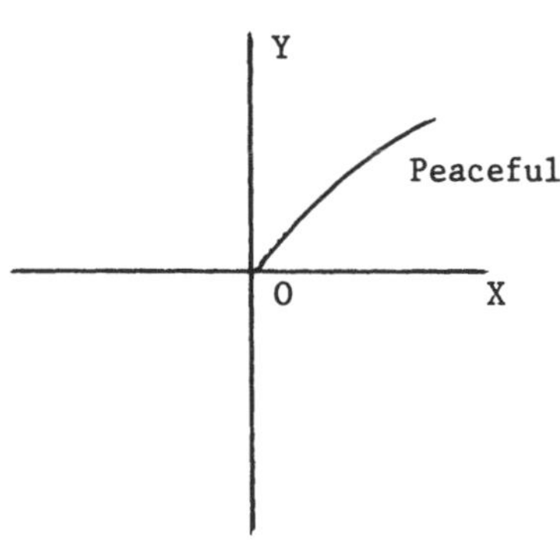

Figure 1.

In an effort to expand the understanding of the language of the horse, a model has been designed to help us deal with the problem of intensity. A basic emotion is graphed on an x-y axis where the y axis indicates intensity and the x axis defines the behavioral expression. At the intersection of x and y, the 0,0 coordinate, the emotion does not exist.

This allows for a more human way to talk about intensity rather than to use words such as more peaceful or less peaceful.

To further characterize the basic emotions, another assumption was made. Each expression must have a counterexpression. Foals are not peaceful all the time and the expression that is peaceful can fade into anger. Peaceful is a state without anger. It is a moment of calm. Anger, on the other hand, lacks calmness as irritation takes over. The model cannot be completed. Once the foal leaves the peaceful state, he enters the other end of the curve. As his irritation increases he moves further into the expression of anger.

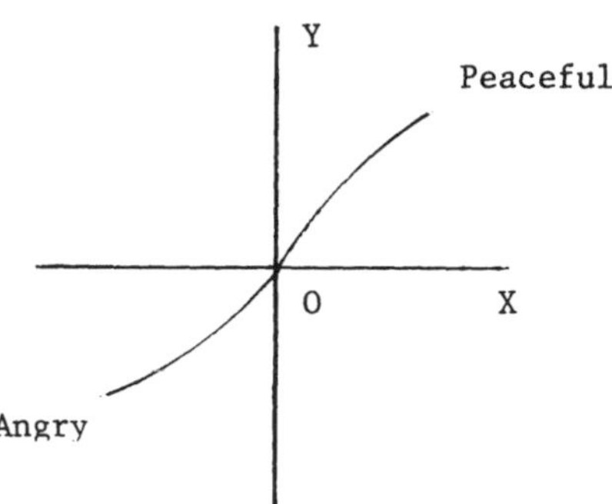

Figure 2.

A model exists for each expression.

The shape of the curve in figure 3 may vary between individuals. Just like humans, some horses are much easier to rile than others. Their curve would have a more intense slope.

This indicates that small changes in intensity bring on angrier expressions. These individuals are quick to fly off the handle and around them one feels as if he were playing with matches in a firecracker factory.

Then there are the super-cool guys and dolls who are determined to see the world as a calm, steady place. Nothing bothers them. Small changes in intensity are hardly noticeable.

It does not make any difference on which end the paired emotions are placed on the graph. The model may be constructed in either way. There is no value judgement placed on the paired expressions. One is not good while the other is bad. All emotions are a part of life and as such have their respected place in the survival of the individual.

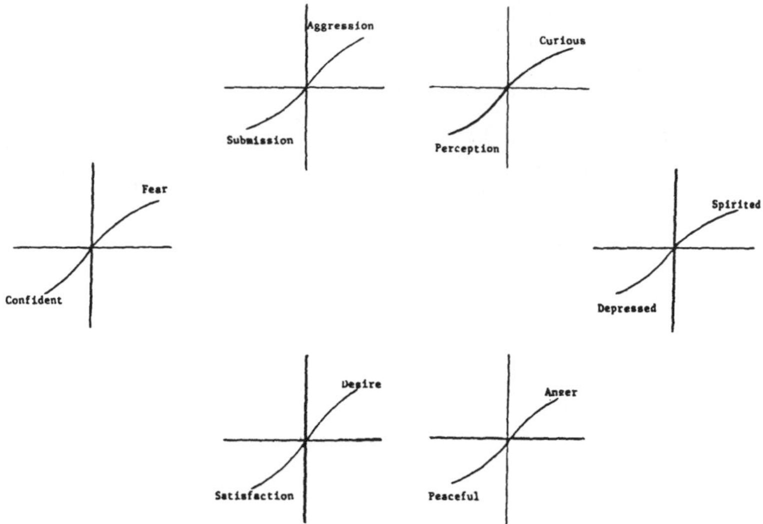

Figure 3.

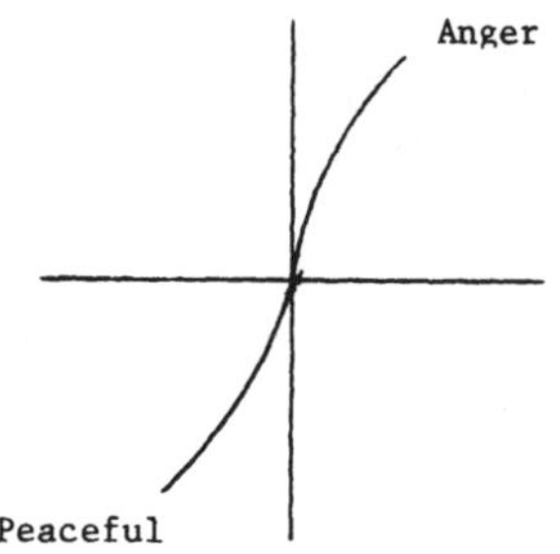

Figure 4.

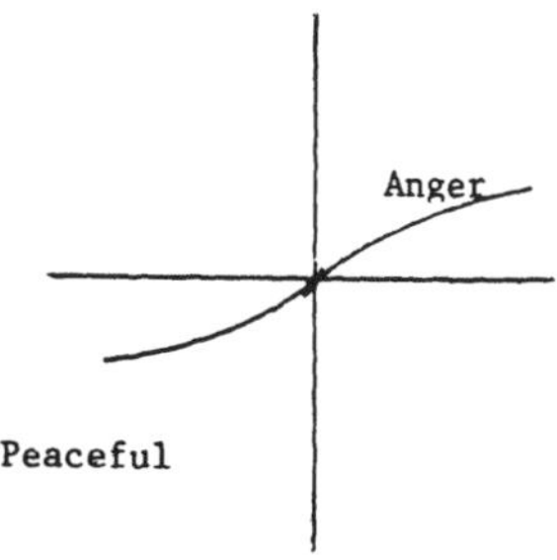

Figure 5.

By manipulating the behavior of the horse, a trainer can control the emotional state of the horse at all times. Ideal communication exists when man can ask the horse to *aggressively* execute a maneuver such as jumping a fence, cutting a cow, or running a race. The horse should feel the *desire* to want to please his master and to *submit* to his *wish.* At the same time, the trainer must *instill* in his charge the *confidence* that the horse can accomplish the goal and the *perception* of exactly what is being asked. In the blink of an eye, the entire mood can be directed by the skillfulness of one educated in the language of the horse to complete *calmness*.

Now, horse and man are one. They think the same thoughts, feel the same expressions, and perform the same jobs. Horsemen throughout the ages have sought to reach this goal. One of the philosophies behind many of the classical schools of Dressage is to capture the various expressions of natural movement of the horse by controlling and directing his energy. However, perhaps the most graphic expression of mastery over the mind and body of the horse can be found in man's mythological past.

As a child I often pondered about the wonderfully strange beast who was part man and part horse. A whimsical creation of man's mind, the centaur seems to express in a picture what thousands of words cannot explain. Two animals united as one. Direction and thought came from the human half, the action came from the horse--a team. Did our ancient ancestors sense the oneness we seek and design the centaur to speak of it?

20
BODY ENGLISH

James P. McCall,
L. R. McCall

Would you like to be able to walk into a pen with a horse and completely control his actions simply by moving your body? You can - once you understand the language of movement - or "body English" as it is commonly called.

Body English is based upon the natural movements and positions through which the dominant members of the herd control subordinates. Aggressive gestures, infringement upon personal space at specific angles, impulsion and blocking are the skills needed for the foundation. Mastery of the body language, however, requires the ability to move like a horse and the acquisition of a special sense called "feel."

One of the rewards for learning this equine language of movement is "free lounging." Free lounging works best in an enclosed area such as a breaking pen or round ring where the colt can be turned completely loose. By use of body English the horse is asked to perform all the preliminary movements that form the foundation for his training. An obvious starting point is in the control of direction and speed. Stopping and backing follow quickly. Turns into the wall and toward the center of the arena can be taught also. All of these basic maneuvers can be achieved during the first session. The natural language of the horse, which body English imitates, is easily interpreted and most naive horses have very little trouble in decifering the messages given by a "two-legged horse." Within just a few sessions, subtle cues can be expressed as the horse becomes more in tune with the manner in which the cues are given. There are no limits to this kind of training--even figure 8s, flying changes, and 360° turns. Whatever you are capable of communicating through movement, the horse can learn to respond.

Free lounging helps to get the horse into condition to withstand further training and develops lines of communication between horse and trainer. The horse becomes sensitized to respond to the body positions of his trainer - a habit that will come in handy when we mount and continue the colt's schooling from a position where we cannot be seen, but must be felt.

Learning body English begins with an understanding of the concept of personal space. There is an area around the body of most mammals that the individual claims as his own. The dimension of this space varies both among species and different members of the same species. The personal space of a horse is a circular area that radiates about 12 feet in all directions. The actual diameter of an individual's personal space is related to age and position within the hierarchy. The more aggressive the animal, the larger area he protects. Oftentimes this area can be defined by watching a submissive horse pass by a dominant horse. He will move in an arc around the higher ranking individual, thus defining the personal space of the dominant horse. This is an important facet of equine behavior--horses tend to move in circles and when threatened, they move in an arc around the aggressor.

The elements of free lounging are the aggressor, the trainer, the circle--a natural response--aided by walls to define the arc and the personal space of the horse which can be moved on to create the natural reactions and movements.

However, the horse is not the only animal in the ring with a personal space. Man, also, has a very sensitive personal space. Just like horses, the dimensions of our personal space depend upon our age, sex, innate level of aggression, and our mood at the moment. Surrounded by a hostile crowd armed with sticks and knives, our personal space swells in an effort to buffer the attack. In a quiet moment with someone very dear to you, the area softens to allow closeness between two individuals.

To free lounge a horse, you must be in tune with the range of energy that your subconscious projects into your personal space. You cannot ask a horse that respects no human to respect you with the expression of a submissive personal space. Likewise, a horse will not respond to the cue to come in if your personal space is swollen up, ready for an attack.

The horse is very sensitive to the area that you are ready to protect. It is part of the world in which he grew up. The flick of an ear, the swish of a tail, and the glare of an eye were all he needed to know exactly how close to push upon another member of his herd. To free lounge, your movements, actions, mood, and energy must match perfectly at all times. Then the slight movement of a foot, the twist of the shoulders, the drop of your head, the swish of an hip can control the movement of your horse.

21
BREAKING WITHOUT FORCE

James P. McCall,
L. R. McCall

Getting up on a totally naive 2-year-old who stands before you completely unrestrained by any tack requires that you have effective communication skills and the confidence to use them to teach the horse without frightening him. Respect and mutual trust in the relationship that has been built between man and horse are absolutely necessary for breaking to proceed without trauma.

In spite of the willingness of horses to assume a role in the transportation of man, man's transition from the ground to the horse's back has long been a difficult training point for the horse. By understanding the natural fears of the horse and patiently showing him that your actions do not relate to the instincts of his primordial past, within 30 min you can gradually make the move up onto his back without the use of any restraint. As the horse understands what you want to do, he will willingly submit to your desires. Instead of the fear-raising, high-energy jolt that mounting can be, the continuation of teaching and showing furthers the philosophy of oneness as you leave the ground to assume your position upon the back. Words, unfortunately, are inadequate to describe this process. Although the mechanics of the technique can be discussed, the timing and feel for the horse can never be explained. Each horse is an individual and must be taught as such. The mood of trust and serenity between the horse and his trainer cannot be captured in words; nor can the timing for each successive step as the horse accepts each point. These serious limitations strip the art from training and make it seem mechanical and textbook-like. It is not.

Yet the first step to achieving mastery over any creative endeavor begins with the foundation--the mechanics of the art, the notes of music, the strokes for painting, the basic positions for dance...and the steps for mounting.

The session begins with a brief longeing warm-up. You want to take the edge off the energy that the colt brings into the ring. Mounting is a low energy process that the horse will find difficult to concentrate on if he has too much energy to burn. The horse should be longed until he is ready to work slow and steady on the rail. Then the horse is called into the center of the area. After a moment of reward for his longeing, you step to his side and press your body against his. The arms are laid across his back. A simple maneuver; nevertheless, it is the first introduction to pressure upon his back. Should the colt try to move away from the weight, he is permitted. You simply lift your arms and he will walk out from under them. The process then starts all over: the colt is asked to return, to accept the pressure of your body on his side, and finally the weight of your arms on his back.

Once this pressure is accepted, let your arms hang over to his off side and gently rub his side with the palms of your hands. Do not use your fingers to scratch his sides. The palm is a sensation that the colt can relate to. It has a similar feel to the touch of another horse. The horse is now being touched on both sides of his body, simultaneously, for the first time. Soon, instead of body and arms draped over his back, you will sit upon his back and your legs will touch his sides.

The next move is to acquaint the horse with what it is going to feel like when you make the move from standing on the ground to sitting on his back. The energy and push needed to mount will spook the green colt unless he is ready for it. This is done in stages. The first step is to hop gently against the horse, increasing the amounts of pressure on his back, but never really leaving the ground. In doing this, be sure to watch out for elbows and fingers! The response of most colts is just to turn around and see what's happening. However, if your youngster decides to move away while he thinks about this new antic, let him. At any time during this session, the horse is allowed to move off. Strangely, nearly all colts will turn right around and come back to you before going more than four or five steps.

Once the gentle hopping against the side of the horse is accepted, it is time to leave the ground with successively bigger jumps. Give time between each larger jumping effort for the horse to understand that he is not being threatened.

Bellying comes next. To get from the ground to a position lying across the horse's back without a running start and without digging into his back takes some athletic ability along with the right body type. Most women find that they cannot accomplish this feat with finesse and flow without the aid of a leg-up. Whoever lends a hand must be sure not to disrupt the relationship and the mood of the session. The helper must be able to lift you smoothly and slowly over the back. Should the colt spook, slide off and begin again. Occasionally, the horse will calmly take a step or two under the weight, then stop. In these instances, you can remain bellied. But, most of the time, the horse won't move because he is uneasy about carrying the extra load. He doesn't know whether your weight will allow him to move as he always has. He is unsure about his balance and ability to maneuver. So rub him and talk to him to reassure him and keep his attention upon you. Let him know he has nothing to fear. When he relaxes, get off.

Before going any further, the colt needs a few minutes to gather himself and to fully realize that nothing bad has happened to him. If you have handled the training right so far, the colt has not been scared. In fact, he serves a reward for his patience and understanding, perhaps a scratch on his favorite "itchy place," or a rub on the withers. The lesson is not over yet, so be sparing with the praise. Be sure that the horse understands that this is just a brief intermission and not the end of the session.

Without rushing, return to the bellying position. A good job done so far will ensure that the colt will be ready for what is going to happen next. Slowly...lift your right leg and gingerly swing it over the croup, being sure not to hit the horse. Barring any mistakes, you can ease yourself up into a sitting position. If the colt becomes excited, swing your right leg over the withers and jump off. To try and ride would require that you hold on with your legs--a response that would serve to intensify your colt's fear and further destroy the harmony of the relationship. Begin again and repeat the procedure until the colt accepts the position without fear.

Once astride the colt, sit squarely on his back with your legs loosely at his side. In this relaxed position, reward the horse and remain upon his back until he relaxes, usually about 2 minutes. To dismount slowly reverse the steps and slide gently to the ground.

The accomplishment of the day's lesson has been a test of learning for both you and the horse. The respect and trust that began to grow will serve as the foundation for further training. Soon the two of you will be walking, trotting, stopping, backing, and turning without the use of man-made devices. Communication will consist of cues built upon the natural response of the horse to different balance and pressure points upon his back. Training will continue to be a man and a horse seeking ways to communicate in order that they might move as one.

22
SOCIAL HIERARCHY

James P. McCall,
L. R. McCall

One of the attributes that sets mammals apart from the rest of the animal world is that their young are born alive and in need of a period of mothering before they can fend for themselves. Evolution has created a bond between parent and offspring that allows this necessary protection and caring to take place. It is during this time that many animals learn the needed tools of survival.

Disturbance of this normal period of mothering by removal of the maternal influence can produce maladjusted animals. Early weaning studies in some mammalian species have shown that such individuals mature physically but lack the proper socialization to become normal adults. For examples, females that do not spend adequate time with their mothers have a greater chance of being poor mothers than normally reared offspring. On the whole, early weaned mammals tend to be less adaptive, less able to fend for themselves, and less able to associate properly with others of their kind.

The world of the horse is made up of a social order or dominance hierarchy. If the horse is not allowed adequate time to learn the necessary social graces from his dam, can he properly function within the structure of the herd?

The traits or behaviors that determine the structure of the pecking order in the horse have been determined by the Law of Nature, the survival of the fittest. Five thousand years of domestication have not changed millions of years of evolution.

Any time two or more horses are placed together they form a herd wherein the stronger will rule the weaker. One of the main characteristics of the dominant individual is a high level of aggression. A threatening gesture from him or her usually convinces all subordinates to comply.

The natural order of things declares that the herd will be ruled by the most dominant and aggressive member of the group, which is the stallion in most cases in the wild. The stallion has come to this position by proving that he is the strongest and most wily male in the area. Nature's reward is to provide him with the opportunity to pass on his genetic makeup through the offspring of the mares in his band.

The mares within the herd have a social order, or ranking. The top mare, or boss mare, rules the entire female herd and is usually submissive only to the stallion. She is the leader when the herd changes locations. Occasionally there may be a role reversal in that the boss mare may dominate the stallion, but this is not likely to occur except in man-made pasture-breeding herds or in the wild where there is a limited number of bachelor stallions to vie for the position of herd stallion.

It is possible that an aged boss mare may take on the role of the matriarch. When this happens, she passes on to her first-in-command, a younger dominant mare, the power to rule. The matriarch then has little interaction with the other mares but her authority is never questioned. She is never encroached on nor does anyone threaten her. This situation is unusual, but it happens.

A mare's position is fairly constant throughout her adult life. There seem to be two critical ages, one at each end of her life, that control her rise and fall in the hierarchy. Sometime between the age of three and five most mares will move up in the hierarchy from the lower end of the pecking order where most of the young and old horses are found.

Battling each successively higher horse for dominance, the mare will reach the limitations of her aggression and find the niche where she belongs. She will maintain this relative position until around age 18 when she may begin to slip back down through the hierarchy of the herd.

There are a few environmental elements that can change the position of a mare. One such factor is a deprived nutritional condition. When the entire herd is under the same nutritional plane, a loss in condition is shared by all members. But if one horse should be in relatively poorer condition while the others remain in good flesh, then that horse may lose position.

Serious illness may also lower the mare's position, although the influence of sickness is difficult to judge. One specific case showed that a 15-year-old thoroughbred mare, which was number three in the social order, dropped only one place during the day she was dying of pneumonia.

When a mare is placed in an existing herd, she is normally positioned at the lower end of the pecking order. Over the course of a year or so, she gradually works herself up to her proper position. On the other hand, a mare who has achieved her proper position may be removed from the herd for a time only to return and immediately assume the same position.

Within the herd are three basic types of mares: alphas, betas, and zetas. Alpha mares are the most dominant. Given the opportunity, they interact very little with each other and receive few if any threats from beta and zeta mares. These mares have the first choices for available grazing lands, shelter, and other limited resources that they need for their comfort and survival.

Beta mares get the next choice of the remaining possibilities, while the zetas get the leftovers. As a group, however, the beta mares are involved in a lot of bickering and fighting in an apparent attempt to maintain their ranked order. The zetas, the lowest group, spend most of their time being chased away by higher ranking mares and then seeking additional resources to provide for their survival.

Dominant mares have a better chance for survival both for themselves and their offspring. They have first choice of water, feed, and the attention of the stallion. In general, they are more adaptable, better able to cope with different environmental situations, and better able to fend for themselves and their young. This increases the chance that their offspring will survive and pass on the genes that helped make them dominant members of the equine herd.

As the picture of the natural equine world becomes clearer, much useful information is gained. Insight into the role of the individual within the herd provides practical knowledge that can be used to make better decisions about the production and management of horses.

For example, certain types of problems can arise when horses are fed in a group-feeding situation. The boss mare will claim her feed tub and remain at that site until she chooses to leave. Except for the time she spends warning other passing mares not to bother her, she will be eating. The rest of the mares will be participating in a situation that resembles musical chairs. Forced from their tubs by more dominant mares, they move around to claim the tub of an even less dominant mare who, in turn, will circulate. This results in the more submissive mares spending more time walking than eating, whereas the more dominant mares will be run off less often. Many times this leads to the top mares getting more feed than they need while the mares at the bottom do not get sufficient amounts. A simple solution lies in increasing the number of feed tubs and placing them more than 20 feet apart so that the more aggressive mares cannot lay claim to more than one feeder.

Even in the pasture the social order will influence which mares get the choice grass--and, if the pasture is limited, which mares get any grass at all. To have a more equal distribution of the natural resources, it may be necessary to divide a herd by placing all the dominant, highly competitive mares together and leave the less aggressive mares to compete among themselves. Surprisingly, this may be a necessary step when there are as few as 18 mares on 150 acres of good grass land.

Another place that social order influences management decisions is in a group-teasing procedure. A stallion is placed in a teasing cage in the center of an enclosed pen. The mares are run into the area and the ones that are in heat will approach the stallion to tease. Managerially this allows for a maximum number of mares to be teased in the shortest period of time. Yet, for this system to be totally effective, the manager must have a thorough understanding of the social structure of the mare hierarchy. The more dominant in-heat mares will vie for the attention of the stallion and may not allow the submissive mares to come near. If the dominant mares do not leave the cage, they may have to be removed so that the other mares will approach. Oftentimes a submissive, timid mare will not react to teasing by an aggressive stallion. In fact, she may be in heat and still completely reject his presence. The same mare, when presented to a more subdued stallion, will tease properly and show the normal signs of estrus.

At one large farm in California, this knowledge led to the design of a teasing system in which mares were run by a series of chutes holding three different stallions with varying levels of aggressiveness. The mares were permitted to tease to the stallion of their choice. By allowing for individual variations, this method has proven highly successful in determining when each mare is ready to breed. It is particularly useful for shy, submissive mares.

Evidence is beginning to accumulate that there might be a correlation between the aggressive/submissive nature of an individual horse and her performance potential. One of the questions that arises from this new supposition is whether aggression in the race horse makes for a faster animal. Can the aggressiveness of the individual be manipulated to produce the desire to be first across the finish line as well as first at the feed tub?

There are a number of thoroughbred breeders who think so. As their young race horses run and play in the fields, the more dominant yearlings go to the front and use threatening gestures to keep their submissive playmates from passing. Does this have any future value in determining which yearlings have the best chance of being good race horses? Unfortunately, it is still too early to know. However, many of the great sires throughout thoroughbred history were reported to have a high level of aggression - St. Simon, Nasrullah, Fair Play, Hastings, and the more contemporary Native Dancer. It is believed that these stallions passed on this trait to many of their offspring. Does this indicate that disposition may, at least in part, be inherited? If so, do the sire and dam contribute equally or does one play a larger role? Originally, a mixed herd of thoroughbred and quarter horse mares was used to provide data to furnish insight into these questions, but upon ranking the group, an interesting problem arose. Instead of the mares being randomized throughout the dominance hierarchy, the breeds tended to segregate with a majority of the thoroughbred mares at the top and the quarter horse mares at the bottom.

Another unusual finding was that the quarter horse mares, although apparently lower in aggression than the alpha and beta thoroughbreds, did not take on the characteristics of zeta mares. The highest ranking quarter horse behaved as an alpha within the quarter horse group--and the betas and zetas maintained their respective rankings. These possible breed differences led to the removal of the quarter horse mares.

The thoroughbred mares were then all bred to the same thoroughbred stallion in an attempt to equalize any effect the stallion would have on the disposition of the foal. Upon reaching one year of age, the foals of these mares were ranked among themselves and each foal's position compared with the rank of his dam. The results showed that the foal's position was highly correlated to his dam's place in the mare pecking order. Dominant mares produced dominant foals, submissive mares had submissive foals (when the stallion influence was constant).

What was the dominant female providing her young? Was it some physical attribute such as weight, height, feed efficiency, or growth per day? The data said no.

Perhaps, then, it was dominant behavior that was learned during the critical period from birth until weaning. To weigh the effect that premature weaning might have on the normal socialization of the foal, a new foal crop was divided into two groups. One group was weaned at two months and the other at five months. Each group consisted of an equal number of foals from alpha, beta, and zeta mares.

Group 1 Weaning age 5 months	Group 2 Weaning age 2 months
Foal of mare ranked #1	Foal of mare ranked #2
Foal of mare ranked #3	Foal of mare ranked #4
Foal of mare ranked #5	Foal of mare ranked #6
Foal of mare ranked #7	Foal of mare ranked #8

At one year of age, both groups of foals were ranked together. All of the late-weaned foals were more successful in dominating control over a feed tub than were their early--weaned counterparts.

Rank of foal	Age at weaning
1	5 months
2	5 months
3	5 months
4	5 months
5	2 months
6	2 months
7	2 months
8	2 months

Closer examination within each group showed another interesting correlation. The ranked order of the foals within each group closely approximated the ranking order of their dams.

Group 1		Group 2	
Rank of mare	Rank of foal	Rank of mare	Rank of foal
4	1	2	5
5	2	3	6
7	3	6	7
1	4	8	8

In group 1, with the exception of the foal of the number 1 mare, which was badly injured several times as a weanling, there was a positive correlation in the dominance of the mare over certain individuals and a dominance of her foal over the foals of the submissive mares. Within this group Mare No. 4 was dominant over No. 5 and No. 7, but not over No. 1. Likewise, her foal was also dominant over the foals of No. 5, No. 7, and the injured No. 1. The correlation was even more complete in Group 2, as there was not an exception. Thus, the ranked order of the mares' foals was much like that of the mares. Weaning age seems to be able to change the ranked order of the offspring. Zeta mare No. 7s foal was the third-ranked individual in the yearling herd although three of the four individuals that were supposed to dominate him were early weaned and a fourth (foal of the No. 1 mare) was injured.

Does this preliminary data indicate that along with a genetic contribution, learning is involved in the preparation of the foal to meet the challenges of dominance and aggression? Can we then determine through selective breeding and, perhaps, weaning age, the level of aggression in the horse? Does breeding our alpha mares to aggressive stallions give the best odds for producing stake horses or horses for other aggressive events such as cutting?

The answers to these questions are still unknowns. Research is currently underway to help unravel some of the mystery.

23

LEARNING ABILITY: THE LIMITING FACTOR IN HORSE TRAINING

James C. Heird

Although horse behavior and training sounds like a rather simple topic, it can be very complicated. Horses are unique among domestic animals in that their value is increased with training. Furthermore, they are worth little if they cannot be trained. The horse's inate behavior affects how he will learn or accept training. Training, in fact, is thought of as simply a "modification of behavior." Through the study of psychology, much as been learned in recent years about behavior. We know that all behavior, either desired or undesired, is controlled by stimuli and reinforcement.

The specific part of behavior controlled by stimuli and reinforcement is known as response. In training, we try to get the horse to make the desired response. One of the common mistakes we make in training is to consider a stop, spin, or jump as a response. Actually, these and similar feats are the result of several responses grouped together to form a complex movement. It is extremely important, therefore, that we realize that all major movements are the result of a series of small responses. These small responses are the result of careful training every time the horse is handled.

Responses are the result of stimulation. In training, we know stimuli as cues. If possible, we like to begin training with those cues that are said to be unconditioned. These are the stimuli that do not have to be learned. In training, there are few useful stimuli that are naturally occurring. However, the horse is exposed to many experiences during his life that serve as unconditioned stimuli. One of the problems associated with unconditioned stimuli is that many trainers start their program by frightening the horse. Fear is an unconditioned stimuli. For billions of years, the horse has been a creature of flight. He runs or flees from the things that scare him. If the natural response to a stimuli is flight, training becomes much harder.

Most stimuli used in training are conditioned stimuli. These are cues that are learned. We like to begin training with the simplest cue possible. Then, as training progresses, we can graduate to more and more complicated cues that result in a trained horse. However, training is a step-by-step process that must progress from the simple to the most complex.

Most tests of intelligence indicate that the horse is not an animal of superior intellect. Therefore, the training program must begin with the assumption that we are working with an animal that is not considered to be smart. If we keep this in mind, we can make training much easier. All cues must be specific. The horse must be able to distinguish betwen any cues presented. It takes a good horseman to be able to give distinguishable cues. This explains why some people train horses better than others. A good trainer is a good athlete who has exceptional balance dexterity. In addition, the good horseman sits quietly so that he never inadvertently cues the horse.

Sometimes it becomes necessary to strengthen the response to certain stimuli. Those events that are capable of strengthening a response are known as reinforcers. Learning is impossible without some kind of reinforcement. There are two types of reinforcement similar to the two types of stimuli. Primary reinforcement is reinforcement that does not have to be learned. There are few primary reinforcers used in horse training. Feed is the possible exception. Reinforcement that has to be learned is known as secondary reinforcement. These are numerous and are used extensively in training. In fact, most training is strengthened by secondary reinforcers rather than by primary reinforcers. It should be obvious that the horse will learn more quickly with primary reinforcement than with secondary reinforcement. However, it is sometimes impractical to use primary reinforcers.

All reinforcement, whether primary or secondary, is classified as either positive or negative. Since most trainers tend to work from the standpoint of negative reinforcement, we should deal with this type of reinforcement first.

Negative reinforcement is punishment. Punishment is defined as that which is adverse in nature and would be avoided if given the opportunity. Most trainers tend to train with the idea that if the horse does not do something correctly, he will be punished. Basically, most horsemen fail to understand the relationship of positive and negative reinforcement in training. Punishment should be used correctly. The severity of the punishment depends upon the severity of the offense. Remember, punishment only suppresses a habit, it does not eliminate it. Therefore, punishment must coincide with the offense. Punishment incorrectly administered or given excessively can cause behavior worse than the offense being punished. Remember, the horse's tendency is to flee from what scares him. Furthermore, the over-use of punishment causes the horse to lose sensitivity. If he is continually severely punished for insignificant acts or misbehavior, the trainer loses the range of reinforcement available to him. If the horse develops a dangerous habit that requires severe punishment to suppress, there is no punishment available that will affect the animal.

Any time punishment is used, an alternate response should be available. This alternate response should be the one that was desired in the beginning. The good trainers are those who are aware of the response desired and, when they use punishment, they allow the horse the opportunity to make the correct response.

As stated previously, reinforcement can also be positive. Positive reinforcement is known as reward training. Reward training is defined as a procedure whereby the animal receives a reward for the correct reponse. Reward training is not used as much as it should be in horses. Rewards for positive reinforcement can be of several types: for example, a pat on the neck, rest, or food.

Horses learn faster with negative reinforcement; however, they retain material longer with reward training. In addition, under stressful situations the horse will be more dependable with the use of reward training.

Avoidance training is also used in horse training. Avoidance training is defined as response preventing adverse reinforcement. Simply stated, this means that the horse learns that for certain acts he will receive punishment. If he does not do those acts, then he will not receive the adverse reinforcement.

The truly great trainers combine reward and punishment with avoidance training. One type of reinforcement will not work as well as the combined use of all three.

Regardless of the type of reinforcement used, it must be contingent. All punishment or reward must be administered quickly to make the horse aware of his actions, both good and bad. This is especially true as we remember the level of intelligence of the horse.

There are some other aspects of behavior and training that affect learning. Athletic ability is one of these. The less physical effort required of a horse to do a maneuver, the easier it is for the horse to learn. The more effort required, the harder it is to learn. That is why some horses seem to learn quickly the simple training maneuvers, but they become impossible "hard-heads" as we ask for more difficult responses. This is basically why some breeds are superior at certain events.

Regardless of the amount of knowledge possessed by a trainer or the athletic ability of the horse, very little learning can take place unless the horse has the proper disposition. No trainer likes to work with a sullen or despondent animal. In addition, some horses are mean and have no desire for human associations.

Another aspect not mentioned up to this point is the effect of early experience upon learning. Studies at Texas Tech indicate that early handling affects learning. Yearlings tested were divided into three groups based upon early handling. Group one had no handling. Group two was handled intermediately. Group three was raised as pets and handled extensively. It was found that the intermediate group learned the quickest. Group one, the nonhandled group, achieved no learning at all at the end of 20 days. Group two had by far the best learning curve. Group three, the very gentle, excessively handled group, learned at a much slower rate than did the intermediately handled group. These results are contrary to the results expected by most horsemen. The intermediate group learned the most because they were gentle enough not to be afraid but not so gentle that they were bored. The extensively handled group seemed not to be challenged by the simple maze that was used to test learning. These results would indicate that we might need to change our training practices based upon the amount of handling a horse has prior to the start of training.

Finally, something that has not been mentioned is intelligence. Some horses are smarter than others. Genetic as well as experience differences influence intelligence. Even full brothers and sisters may differ in intelligence. The good trainer recognizes this and works accordingly.

In summary, problems connected with training can be answered if we remember two simple statements:

> The trainer has to be smarter than the horse.
> The trainer has to have common sense.

Those two statements are the root to the answer of all problems connected with training.

The only advantage that we as trainers have over the horse is a superior ability to reason and think. If we are smart, we try to make learning easier by varying our training methods. Common sense will let us know that in training we have to use our minds. The horse is stronger, faster, and quicker than any human. If a trainer loses his ability to think through either anger or poor control of his temper, he puts himself on the intellectual level of the horse. If this happens, the horse is superior. Training rules are not written in stone, thus we must learn to think and use the method that will get the best results.

Part 6

ANIMAL PSYCHOLOGY, ENVIRONMENT, AND WELFARE

24

LIVESTOCK PSYCHOLOGY AND HANDLING-FACILITY DESIGN

Temple Grandin

Handling your cattle and sheep will be much easier if you learn a little livestock psychology. Many people do not realize that cattle and sheep have panoramic vision and they can see all around themselves without turning their heads (Prince, 1977; McFarlane, 1976). Sheep with heavy fleeces would have a more restricted visual field depending on the amount of wool on their head and neck. Both cattle and sheep depend heavily on their vision and are easily motivated by fear (Kilgour, 1971). Livestock are sensitive to harsh contrasts of light and dark around loading chutes, scales, and work areas. "Illumination should be even and there should be no sudden discontinuity in the floor level or texture" (Lynch and Alexander, 1973).

Solid shades should be used over the working, loading, and scale areas (Grandin, 1981). Slatted shades are fine for areas where the animals live and feel familiar. However, when the animals come into the handling areas they are often nervous. The zebra stripe pattern cast by the slatted shades constructed from snow fence or corrugated sheets suspended on cables will cause balking. The pattern of alternating light and dark has the same effect as building a cattle guard in the middle of the facility. Contrasts of light and dark have such a deterrent effect on cattle that in Oregon lines are painted across the highway to take the place of expensive steel cattle guards.

Shadowy stripes will cause balking problems with sheep. A single-file chute for sorting sheep should be oriented so that the sun does not form a shadow down the middle of the chute. The worst possible situation for sheep is to have half the floor of the chute in the shade and the other half in the sunlight. In shearing sheds and sheep holding areas, the wooden slats on the floor should face so that the sheep walk across the slats instead of in the same direction as the slats (Hutson, 1981). If you get down on your hands and knees and look at the floor, the floor appears more solid if you move across the slats. The floor should also be constructed to prevent sunlight from shining up through the slats.

A single shadow that falls across a scale or loading chute can disrupt handling. The lead animal will often balk and refuse to cross the shadow. If you are having problems with animals balking at one place, a shadow is a likely cause. Balking can also be caused by a small bright spot formed by the sun's rays coming through a hole in a roof. Patching the hole will often solve the problem. Handlers themselves should be cautious about causing shadows. Figure 1 illustrates a shadow that was formed when the handler waved at the cattle. The animals refused to approach the shadow of the waving handler cast at the entrance to the single-file chute.

Figure 1. The handler's shadow cast on the entrance to the single-file chute caused the cattle to balk. This is just one of the many kinds of shadows which can cause balking problems in your cattle handling facility.

APPROACH LIGHT

Both cattle and sheep have a tendency to move towards the light. If you ever have to load livestock at night, it is strongly recommended that frosted lamps that do not glare in the animals face be positioned inside of the truck (Grandin, 1979). However, loading chutes and squeeze chutes should face either north or south; livestock will balk if they have to look directly into the sun.

Sometimes it is difficult to persuade cattle or sheep to enter a roofed working area. Persuading the animals to enter a dark, single-file chute from an outdoor crowding pen in bright sunlight is often difficult. Cattle are more easily driven into a shaded area from an outdoor pen if they are first lined up in single file.

Many people make the mistake of placing the single-file chute and squeeze chute entirely inside a building and the crowding pen outside. Balking will be reduced if the single file chute is extended 10 to 15 feet outside the building. The animals will enter more easily if they are lined up single file before they enter the dark building. The wall of the building should NEVER be placed at the junction between the single file chute and the crowding pen. Either cover up the entire squeeze chute and crowding pen area or extend the single file chute beyond the building. If you have just a shade over your working area, make sure that the shadow of the shade does not fall on the junction between the single file chute and the crowding pen.

PREVENT BALKING

Drain grates in the middle of the floor will make both sheep and cattle balk because the animals will often refuse to walk over them. A good drainage design is to slope the concrete floor in the squeeze chute area toward an open drainage ditch located outside the fences. The open drainage ditch outside the fences needs no cover and so it is easier to clean.

Animals will also balk if they see a moving or flapping object. A coat flung over a chute fence or the shiny reflection off a car bumper will cause balking. You should walk through your chutes and view them from a cow's eye level before moving or loading animals. You will be surprised at the things you may see. When cattle and sheep are being worked, the handlers should stand back away from the headgate so that approaching animals cannot see them with their wide angle vision. The installation of shields for people to hide behind can facilitate the movement of livestock (Kilgour, 1971; Freeman, 1975).

Problems with balking tend to come in bunches; when one animal balks, the tendency to balk seems to spread to the next animals in line (Grandin, 1980). When an animal is being moved through a single-file chute, the animal must never be prodded until it has a place to go. Once it has balked, it will continue balking. The handler should wait until the tailgate on the squeeze chute is open before prodding the next animal (Grandin, 1976). A plastic garbage bag attached to a broom handle is a good tool for moving cattle in pens. The cattle move away from the rustling plastic. When livestock are being moved, well-trained dogs are recommended for open areas and large pens. Once the animals are confined in the crowding pen and single-file chute, dogs should not be allowed near the fences where they still can bite at the cattle or sheep.

SOLID CHUTE SIDES

For both cattle and sheep the sides of the single-file chute, loading chute, and crowding pen should be solid. Solid sides prevent the animals from seeing people, cars, and other distractions outside the chute. A study with sheep showed that they moved more rapidly through a single-file chute that had solid sides (Hutson and Hitchcock, 1978). The principle of using solid sides is like putting blinkers on the harness horse. The blinkers prevent the horse from seeing distractions with his wide-angle vision. Cattle and sheep in a handling facility should be able to see only one pathway of escape--this is extremely important. They should be able to see other animals moving in front of them down the chute, when sheep are being sorted, the approaching animals should be able to see the previously sorted sheep through the end of the sorting chute.

Livestock will balk if a chute appears to be a dead end (Brockway, 1975; Hutson, 1980). Sliding and one-way gates in the single-file chute must be constructed so that your animals can see through them, otherwise the animals will balk (figure 2). The sides of the single-file chute and the crowding pen should be solid. The crowding-pen gate also should be solid so that animals cannot see through and will head for the entrance to the single-file chute (Rider, 1974). Mirrors could be used to attract sheep into pens and other areas that appear to be a dead end. The sheep are attracted to the image of sheep in the mirror (Franklin & Hutson, 1982).

HERD BEHAVIOR

All species of livestock will follow the leader and this instinct is strong in both cattle and sheep (Ewbank, 1961). Many people make the mistake of building the single-file chute to the squeeze too short. The chute should be long enough to take advantage of the animal's tendency to follow the leader. The minimum length for the single-file chute is 20 ft. In larger facilities 30 to 50 lineal ft is recommended.

Cattle and sheep are herd animals and, if isolated, can become agitated and stressed. This is especially a problem with Brahman-type cattle. An animal left alone in the crowding pen after the other animals have entered the single file chute, may attempt to jump the fence to rejoin its herdmates. A lone steer or cow may become agitated and charge the handler. A large portion of the serious handler injuries occur when a steer or cow, separated from its herd-mates, refuses to walk up the single file chute. When a lone animal refuses to move, the handler should release it from the crowding pen and bring it back with another group of cattle.

Figure 2. The single-file chute to the squeeze should have solid sides to prevent the cattle from seeing distractions outside the fence. Sliding gates in the single-file chute must be constructed from bars so that the cattle can see through them. Solid sliding or one-way gates will cause balking.

EFFECTS OF SLOPE AND WIND

To prevent livestock from piling up against the back gate in the crowding pen, the floor of the pen must be level. A 10° slope in the crowding pen will cause the animals to pile and fall down against the crowding gate. A small 1/4 in. to 1/8 in. slope per foot for drainage will not cause a handling problem. Livestock move more easily uphill than down, but they move most easily on a flat surface (Hitchcock and Hutson, 1979).

Research by Hutson and Mourik (1982) indicates that sheep will move more easily when they are heading into the wind. Heading into the wind can stimulate sheep to start moving along a chute.

WHY A CURVED CHUTE WORKS

A curved chute works better than a straight chute for two reasons. First it prevents the animal from seeing the truck, the squeeze chute, or people until it is almost in the truck or squeeze chute. A curved chute also takes advantage of the animal's natural tendency to circle around the handler (Grandin, 1979). When you enter a pen of cattle or sheep you have probably noticed that the animals will turn and face you, but maintain a safe distance (figure 3). As you move through the pen, the animals will keep looking at you and circle around you as you move. A curved chute takes advantage of this natural circling behavior.

Figure 3. When you walk through a pasture the cows will turn and look at you. They will circle around you as you move about the pasture. Curved chutes take advantage of the cow's circling behavior.

Cattle can be driven most efficiently if the handler is situated at a 45° to 60° angle perpendicular to the animal's shoulder (Williams, 1978) (figure 4). A well-designed, curved single-file chute has a catwalk for the handler to use along the inner radius. The handler should always work along the inner radius. The curved chute forces the handler to stand at the best angle and lets the animals circle around him. The solid sides block out visual distractions except for the handler on the catwalk.

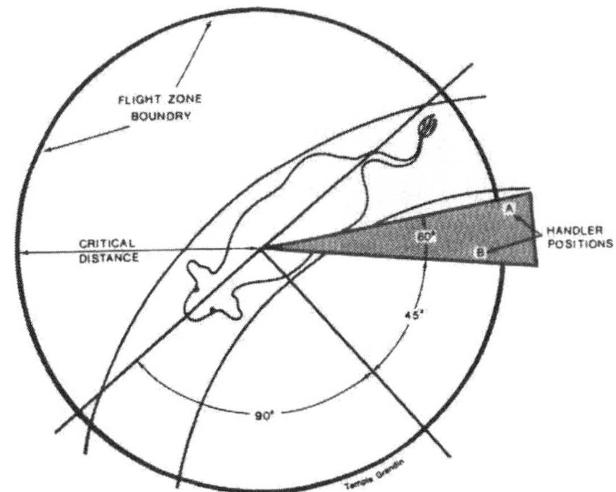

Figure 4. The shaded area shows the best position for moving an animal. To make the cow move forward the handler moves into Position B which is just inside the boundary of the flight zone. The handler should retreat to Position A if he wants the animal to stop. The solid curved lines indicate the location of the curved single-file chute.

The catwalk should run alongside of the chute and NEVER be placed overhead. The distance from the catwalk platform to the top of the chute fence should be 42 inches. This brings the top of the fence to belt-buckle height on the average person.

Figures 5 and 6 illustrate curved facilities for handling cattle and sheep. Curved designs are recommended by both Grandin (1980) and Barber (1977).

FLIGHT DISTANCE

When a person penetrates an animal's flight zone, the animal will move away. If the handler penetrates the flight zone too deeply, the animal will either turn back and run past him or break and run away. Kilgour (1971) found that when the flight zone of bulls was invaded by a mechanical trolley, the bulls would move away and keep a constant distance between themselves and the trolley. When the trolley got too close the bulls bolted past it. The best place for the handler to work is on the edge of the flight zone. This will cause the animals to move away in an orderly manner. The animals will stop moving when the handler retreats from the flight zone.

The size of the flight zone varies depending on the tameness or wildness of the animal. The flight zone of range cows may be as much as 300 ft whereas the flight zone

Figure 5. Cattle handling facility utilizing a curved single-file chute, round crowding pen, and wide curved lane. Up to 600 cattle per hour can be moved through the dip vat with only three people. The handlers work along the inner radius of the single-file chute and the wide curved lane (designed by Temple Grandin).

Figure 6. Sheep handling and sorting facility with a curved bugle crowding pen. The inner radius is solid to prevent the sheep from seeing the handler standing at the sorting gates (designed by Adrian Barber, Australia).

of feedlot cattle may be only 5 to 25 ft (Grandin, 1978). Extremely tame cattle or sheep are often difficult to drive because they no longer have a flight zone.

Many people make the mistake of getting too close to the cattle when they are driving them down an alley or putting them in a crowding pen. Getting too close makes cattle feel cornered. If the cattle attempt to turn back, the handler should back up and retreat to remove himself from the animal's flight zone instead, of moving in closer.

Cattle will often rear up and get excited while waiting in the single-file chute. The most common cause of this problem is the handler leaning over the single file chute and deeply penetrating the animal's flight zone. The cattle will usually settle down if the handler backs up.

When sheep are being handled in a confined area, pile-ups can occur if their flight zone is deeply penetrated. This is why dogs should not be used in the crowding pen or the single-file chute, because a dog, in a confined area, deeply penetrates the flight zone and the sheep have no place for escape. Dogs are recommended only for open areas and larger pens where there is room for the sheep to move away. During handling, minimize yelling and screaming so as to avoid enlarging the size of the animal's flight zone.

BREED DIFFERENCES

The breed of the cattle or sheep can affect the way it reacts to handling. Cattle with Brahman blood are more excitable and may be harder to handle than the English breeds. When Brahman or Brahman-cross cattle are being handled, it is important to keep them as calm as possible and to limit use of electric prods. Brahman and Brahman-cross cattle can become excited; they are difficult to block at gates (Tulloh, 1961) and prone to ram into fences. With this type of cattle it is especially important to use substantial fencing. If thin rods are used for fencing, a wide belly rail should be installed to present a visual barrier. Angus cattle tend to be more nervous than Herefords (Tulloh, 1961). Holstein cattle tend to move slowly (Grandin, 1980). Brahman cattle tend to stay together in a more cohesive mob than English cattle.

Brahman and Brahman-cross cattle can become so disturbed that they will lie down and become immobile, especially if they have been prodded repeatedly with an electric prod (Fraser, 1960). When a Brahman or Brahman-cross animal lies down, it must be left alone for about five minutes or it may go into shock and die. This problem rarely occurs in English cattle or European cattle such as Charolais.

There are distinct differences in the way various breeds of sheep react during handling (Shupe, 1978; Whately et al., 1974). Rambouillet sheep tend to bunch tightly together and remain in a group; crossbred Finn sheep tend to turn, face the handler, and maintain visual contact. If the handler penetrates the collective flight zone of a group of Finn sheep, they will turn and run past the handler.

Cheviots and Perendales are the easiest to drive into a crowding pen; the Romney, Merino-Romney cross, and the Dorset-Romney are the most difficult. The Romney tends to follow the leader but it is easily led into blind corners. Cheviots have a strong instinct to maintain visual contact with the handler and to display more independent movements than other breeds.

DARK BOX AI CHUTE

For improved conception rates, cows should be handled gently for AI and not allowed to become agitated or overheated. The chute used for AI should not be the same chute used for branding, dehorning, or injections. The cow should not associate the AI chute with pain. Cows can be easily restrained for AI or pregnancy testing in a dark box chute that has no headgate or squeeze (Parsons & Helphinstine, 1969; Swan, 1975). Even the wildest cow can be restrained with a minimum of excitement. The dark box chute can be easily constructed from plywood or steel. It has solid sides, top, and front. When the cow is inside the box, she is inside a quiet, snug, dark enclosure. A chain is latched behind her rump to keep her in. After insemination the cow is released through a gate in either the front or the side of the dark box. If wild cows are being handled, an extra long dark box can be constructed. A tame cow that is not in

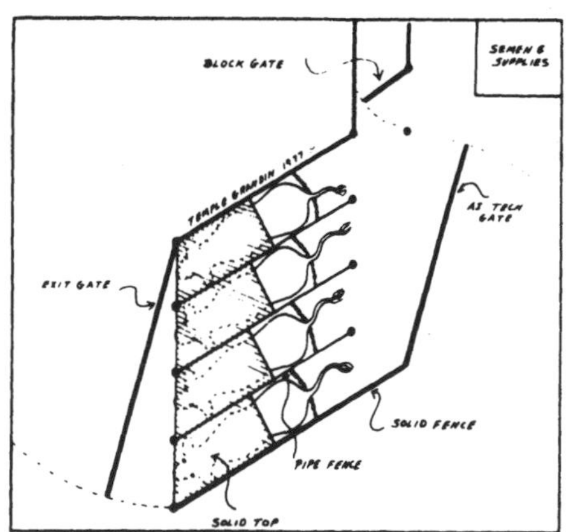

Figure 7. Chutes for A.I. can be laid out in a herringbone design. The two outer fences and the grates should be solid. The inner partition in between the cows should be constructed from bars. Cows will stay calmer if they know they have company.

heat is used as a pacifier and is placed in the chute in front of the cow to be bred. Even a wild cow will stand quietly and place her head on the pacifier cow's rump. After breeding, the cow is allowed to exit through a side gate, while the pacifier cow remains in the chute.

If a large number of cows have to be pregnancy checked or inseminated, two to six AI chutes can be laid out in a herringbone pattern (figure 7). This design is recommended by McFarlane (1976) from South Africa. The chutes are set on a 60° angle. They are built like regular dark box AI chutes except that the partitions inbetween the cows are constructed from open bars so the cows can see each other. The cows will stand more quietly if they have company. The two outer fences should be solid. If the cows are reluctant to enter the dark box, a small 6 in. by 12 in. window can be cut in the solid front gate in front of each cow.

LOADING CHUTE DESIGN

Loading chutes should be equipped with telescoping side panels and a self-aligning dock bumper. These devices will help prevent foot and leg injuries caused by an animal stepping down between the truck and the chute. The side panels will prevent animals from jumping out the gap between the chute and the truck.

A well-designed loading ramp has a level landing at the top. This provides the animals with a level surface to walk on when they first get off the truck. The landing should be at least 5 feet wide for cattle. Many animals are injured on ramps that are too steep. The slope of a permanently installed cattle ramp should not exceed 20°. The slope of a portable or adjustable chute should not exceed 25° (Grandin, 1979). Steeper ramps may be used for loading sheep but they are NOT recommended for unloading. Sheep will move up a steep ramp readily.

If you build your ramp out of concrete, stairsteps are strongly recommended. For cattle the steps should have a 3.5 to 4 in. rise and a 12 in. tread width. The surface of the steps should be rough to provide good footing. For sheep the steps should have a 2 in. rise and a 10 in. tread width.

On adjustable or wooden ramps, the cleats should be spaced 8 in. apart from the edge of one cleat to the edge of the next cleat (Mayes, 1978). The cleats should be 1 1/2 to 2 in. high for cattle and 1 in. by 1 in. for sheep.

Chutes for both loading and unloading cattle should have solid sides and a gradual curve (figure 8). If the curve is too sharp, the chute will look like a dead end when the animals are being unloaded. A curved single-file chute is most efficient for forcing cattle to enter a truck or a squeeze chute. A chute used for loading and unloading cattle should have an inside radius of 12 ft to 17 ft, the bigger radius is the best. A loading chute for cattle should be 30 in. wide and no wider. The largest bulls will fit through a 30 in. wide chute. If the chute is going to be used exclusively for calves, it should be 20 to 24 in. wide.

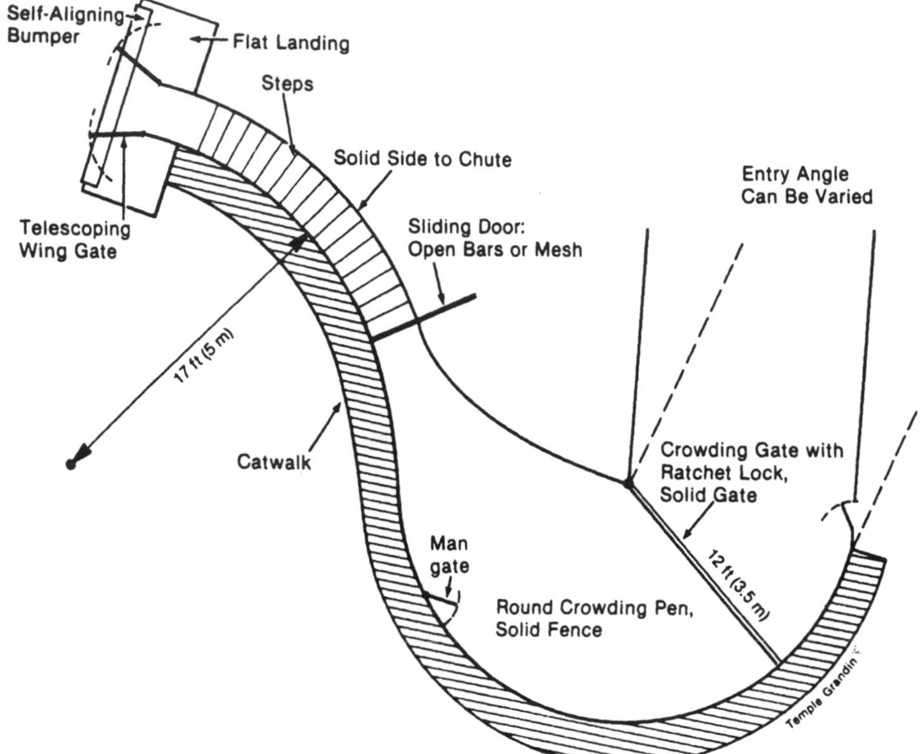

Figure 8. Curved loading chute with a round crowding pen. The sides of the chute are solid.

In auctions and meat packing plants where a chute is used <u>to unload only</u>, a wide straight chute should be used. This provides the animals with a clear path to freedom. These chutes can be 6 to 10 ft wide. A wide, straight chute should not be used for loading cattle.

SHEEP LOADING

Since most trucks have a 30 in. wide door, a good chute design for sheep is a 30 in. wide ramp that enables two sheep to walk up side by side. If a single-file chute is used, it should be 17 in. to 18 in. wide. The chute should be designed so that the animals walk up either in a single file or two abreast. Don't build a chute that is one and one-half animals wide--this creates jamming problems.

A wide ramp is recommended for loading sheep into shearing sheds or onto trucks that can be opened up the full width of the vehicle. In Australia sheep moved easily up ramps 8 to 10 ft wide when loading sheep onto ships for shipment to the Middle East (figure 9). The entrance ramp into a raised shearing shed should be 8 ft to 10 ft wide (Simpson, 1979).

Figure 9. A wide ramp is used to load sheep onto a ship in Australia. Once the flow of sheep was started the animals moved easily up the ramp.

REFERENCES

Barber, A. 1977. Bugle sheep yards. Fact Sheet, Dept. of Agriculture and Fisheries South Australia, Adelaide, Australia.

Brockway, B. 1975. Planning a sheep handling unit. Farm Buildings Center, Nat. Agr. Center, Kenilworth, Warickshire, England.

Ewbank, R. 1961. The behavior of cattle in crises. Vet. Rec. 73:853.

Franklin, J. R., G. D. Hutson. 1982. Experiments on attracting sheep to move along a laneway. III Visual Stimuli, Appl. Animal Ethology 8:457,

Fraser, A. F. 1960. Spontaneously occurring forms of "tonic immobility" in farm animals. Canad. J. Comp. Med 24:330.

Freeman, R. B. 1975. Functional planning of a shearing shed. Pastoral Review 85:9.

Grandin, T. 1981. Innovative cattle handling facilities. In: M.E. Ensminger (Ed.). Beef Cattle Science Handbook, 18:117. Agriservices Foundation, Clovis, Calif.

Grandin, T. 1980. Livestock behavior as related to handling facilities design. Int. J. Stud. Animal Problems 1:33 etc.

Grandin, T. 1979. Understanding animal psychology facilitates handling livestock. Vet. Med. and Small Animal Clinician 74:697.

Grandin, T. 1978. Observations of the spatial relationships between people and cattle during handling. Proc. Western Sec. Amer. Soc. of Animal Sci. 29:76.

Grandin, T. 1976. Practical pointers on handling cattle in squeeze chutes, alleys, and crowding pens. In: M.E. Ensminger (Ed.). Beef Cattle Science Handbook 13:228.

Hitchcock, D. K., G. D. Hutson. 1979. The movement of sheep on inclines. Australian J. Exp. Agr. and Animal Husbandry 19:176.

Hutson, G. D., S. C. van Mourik. 1982. Effect of artificial wind on sheep movement along indoor races. Australian J. Exp. Agr. and Animal Husbandry 22:163.

Hutson, G. D. 1981. Sheep movement on slatted floors. Australian J. Exp. Agr. and Animal Husbandry 21:474.

Hutson, G. D. 1980. The effect of previous experience on sheep movement through yards. Appl. Animal Ethology 6:233.

Hutson, G. D. and D. K Hitchock. 1978. The movement of sheep around corners. Appl. Animal Ethology 4:349.

Kilgour, R. 1971. Animal handling in works, pertinent behavior studies. 13th Meat Industry, Res. Conf. Hamilton, New Zealand. pp 9-12.

Lynch, J. J. and G. Alexander. 1973. The Pastoral Industries of Australia. pp 371. Sydney University Press, Sydney, Australia.

Mayes, H. F. 1978. Design criteria for livestock loading chutes. Technical Paper No. 78-6014, Amer. Soc. Agr. Eng. St. Joseph, Michigan.

McFarlane, I. 1976. Rationale in the design of housing and handling facilities. In: M. E. Ensminger (Ed.). Beef Cattle Science Handbook 13:223.

Parsons, R. A. and W. N. Helphinstine. 1969. Rambo AI breeding chute for beef cattle. One-Sheet-Answers, University of California Agricultural Extension Service, Davis, California.

Rider, A., A. F. Butchbaker and S. Harp. 1974. Beef working, sorting, and loading facilities. Technical Paper No. 74-4523, Amer. Soc. Agr. Eng. St. Joseph, Michigan.

Shupe, W. L. 1978. Transporting sheep to pastures and markets. Technical Paper No. 78-6008, Amer. Soc. Agr. Eng. St. Joseph, Michigan.

Simpson, I. 1979. Building a modern shearing shed. Division of Animal Industry Bulletin A3.7.1. New South Wales Dept. of Agr., Australia.

Swan, R. 1975. About AI facilities. New Mexico Stockman. Feb., pp 24-25.

Tulloh, N. M. 1961. Behavior of cattle in yards: II. A study of temperament. Animal Behavior 9:25.

Whately, J., R. Kilgour and D. C. Dalton. 1974. Behavior of hill country sheep breeds during farming routines. New Zealand Soc. Animal Production 34:28.

Williams, C. 1978. Livestock consultant, personal communication.

25
MEASURING AN ANIMAL'S ENVIRONMENT

Stanley E. Curtis

ASSESSING ANIMAL ENVIRONMENTS

Animal environments are characterized according to problems suspected of being associated with environmental stress on the animals and the effectiveness of control measures. Most of the elements of animal environments can be measured but, interpreting the results in terms of animal well-being, facility operation, and production economics often remains a dilemma because of the interaction of environmental factors. The effect of one stressful factor on an animal quite often depends on the nature of the rest of the environment.

ANIMAL ENVIRONMENT PROBLEMS

Troubleshooters often must engage in trial-and-error to identify animal-environment problems, but there are several points to be kept in mind:

- The environment results from all external conditions that the animal experiences, so all elements must be considered. Those which cannot be measured or controlled readily might influence animal health and performance nonetheless, so even they must be considered so far as is possible.
- The environmental complex acts as a whole on the animal, so interactions must be kept in mind. The combined effects of two or more environmental components may be difficult to evaluate, but they must be considered.
- Time and space affect the environmental factors. Environmental variables should be measured where the animal experiences them--in its micro--environment--taking into account the lack of spatial uniformity that occurs in all facets of the surroundings. Most important are vertical stratification of various parameters of the thermal environment and horizontal and vertical

variation in airflow due to design or mode of operation of the ventilation system.

- Environmental elements change with time at a given place. Because weather and facility occupancy vary with time, control requirements do also. Thus, environmental assessments and control schemes must take daily and seasonal environmental cycles into account. Most animals adjust to environmental cycles readily as long as extremes of the excursions are not unduly stressful.
- The rate of environmental change is critical. Abrupt environmental changes tend to be more stressful than those occurring over a longer period. For example, preconditioning young animals to a cool environment before moving them from a warm, closed house to a cool, open one during cold weather reduces the stress. It is sometimes difficult to identify a single index of environmental stress or even an adequate multiple index or combination of indices. For example, daily temperature range per se may be of little consequence as long as the day's maximum and minimum values do not exceed or fall below respective trigger levels. Likewise, a certain rate of temperature change might be stressful if it occurs at extremes of temperature, but not within more moderate ranges. With modern statistical techniques, it is possible to develop many environmental indices, with relative ease but it is a very difficult job to determine their respective significances in terms of the environment's impingements on the animal.
- Animals modify their own environments by giving off heat, water vapor, urine and feces, disease causing microbes, and others. The animals' own processes help determine the nature of their microenvironment. Changes in age or number of animals in a facility alter these impacts and, therefore, the control measures required.
- Anthropomorphism is a common pitfall in the assessment of animal environments. A comfortable environment for a human is not necessary for an animal. Animals send signals of discomfort or uneasiness to alert caretakers. These behavioral indications are always useful signs that the environment could be improved. In some cases, the way animals behave is the only clue that stress is present.

MEASURING ENVIRONMENTAL FACTORS

A well-planned and organized approach to environmental-measurement programs is important. The means employed will depend on the amount of accuracy, the extent of detail required and the effort devoted to the actual work of measurement. Insights into sampling theory, as well as descriptions of some of the instruments and techniques that have proved especially applicable to measuring outdoor and indoor animal environments, are provided in the following sections.

Sampling Theory

Environmental assessments are based on interpretations of measurements of pertinent variables. Hence, the observations must represent the situation faithfully. To ensure this, environmental sampling programs must be planned carefully. The main reasons for this have been alluded to already: most environmental elements vary over time and space. Some of these variations are regular and predictable, others are not. The times and the places the environment is measured determine whether the resultant information reflects the character of the environment well enough to be of use. Of course, it is possible to gather more data than necessary, too.

Time considerations. To estimate the average impingement of an environmental factor over a period of hours or days, or to learn about excursions of these values over time, a continuous sample is needed. In short, continuous sampling from more or less permanent instrument stations, usually coupled with recording equipment, provides the data needed for detailed analyses of animal environments.

The most important consideration in regard to time is the length of the observation period. It should be a multiple of a well-established environmental cycle. In studying an animal facility in a temperate climate, for instance, seasonal periodicities must be accounted for to appreciate the facility's nature all year long. On the other hand, a particular problem may be limited to one season, in which case the observation period would be a multiple of the day, to account for diurnal cycles in meteorologic phenomena.

Of course, there is considerable variation among years and even among days within seasons. For example the number of cycles--the number of years or days--that should be observed to give meaningful results depends partly on the nature of this variation already known to occur in the facility's locale. Furthermore, interpretation of any results should include an historical perspective. For example, if observed values for air temperature are at the lower end of the acceptable range during a winter known to be relatively mild by local standards, the problem of a too-cold animal micro-environment might well be encountered during a more nearly normal winter.

The frequency of observations needed within a sampling period is another important decision. In general, measurements should be made as frequently as feasible, especially if automatic recording equipment is being used. Then, after observations have been completed, key periods of environmental extremes or change can be evaluated in detail, while less interesting periods can be ignored. Also, when observations are made too infrequently, errors can be made in characterizing both the ranges and the average values of the variables.

Time-averaging can obscure extremes of environmental factors that may trigger animal responses, hence it must be done only when warranted. For example, effective environmental temperature in an outdoor environment might range from -10° to 20°C one day, from 0° to 10°C another. Average temperature might be around 5°C on both days but the nature of the animals' thermoregulatory reactions would be different on these days, and thus averaging the environmental data over a day could lead to misleading impressions.

Nonlinearity in interactions among environmental factors causes additional difficulties so far as the time-averaging of environmental measurements is concerned. In other words, effects of combined factors must be calculated carefully, even when their relations for steady-state conditions are well-known. Take as an example the case of the wind-chill index (table 1). The average of the wind-chill indices for the three sets of conditions in the table is 1.6×10^3 Kcal m^{-2} hr^{-1}, while the wind-chill index for the average of the three conditions (namely, temperature -34°C and wind speed 9 m sec^{-1}) is 2.1×10^3 kcal m^{-2} hr^{-1}.

TABLE 1. AN EXAMPLE OF NONLINEAR INTERACTION AMONG ENVIRONMENTAL VARIABLES: WIND-CHILL INDEX

	Condition		
	A	B	C
Temperature (°C)	-18	-34	-51
Wind speed (m sec^{-1})	18	9	0
Wind-chill index (kcal m^{-2} hr^{-1})	1.8 x 10	2.1 x 10	.8 x 10

Time lags between environmental occurrences and animal responses also must be recognized--taking them into account could improve the probability of defining a connection between animal and environment.

Another approach is discontinuous sampling, sometimes called spot- or grab-sampling. This usually involves more portable equipment and is aimed at gaining information over short periods, such as hours. Discontinuous sampling is best suited to determining extremes in an environmental element when the basic nature of the variation of the factor is known. For example, air temperature might be measured only in the early afternoon on relatively hot days to gain information on upper values of air temperature, or concentrations of air pollutants might be measured in closed animal houses on relatively cold days when ventilation rate is relatively low.

Discontinuous sampling has several advantages. It requires less time and often less equipment. Further, fewer data are generated, so data-processing equipment requirements are less than when continuous sampling is practiced. Finally, equipment portability often facilitates economical use of a single instrument at many locations in a facility to learn more about environmental variation over space than might be feasible otherwise.

Space considerations. Two prime considerations should determine where environmental measurements are to be taken. In the first place, environmental factors generally should be measured in the immediate surroundings of the animals. For example, most elements of the environment vary consistently with distance above the floor, but most animals reside at discrete heights within a facility. Thus, the height at which the environment is measured is crucial if the measured values are to reflect the conditions to which the animals are being exposed.

In animal facilities it is sometimes tempting to sample the environment in an alleyway or some other place where the instruments will be relatively safe from animal damage. For the most part, these temptations should be resisted. The animals affect their own surroundings so greatly that most variables differ even from animal microenvironments to nearby areas where the animals are not permitted.

Second, the environmental and animal features known to affect the variable under study should be clearly in mind when the measurement sites are chosen. Major items include heat sources: air inlets and outlets; orientation of the facility to winds, the sun, and other structures; location of mechanical services, such as feed-delivery systems; and animal size and population density.

There also is the substantial problem of the effect of the measurement instrument itself, and its protective hardware, on the environment. Some equipment stations obstruct airflow, for example, and the air samplers may be drawing air from different heights and extracting components from the air around the sampler, thereby modifying it.

Instrument Choice

Dozens of instruments for measuring environmental factors are available on the commercial market today. The choice of instrument or set of instruments to be used in a given environmental-measurement program is based on several considerations: (1) the kind of information needed, (2) the relative efficiencies of the various instruments and their reliabilities under field conditions, (3) ease of use, cost, and availability, and (4) personal choice--often based on past experiences--is an important point.

These instruments have accompanying instructions. If the manufacturer's written advice proves inadequate, get in touch with a technical representative of the manufacturing firm directly; most will assist customers by mail or telephone with problems in specific applications of their product.

In the sections that follow, some instruments commonly used in measuring environmental factors in animal facilities are described in brief detail.

MEASURING AIR TEMPERATURE

Liquid-In-Glass Thermometers

A variety of commercially available mercury- or alcohol-in-glass thermometers are used widely to measure air temperature. As temperature rises, the liquid expands to occupy more of the capillary tube in which it is held. Of course, as in all thermometry, the temperature registered is actually that of the thermometer, not necessarily that of the environment, so factors apart from the temperature of the surrounding air that affect thermometer temperature lead to errors. Chief among these are solar and thermal radiations and air movement. Precautions must be taken to shield the thermometer against them. Furthermore, the thermometer may be placed in a spot where the air is stagnant or otherwise unrepresentative of the general area.

Lag time. When environmental temperature changes, the value required for a thermometer to reduce the difference between registered temperature and actual air temperature to 36.8% of the original difference is called the lag time of the thermometer. Lag time for most mercury-in-glass thermometers is around 1 min, while that for alcohol-in-glass is roughly 1.5 min. Some electronic thermometers have much shorter lag times. Especially in discontinuous sampling, lag time is an important consideration because a measurement can be made more quickly with the shorter lag time.

Maximum-minimum thermometer. A thermometer designed to register the maximum and minimum temperatures experienced during the measurement period often has been used in animal facilities. It consists of a U-tube with bulbs at both ends. One side of the U serves as the scale for maximum temperature, the other for minimum. The bulb on the maximum side serves as a safety reservoir and is partially filled with a liquid such as creosote solution. That on the other side is completely filled with the liquid. Between these two portions of liquid, in the bottom of the U-tube, is mercury. As the thermometer becomes warmer, the liquid in the filled bulb expands, pushing the mercury up on the maximum-scale side (the opposite side). Atop the mercury on both sides is an iron index (a sliver of iron), and as the mercury moves up the maximum side it pushes the index ahead of it. As the thermometer cools, the mercury retracts, but the index remains at its highest point due to friction with the inside of the U-tube that can be overcome by the mercury, but not by the other liquid. Of course, as the mercury column retracts upon cooling, it pushes the minimum-temperature side's index ahead of it, so this one registers the lowest temperature of the measurement period on the maximum-temperature scale, which is upside-down. When maximum and minimum temperature have been observed, the thermometer is reset by replacing the indices atop respective mercury meniscuses by means of a magnet.

Although maximum and minimum temperature may be all the information needed in some situations, and despite the fact it is relatively inexpensive and straightforward in design, this instrument has serious drawbacks. Chief among them are the tendencies for the mercury column to become separated or broken and for the indices to become permanently lodged in the U-tube.

Bimetallic-strip thermometer. The sensor of a bimetallic-trip thermometer comprises a sheet of each of two metals having dissimilar coefficients of linear thermal expansion, which are joined along their faces. When such a strip's temperature changes, it bends. Lag time is relatively short--around 10 seconds.

Thermograph. The bending movement noted above for the bimetallic strip is usually magnified by shaping the strip appropriately--and in a thermograph the movement is recorded by affixing a pen to its free end and applying this pen to a piece of graph paper on a drum rotated by a clockworks. By this relatively inexpensive means, a permanent record of temperature variation is made. Further, it requires no electrical supply. Of course, a thermograph is so large that it affects the microenvironment it is used in and, because of its construction, is prone to error due to radiation and stagnant air pockets.

Calibration of a thermograph is a critical matter. The recording element is very sensitive to physical shock, which often occurs when the instrument is moved from place to place, thus, after the instrument is moved, it is absolutely necessary to calibrate a thermograph by adjusting the recording pen several days in a row, preferably near the times of the daily high and low temperatures. An artificially ventilated psychrometer is commonly used as the standard instrument for calibration of thermographs.

Electric thermometers

There are two general kinds of electric thermometers. One kind depends on the principle that as the temperature of a substance changes, so does its electrical resistance. The other kind--thermocouple thermometry--depends on the principle that when wires of two specific metals are joined at both ends, and when these two junctions are kept at two temperatures, an electromotive force is generated in that circuit. Voltage in such a circuit, when measured with a potentiometer, is directly proportional to the temperature difference between two junctions.

Thermistor. Certain semiconducting materials have negative temperature coefficients of electrical resistance: as temperature rises, resistance decreases. Resistance in the circuit to which a current has been applied is measured by an indicating unit. These sensing elements are called thermistors--parts of the electric thermometers most applicable to air thermometry in animal environments.

Thermistor probes, indicating units, and recording units are commercially available in a wide range of models. At one time, thermistor systems were notoriously unstable, but nowadays stable, calibrated probes are on the market and in recent years they increasingly have become the sensors of choice for routine assessment of animal environments.

Thermocouples. A variety of combinations of dissimilar wires are used for the two sides of the circuit in thermocouple thermometry. Copper and constantan are frequently chosen. One of the thermocouple junctions (the reference junction) is held at a constant or known temperature so changes in electromotive force measured reflect changes in the temperature of the measurement junction, which is placed in the environment to be monitored. The voltage generated can be used to drive a millivolt recorder or registered on a millivoltmeter.

In general, use of thermocouples for air thermometry in animal facilities has some disadvantages. The needed equipment, especially the constant-temperature bath for the reference junction, is relatively cumbersome, the physical integrity of the measurement and reference junctions critical and sometimes difficult to maintain, and careful calibration of the thermocouples very important.

Metal-resistance thermometer. In metals, as temperature rises, so does electrical resistance. Small-diameter platinum wire wound on a support having a small thermal expansion coefficient is a frequent choice. A small current is introduced into the wire, and resistance of the whole circuit is measured using a Wheatstone bridge circuit. The most sensitive resistance thermometers tend to be fragile, and for this reason alone they are of limited use in animal environments. In addition, the measuring equipment is relatively expensive.

MEASURING AIR MOISTURE

The measurement of water vapor in air is called psychrometry or hygrometry. Several principles have been used to measure air moisture, and three have been applied widely in quantifying animal environments.

Hair Hygrometer

Hair is hygroscopic, and its length is related directly with the amount of water it contains. Further, there is a nonlinear, direct correspondence between length of hairs and the relative humidity of the air surrounding them. Hair hygrometry is most accurate when relative humidity ranges between 20% and 80%. This principle has been employed extensively in hygrometry in animal facilities, partly because hair hygrometers require no electrical supply and they are affected little by other factors.

Hygrograph. Elongation and shortening of hair bundles can be magnified by an appropriate level system and transformed into movement of an arm holding a pen that is applied to graph paper on a rotating drum, giving a record of changes in relative humidity. Just as for the thermograph, hygrographs must be calibrated carefully over a period of several days after they have been moved to a new location before reliable measurements can be made.

Psychrometers

Psychrometers are a class of instrument by means of which the air's moisture content or relative humidity can be estimated indirectly. They use both a wet-bulb thermometer and a dry-bulb, and their measurement principle is based on the thermodynamic relation between the air's moisture content and wet-bulb temperature. (Wet-bulb temperature is affected by dry-bulb temperature and air pressure.) Once dry-bulb and wet-bulb temperatures of the air are known, the air's moisture content and relative humidity can be estimated from a psychrometric chart.

The dry-bulb thermometer can be of any type, while the wet-bulb thermometer is a similar instrument having a water-saturated wick closely surrounding the bulb or sensing the element. As the psychrometer is operated, evaporation from the wick occurs, and the temperature of the wet bulb is depressed. Of course, the drier the air, the greater the evaporation, and the greater the wet-bulb depression.

To give an accurate estimate of wet-bulb temperature, the thermometers must be ventilated adequately; maximum cooling of the wet bulb does not occur in still air. Also, temperature readings must not be taken until equilibrium has occurred. Other sources of error when using any psychrometer are heat conduction down the wet-bulb thermometer (the wick is ordinarily extended up the stem), receipt of solar and thermal radiation (the latter even from the operator of the instrument), and wicks that are too thick or dirty (one way to minimize mineral crust is to wet the wick with distilled water only).

Wet-bulb temperature should always be read before that of the dry-bulb, as it will begin to rise as soon as ventilation ceases. It is also good practice to repeat the measurement several times to make sure the lowest wet-bulb temperature has been attained.

Sling psychrometer. The sling psychrometer, once the standard instrument for spot-sampling psychrometry, consists of dry- and wet-bulb liquid-in-glass thermometers in a frame that can be revolved around a handle. The thermometers are whirled--usually for a minute or more--to permit wet-bulb depression to occur. Larger sling psychrometers must be revolved at least two times per second to provide sufficient ventilation, and smaller ones five times. The movement of the operator mixes the air in the region.

Artificially ventilated psychrometer. Various models of psychrometer are now available in which air is drawn artificially past the temperature sensors. One popular model employs dry-cell batteries that supply a small fan, which pulls air at speeds up to 5 m sec^{-1} past the bulbs. Of course, this kind of instrument may draw air from as far away as 1 foot, hence it is not applicable to some microenvironmental measurements. Still, it is reliable, rugged, and portable.

Electrical Conductivity of Hygroscopic Materials

Salts such as lithium chloride are hygroscopic, and their electrical conductivity increases--thus, their resistance decreases--with increasing water content.

Electric hygrometer. Commercially available sensing units usually involve a film of lithium chloride on a nonconducting frame through which an electrical current is passed for the purpose of measuring electrical resistance changes. The logarithms of the resistance and the atmospheric humidity parameters are inversely related.

Dew cell. When a film of lithium chloride is applied to a heating-element frame, and the temperature is so high the salt is dry, the salt is also highly resistant to conducting electricity. The electrical circuitry of a dew-cell apparatus is designed so that when the salt is a conductor, the element is being heated, but the heating stops when the salt becomes warm enough to become dry. Thus, the lithium-chloride film is kept more or less at the same temperature and dry at all times. This equilibrium salt temperature is related directly with the air's dew-point temperature.

Electric hygrometers and dew cells make it possible to monitor air humidity continuously, but in animal environments they often become dirty, and this can lead to errors. Further, they remain operational for periods of only a few months under the best of conditions.

MEASURING AIR MOVEMENT

Drafts, stagnant spaces, and inadequate removal of moisture or noxious gases in animal houses are among the common symptoms of improper design or operation of a ventilation system. Air speed and distribution throughout an animal facility must be known if ventilation problems are to be remedied.

Anemometers

Several kinds of instruments to measure air speed are available. Each is best suited to a particular application in animal-environment measurement.

Pitot-tube measurement. When air moves into or across the mouth of an open tube, the air pressure in the tube changes; it increases in the former case, decreases in the latter. Such pressure changes are proportional to air speed and serve as the basis of pitot-tube anemometry.

The most common pitot-tube anemometer used today is the Velometer, a rugged and portable instrument that gives a direct reading of air speed and comes supplied with a variety of probes for different velocity ranges. This instrument is most adaptable to measuring air velocity at inlets, in areas of strong drafts, and in air ducts.

Because of its very nature, the pitot-tube is extremely directional; it is sensitive to air movement in one direction. Large errors can result when a probe is not properly oriented.

Hot-wire anemometer. Several modes of hot-wire anemometer are on the market. Some are directional and therefore applicable to measuring air speed in ducts and at inlets; others are more nearly omnidirectional. They operate on the principle that as air passes across a fine platinum or nickel wire that has been heated electrically, the wire tends to cool by an amount proportional to air speed. This instrument is designed so current flowing through the wire automatically changes so as to keep wire temperature nearly constant. The amount of current required to achieve this is thus related directly with air velocity.

Hot-wire anemometers are relatively sensitive and thus especially applicable to situations--such as animal microenvironments--where velocity can be as low as .5 cm sec^{-1}. Another advantage is that some designs are very portable and fairy rugged, and so small they do not interfere much with the environment. They also respond very quickly to changes in air speed. However, the sensing wire can be affected over time by atmospheric pollutants, and for this and other reasons frequent calibration is necessary. Further, the sensing wire itself is exceedingly fragile and subject to damage. The hot-wire anemometer cannot be used outdoors during rainy periods or in any environments where water can reach the sensing wire.

Rotating-vane anemometer. There are various kinds of anemometers employing a lightweight propeller with eight or more blades. All are highly directional, and for applications where air direction changes greatly they are commonly attached to a vane device that keeps them aimed into the main flow of the air.

A variety of physical and electronic metering devices are used in conjunction with these anemometers. They are most useful at high air speeds, and thus are most applicable in outdoor settings. Indoor models are relatively fragile and subject to damage by corrosive gasses.

Cup anemometer. The device used to measure air velocity outdoors consists of several cups, each connected to a rod radiating from a rotor. A revolution of the assembly is directly related to air velocity. Cup anemometers are not suitable for measuring low air speeds. They are most applicable to estimating the average speed of the wind during periods of at least several hours.

Kata thermometer. A simple instrument was developed over a century ago to estimate the cooling power of the air. The kata thermometer is capable of measuring very low air velocities. It is simply an alcohol-in-glass thermometer with the stem marked at the 37.5° and 35°C levels.

The thermometer is warmed in a water bath to around 40°C, removed from the water and wiped dry, and placed in the environment to be measured. The time required for temperature to fall from 37.5° to 35°C is determined using a stopwatch, and this value substituted into a formula (which also includes air temperature and an individual instrument-calibration factor supplied by the manufacturer) for calculating air velocity.

Airflow-pattern measurement

Certain visible particles suspended in the air can be used as an aid to tracing how the air is distributed in an animal house. The source of the particulate matter is simply placed at the air inlet or in the microenvironment to be studied, and the course of the pollutant followed visually through the space of interest. Much cigar and pipe smoke has been used quite effectively for this purpose in the past. More reliable sources of larger amounts of visible tracers are now on the market.

In the absence of an anemometer, low and moderate air speeds can be estimated to a first approximation by briefly interrupting the tracer's flow and monitoring the movement of the turbulence so induced with the aid of a measuring tape and a stopwatch.

"Smoke"vials. When a solution of titanium tetrachloride is exposed to air, it gives off copious whitish fumes. Small glass vials of this solution areideal for use in studying airflow patterns in microenvironments or parts of a room. A vial is broken open and simply can be held in the location to be observed. An open vial is commonly placed in a cup, which may be attached to a pole in order to reach certain parts of the room of interest. Several devices are also available to increase the evolution of fumes by passing air over the solution. Titanium tetrachloride fumes in the concentrations used are not toxic to animals.

Talcum aerosolizer. Fine talcum powder is also used as a tracer in small-scale air-distribution studies in animal houses. Special talcum-powder aerosolizers are available commercially. Talcum tends to precipitate faster than does titanium tetrachloride, and it is generally more difficult to generate an adequate aerosol of the powder. For these reasons, it is inferior to titanium tetrachloride for this purpose.

"Smoke" pellets and candles. When long-term or large-scale observations are desired, more tracer might be needed than can be supplied by titanium tetrachloride or talcum aerosols. The commercial market has a wide range of pellets and candles that, when lighted, produce very large amounts of fumes. These are generally set in the way of the incoming air. The fumes of some of these devices are toxic and therefore must not be used in occupied houses.

Static-Pressure Measurement

In negative-pressure or exhaust-ventilation systems, adequate negative-static pressure must be maintained by the fans. Static pressure inside an animal house is usually measured by means of a manometer sloped 1:10 to increase accuracy. One end of the manometer is open to the outside atmosphere, the other to the inside, and the pressure difference is registered by the manometer.

Every animal house employing mechanical ventilation should be equipped with its own static-pressure manometer.

MEASURING SURFACE TEMPERATURE

The temperature of an animal's surface is an important determinant of convective and radiant exchanges of heat.

The temperatures of environmental surfaces likewise play a central role in determining the magnitude and direction of thermal-radiant flux. Accurate measurement of surface temperature is difficult, but there are two ways this can be approached.

Portable radiation thermometer. A variety of battery-powered instruments now available provide rapid measurement of surface temperature by a technique that doesnot involve contact with the surface. The portable radiation thermometer is simply aimed at the surface to be measured, and the surface's temperature is registered on the meter. Models providing different fields of view are available.

This kind of thermometer must be calibrated before every observation by means of a Leslie cube or some other temperature-calibration device. Solar radiation reflected by a surface leads to overestimation of that surface's temperature when measured with this instrument. Also, surfaces in the surroundings substantially cooler than that being measured tend to bias these estimates in the opposite direction. The instrument is reliable, reasonably rugged, and is an excellent means for spot-sampling animal and environmental surface temperatures.

Contact thermometers. Thermometers in a variety of styles involving contact with animal and environmental surfaces have been used sometimes to measure surface temperature. These include thermistor probes in hypodermic needles, contact discs (banjo probes), as well as thermocouples taped or glued to surfaces.

There are serious drawbacks to contact thermometers. First, the contact must be flawless--otherwise insulative air pockets, even very small ones, between thermometer and surface will introduce measurement errors--and this is rarely achieved on either animal or environmental surfaces.

There are other problems of introducing artifacts with animal surfaces in particular. It is a practical impossibility to achieve the necessary contact with covered areas of an animal's surface without disrupting that cover and altering surface temperature. Even on nude areas, affixing a thermometer to the skin in such a way as to ensure adequate contact alters cutaneous blood flow and surface temperature.

MEASURING SOLAR AND THERMAL RADIATION

Solar Radiation

The total direct and diffuse (sky) solar radiation received by a horizontal surface is measured by an instrument called a pyranometer. Several designs are available on the market. The standard instrument in the United States is the Eppley pyranometer.

Eppley 180° pyranometer. This instrument consists of horizontal concentric silver rings, one black and one white. The glass hemisphere that covers the rings transmits only radiation with wavelengths less than 3.5 em (solar radiation), for which the black and white rings have different absorptivities. The temperature difference between the two rings generates an electromotive force in a thermopile (a series of thermocouple junctions), alternate junctions of which are in thermal contact with respective rings.

The Eppley 180° pyranometer measures direct and diffuse solar radiation received by a horizontal flat surface from the upper hemisphere. When direct sunshine is blocked by a shade, only the sky radiation is measured.

This instrument is fragile and must be sited carefully. Its glass bulb must be cleaned daily.

Thermal Radiation

The rate of incoming thermal radiation is usually estimated as the difference between the rate of total incoming radiation having wavelengths between .1 and 100 em, and the rate of total incoming solar radiation as estimated by an unshaded pyranometer. The rate of total incoming radiation can be measured by any of several kinds of total radiometer.

Beckman and Whitley total radiometer. This instrument is ventilated to minimize errors due to variable convective heat loss and measures the rate of incoming radiation at wavelengths between .1 and 100 em. It essentially consists of three layers of plastic: the top one is painted black and exposed to radiation from the upper hemisphere; the middle one contains a thermopile; and the lower one is shielded to minimize its receipt of radiation. The temperature difference between top and bottom of the middle plate is related directly to the upper plate's rate of radiation absorption.

MEASURING LIGHT INTENSITY

Photovoltaic Meters

Light meter. A selenium photovoltaic cell comprises the basis of most light-intensity meters used by photographers. Illumination of the sensitive layer of selenium by visible radiation (wavelengths from .3 to 7 em set up a flow of current in an appropriate circuit. Photographic light meters are portable and very useful in measuring light intensity in animal environments.

In most commercially available light meters, sensitivity over the visible spectrum is trimmed to resemble that of the human eye; that is, it peaks at a wavelength of about .55 em instead of the .63 em wavelength radiation that drives photoperiodisms in animals.

Illuminometer. A recording hemispherical photometer called the Illuminometer is also on the market. It is particularly suited to stationary installations where light intensity is known to vary over time, such as it does outdoors.

MEASURING SOUND LEVEL

Sound-level meter. Several battery-powered models of sound-level meter are available commercially. All provide a simple, portable, and reliable means of measuring sound level in decibels. The microphones on such instruments are relatively nondirectional, and the operation of sound-level meters is straightforward. Necessary precautions include recognizing the possible presence of obstacles to sound waves; locating the microphone at the observer's side, not between observer and sound source; shielding the microphone from any moving air; and making sure interfering electromagnetic fields from other electrical equipment are accounted for.

MEASURING AIR PRESSURE

Aneroid barometer. Measurement of air pressure in conjunction with animal production is ordinarily accomplished by means of an aneroid barometer--an instrument in which the walls of an evacuated cell move as air pressure changes. The movements are transmitted to a pointer, which indicates air pressure. While not as accurate as a mercury barometer, the aneroid version is nonetheless quite useful in animal work. If moved, it must be recalibrated against a mercury barometer.

Barograph. Movements of an aneroid barometer cell's wall can be recorded on graph paper affixed to a rotating drum when a pen is linked to that wall.

MEASURING AIR POLLUTANTS

Measuring Aerial Gases and Vapors

Colorimetric indicator tubes. Several systems of the same general type are available commercially for the convenient and, when properly used, reasonable accurate measurement of aerial gases and vapors. These consist of an indicator tube and a precision piston or bellows pump operated manually to draw air. The detector tube contains a specific chemical that reacts with the gas or vapor being measured. These small detector tubes are available for all the major gases and some of the vaporous compounds commonly present in animal-house air.

When air is pulled through an indicator tube, the pollutant for which the tube's indicating gel is specific reacts with the chemical, resulting in discoloration. The extent of this change is related to the concentration of the pollutant in the air. One problem with such a measurement system is that gases associated with dust particles--for example, some of the ammonia in dusty air--are filtered out of the air before it reaches the colorimetric indicator. This tend to cause underestimation of the pollutant's concentration in the air.

The pump for this kind of system must be kept leakproof and must be calibrated. Also, the detector tubes must be handled carefully and their predicted shelf lives observed.

Measuring Aerial Dust

High-volume sampler. Several models of dust samplers in wide use draw through a filter made of cellulose paper, glass or plastic fibers, or organic membrane. Dust particles too large to pass the filter are collected on it.

In practice, a filter is dried and tared before sample collection, and the particle-laden filter is dried and weighed again at the end of a sampling period. The difference between the two is an estimate of the mass of the particles collected during the sampling period. This usually is divided by the product of sampling period and average airflow rate to give the concentration of the pollutant. Filters are commonly handled with tweezers and transported outside the laboratory in large, covered petri dishes.

Another critical factor is calibration of airflow rate. most high-volume air samplers are equipped with some sort of airflow meter, but these instruments should be calibrated frequently because errors in this estimate are perpetuated as errors in all concentration estimates.

Particle counting and sizing. Several dozen models of instruments to count airborne particles are on the market. Some are primarily collectors--based on the principle of impaction on a solid surface, impingement in a liquid medium, centrifugation, or settling--and used in conjunction with subsequent visual observation. Others are direct-reading instruments employing optics and electronics.

Impactors are most commonly used for discontinuous sampling in animal environments. Two popular instruments are the Anderson six-stage, stacked-sieve, nonviable sampler, and the four-stage cascade impactor. Both feature impaction of dust particles on pieces of glass, which are then inspected microscopically for counting. As for sizing of the particles, both instruments have several stages designed so that the polluted air is drawn through a series of jets with progressively smaller cross-sections. The result is that relatively large particles are impacted in early stages, smaller ones at later stages. The size ranges monitored by these instruments are pertinent to the site of deposition of the particles within an animal's respiratory tract.

MEASURING AERIAL MICROBES

Qualitative studies of airborne microbes in animal environments have long involved opening a petri dish of culture medium, permitting viable particles to settle out of the air onto the medium's surface. Of course, special media can be used when there is interest in particular kinds of microbes. When the aerial concentration of microbes or microbe-carrying particles must be determined, another method must be used.

<u>All-glass impinger</u>. One method of quantifying airborne microbes is to impinge them in an isotonic solution that can then be diluted appropriately, combined with nutrients, and cultured in preparation for counting. This method is well-adapted to situations where aerial microbic level is high, but has the disadvantage of disintegrating airborne particles containing more than one microbe so that, for instance, an airborne particle that would give rise to one colony in an animal's respiratory tract might give rise to hundreds to be counted in the culture dish.

<u>Andersen six-stage viable sampler</u>. The viable version of the Andersen stacked-sieve sampler holds special culture-medium plates instead of flat pieces of glass as in the nonviable model. Particles are impacted onto the solid medium's surface where colonies can grow and be counted.

The Andersen viable sampler is a very useful instrument for both counting and sizing airborne microbic particles in animal environments. Special media can be used when desirable, and both the size and the number results can be interpreted in terms of the challenge the aerial microbic particles present to the animals' respiratory tracts. On the other hand, because of its relatively high air-sampling rate, the Andersen sampler is less well-adapted to air environments in which microbic populations are very high, as in some closed animal houses during cold weather. In such case, the sampling period may have to be short as 15 seconds, and thus special care must be exercised to ensure accuracy in estimating the volume of air drawn through the instrument during the sampling period.

<u>Andersen disposable two-stage viable sampler</u>. A less expensive device is a disposable-plastic, two-stage sampler fashioned after the original Andersen six-stage model. Commercial petri dishes available in hospital microbiology laboratories and a variety of vacuum sources can be used with this system.

This instrument has a critical orifice providing an air-sampling rate of 1 ft^3 per min when a vacuum of at least 10 in. of mercury is maintained. When operated in this way, the colonies that grow on the upper stage have arisen from particles having an aerodynamic diameter greater than 7 em, and hence they would not have deposited in the lungs of an animal. The particles that are impacted on the lower stage are between 1 and 7 em in diameter, and many of these could have reached the animal's lungs.

Like the Andersen viable sampler, the disposable sampler sometimes must be operated for a short sampling period in commercial animal houses.

Also, the collection efficiency of the Andersen disposable sampler seems to be less than that of the standard version. Despite these drawbacks, the disposable model is an inexpensive, relatively accurate means of estimating the concentration of microbe-bearing particles in the air, and whether they are of such a size as to directly threaten pulmonary health.

REFERENCES

Anonymous. 1972. Air Sampling Instruments for Evaluation of Atmospheric Contaminants. Fourth Ed. Am. Conf. Gov. Indust. Hygienists. Cincinnati.

Curtis, S.E. 1981. Environmental Management in Animal Agriculture. Animal Environment Services, Mahomet, Illinois.

Gates, D.M. 1968. Sensing biological environments with a portable radiation thermometer. Appl. Optics 7:1803.

Hosey, A.D. and C.H. Powell (Eds.). 1967. Industrial noise--a guide to its evaluation and control. Pub. Health Serv. pub. 1572. U.S. Gov. Printing Off., Washington.

Johnstone, M.W. and P.F. Scholes. 1976. Measuring the environment. In: Control of the Animal House Environment. Vol. 7, Laboratory Animal Handbooks. Laboratory Animals, Ltd., London.

Kelly, C.F. and T.E. Bond. 1971. Bioclimatic factors and their measurement. In: A Guide to Environmental Research on Animals. Nat. Acad. Sci., Washington.

Munn, R.E. 1970. Biometerological Methods. Academic Press, New York.

Platt, R.B. and J.F. Griffiths. 1972. Environmental Measurement and Interpretation. Krieger, Huntington, NY.

Powell, C.H. and A.D. Hosey (Eds.). 1965. The industrial environment--its evaluation and control. Pub. Health Serv. Pub. 614, U.S. Gov. Printing Off., Washington.

Schuman, M.M., et al. 1970. Industrial Ventilation. Eleventh Ed. Am. Conf. Gov. Indust. Hygienists, Cincinnati.

Spencer-Gregory, H., and E. Rourke. 1957. Hygrometry. Crosby Lockwood, London.

Stern, A.C. (Ed.). 1976. Measuring, Monitoring, and Surveillance of Air Pollution. Vol. III, Air Pollution. Third Ed. Academic Press, New York.

Tanner, C.B. 1963. Basic instrumentation and measurements for plant environment and micrometerology. Soils Bull. 6, Univ. of Wisconsin, Madison.

Wolfe, H.W., et al. 1959. Sampling Microbiological Aerosols. Pub. Health Serv. Pub. 686. U.S. Gov. Printing Off., Washington.

26
MEASURING ENVIRONMENTAL STRESS IN FARM ANIMALS

Stanley E. Curtis

Relations between agricultural animals and their surroundings always have been important. Those species recruited for domestication generally differ from their wild cousins in that the domesticated animals are adaptable to a wider range of environments than are their cousins (Hale, 1969). Hence, these animals we keep are more amenable to being confined and managed by the humans they serve (Bowman, 1977).

Ecology always has been at the heart of animal production. The shelter aspect of environmental management has been applied for a long time. shepherds kept their flocks in folds at night thousands of years ago. Only with the advent of widespread spacewise and time wise intensiveness in animal agriculture have animal-environmental relations become so important relative to other factors of production. And only with this intensiveness has major environmental modification been possible not to mention economically feasible. Now in addition to increasing the fit of the animals to the environment, we are coming closer to meeting the animals' needs by modifying their environments.

A DIGRESSION

"Measuring Environmental Stress in Farm Animals," calls to mind that which we should keep in mind. Let us examine the last five words of the title first and the first word last.

Environmental Stress in Farm Animals

Stress is of the environment, not of the animals (Fraser, et al., 1975; Curtis, 1981). Nevertheless, we measure stress in the animals, not in the environment. An animal is under stress when it is required to make extreme functional, structural, or behavioral adjustments in order to cope with adverse aspects of its environment. Thus, an environmental complex is stressful only if it makes extreme demands on the animal.

In other words, an environment is not stressful in and of itself; it is stressful only if it puts an animal under stress. And because animals differ in the ways they perceive and respond to the environmental impingements, the very same environment can be stressful to one animal and not to another.

An environmental factor that contributes to the stressful nature of an environment is called a stressor. When we "measure stress in an animal" we really measure the effects of the stress: the changes the stressor causes in the animal (such as the rise in body temperature when the animal is experiencing a net gain of heat from the environment) or the responses the animal invokes in an effort to establish a normal internal state in the face of a stressor (such as the rise in breathing rate when the animal needs to increase its heat-loss rate to bring body temperature back down to the desired point).

Measuring Environmental Stress

Scientists in a wide range of disciplines have been "measuring stress in animals" with increasing frequency over the past century and a half. Almost twenty years ago, The American Physiological Society published an epic tome of some 1056 pages called Adaptation to the Environment (Dill, 1964). The means of measurement have continued to develop as the scientific inquiry in animal ecology has blossomed profusely in the intervening two decades.

Yet the measurement of stress in animals is but the first step in applying ecological knowledge to animal production. The second step is the interpretation of the values. And of the two, the second step is by far the more difficult. In particular, it is necessary to determine where stress leaves off and distress (excessive or unpleasnt stress) begins.

Interpretation of stress parameters and indices is thus the real challenge as we continue to generate more knowledge and endeavor to use more completely what is already known for the purpose of increasing the fit between agricultural animals and their environments. And so it is this interpretation step on which we shall dwell.

STRESS RESPONSES: TRADITIONAL CONCEPTS

It is the unusual moment when an animal--in the wild or on a farm--is not responding to several stressors at once. Stress is the rule, not the exception. And nature has endowed the animals with a marvelous array of reactions to these impingements.

External environment comprises all of the thousands of physical, chemical, and biological factors that surround an animal's body. Each environmental factor varies over space and time. The animal's environment is therefore exceedingly complex.

The animal must maintain a steady state in its internal environment despite fluctuating external conditions. Claude Bernard (1957) said: "All vital mechanisms, however varied they may be, have only one object, that of preserving constant the conditions of life in the internal environment." This is the concept based on negative-feedback control loops that Walter Cannon later called homeostasis. More recently it has been called homeokinesis to emphasize its dynamic, yet consistent, nature.

All sorts of external environmental elements tend to modify corresponding internal environmental elements in an animal. Ultimately, if no homeokinetic mechanism acted, the internal environment would resemble the external, and life would cease.

Homeokinetic Control Loops

The homeokinetic animal attempts to control all aspects of its internal environment via adaptive responses similar in principle to a house's temperature-control system. Neural mechanisms participate in input reception and analysis, decision-making, and effector activation. Neuroendocrine mechanisms link neural and endocrine elements and activate effectors. Endocrine mechanisms take part in neural-endocrine and endocrine-endocrine linkages, as well as effector activation, and in some cases even effector action. These processes occur in specific configurations in the animal's many specific control loops. Muscles and glands are the body's chief effectors. Effector action is usually specific for the particular remedial reaction required.

Nonspecific stress response. In addition to specific stimulus/effector activation loops, Hans Selye (1952) has developed the concept of a nonspecific initial reaction to diverse stimuli. According to this facet of Selye's general adaptation syndrome, the rate of adrenal glucocorticoid secretion increases abruptly following any insult to the body. The teleological reason for this is that glucocorticoids promote mobilization of proteins from tissues. The amino acids liberated in this way can be used either as fuel or for synthesis of other proteins, such as immunoglobulins or scar tissue, that might be crucial at the moment.

This nonspecific reaction no doubt occurs, but specific impingements sooner or later require specific counteractions. Further, for domestic animals, the nonspecific alarm reaction seems to be superfluous, if not counterproductive, whereas insult or injury might interfere with a wild animal's getting food, and thus crucial amino acids, food-getting is ordinarily not a problem for domestic animals. Finally, all productive processes involve protein synthesis, so a high glucocorticoid secretion rate can be detrimental to food-animal performance at least in the short term.

Adaptation: Stress and Strain

An environmental adaptation refers to any functional, structural, or behavioral trait that favors an animal's survival or reproduction in a given environment, especially an extreme or adverse surrounding. Rates of life processes are the criteria used most often to assess adaptation. Adaptation can involve either an increase or a decrease in the rate of a given process.

A strain is any functional, structural, or behavioral reaction to an environmental stimulus. Strains can be adaptive or nonadaptive. Many enhance the chances of survival, but others are seemingly of little consequence.

A stress is any environmental situation--and a stressor any environmental factor--that provokes an adaptive response. A stress might be chronic (gradual and sustained) or acute (abrupt and often profound). Thus, by definition, environmental stress provokes animal strain, or in other words environmental stress provokes a stress response.

Environmental stress occurs when a given animal's environment changes so as to stimulate strain (as when environmental temperature falls below the crucial level) or when the animal itself changes in relation to a given environment (as when shearing reduces a sheep's cold tolerance).

Kinds of adaptation.

There are several categories of environmental adaptation. A given animal represents one stage in a continuum of evolutionary development. An animal's heredity determines the limits of its environmental adaptability. Hence, there are genetic adaptations to environment. Genotypic changes occur naturally due to genetic mutations. Environmental stress theoretically permits mutations having adaptive utility to be realized and ultimately to become fixed in animal populations. Artificial selection pressures for productive traits reduce such natural selection pressures. But individuals selected on the basis of productive performance are at least adequately adapted to the production environment; otherwise, they would neither perform at relatively high levels nor reproduce.

There are also induced adaptations. A given stressful environmental complex provokes various responses depending on the individual animal's current adaptation status, which is determined by heredity and by its life history, as well.

Acclimation is one kind of induced adaptation. It refers to an animal's compensatory alterations due to a single stressor acting alone, usually in an experimental or artificial situation, over days or weeks. A hen in a layer house might acclimate to altered day length, for example.

Acclimatization, on the other hand, refers to reactions over days or weeks to environments where many environmental factors vary at the same time. A ewe at pasture acclimatizes to seasonal variations in day-length in conjunction with variations in other environmental factors.

Finally, an animal may become habituated to certain stimuli when they occur again and again. Sensations and effector responses associated with particular environmental stimuli tend to diminish when these stimuli occur repeatedly. A pig raised near an airport becomes habituated to the roars of jet airplanes, for example.

Level of Adaptation. An animal's environmental adaptation can be analyzed at several levels of organization. At one end of the spectrum, adaptations can take the form of enzyme inductions or of changes in other modifiers of catlyzed biochemical reactions. In the middle are changes associated with adaptive responses to environmental stress in sensory, integrative, and coordinative neural functions, in neuroendocrine and endocrine functions, and in effectors' outputs. At the other end of the range, the animal's behavior often changes in response to environmental stimuli. Malcolm Gordon (1972) said: "There is certainly no logical basis for any claim that understanding the nature of life at one level of organization is more fundamental to overall understanding than comprehension at any other level."

STRESS RESPONSES: PSYCOLOGICAL COMPONENTS

It is now generally recognized that the amount of stress an animal is under depends not only on the intensity and duration of the noxious agent (the traditional concept), but on the animal's ability to modify the effects of the stressor as well (Mason, 1975; Archer, 1979).

Lack of Control

A recent study of stress effects on tumor rejection demonstrated psycological components of stress responses (Visintainer, et al., 1982). Stressors such as mild electrical shock depress an animal's ability to reject certain tumors in experimental settings. In this particular experiment, individually held rats were inoculated with a standard dose of tumor-causing cells and assigned to three treatments: control (no shock), mild shock that could be stopped by pressing a switch (escapable shock), and mild shock that stopped anytime the escapable-shock rate in the trial pressed its switch but over which this rat itself had no control (inescapable shock).

In other words, the amount of physical impingement received by animals in the two shock treatments was the same, but those in one group (escapable shock) could control the duration, while those in the other group (inescapable shock) could not.

Fifty-four percent of the control rates rejected their tumors. Inescapable shock caused so much stress that tumor rejection occurred in only 27% of the rats, while 63% of those subjected to escapable shock rejected their tumors. The conclusion: the low rate of tumor rejection was due not to the shock itself, but to the animal's inability to control this stressor.

Alliesthesia Modification

Central perception ("alliesthesia") of stress intensity depends on the context within which it occurs. Alliesthesia in the form of comfort rating or pleasure rating is affected by the animal's internal state and, hence, by its external surroundings as well.

For example, a thermal stimulus can feel pleasant or unpleasant depending on the body's thermal status. Hypothermic humans find cold stimuli very unpleasant and hot very pleasant, while hyperthermic humans have the opposite perceptions (Cabanca, 1971). Similarly, gastric loading with glucose decreased the human subjects' pleasure rating of the sweet taste of sucrose in a thermoneutral environment (26°C), but this negative alliesthesia due to glucose loading was eliminated when ambient temperature was reduced to 4°C (Russek, et al., 1979).

These findings remind us that "variety is the spice of life" and suggest that "taking the bitter with the sweet" is pleasurable in the long run. Extrapolating the concept of alliesthesia modification to agricultural animals' lives, it would appear that stress of one sort often primes the animal to receive pleasure from some other aspects of its environment.

In any case, the fact that an animal's psychological state can modify its perception of stress makes it all the more difficult to interpret how a specific stressor is affecting a specific animal.

MEASURING AND INTERPRETING STRESS RESPONSES

The scientific literature stores report of hundreds of experiments purported to measure stress in food animals (Hafez, 1968; Hafez, 1975; Johnson, 1976a,b; Stephens, 1980; Craig, 1981; Curtis, 1981). It is a relatively simple task to subject experimental animals to a controlled stressor and measure a resultant change in some physiological, anatomical, or behavioral parameter. Hormonal, cardio-respiratory, and heat-production parameters have been studied most in the past. Behavioral and anatomical changes are being characterized more lately. But an objective index of stress in terms of animal health, performance, and well-being has been elusive.

As Graham Perry (1973) said: "Even marked physiological changes may indicate only that an animal is successfully adapting to its environment--not necessarily that it is succumbing to adversity." And, similarly, Ian Duncan (1981) said: ". . . it should be of no surprise that chickens behave differently in different environments. This may simply demonstrate how adaptable they are." Again: at what point does stress become dis-stressful.

As for methodology, it is very difficult to study the effects of specific supposed stressors on an animal without introducing artifacts due to the stressfulness of the investigative techniques themselves (Adler, 1976). This is especially so in real or simulated production situations. What is the baseline adrenal-glucocorticoid secretion rate of an animal? Will it ever be certain beyond a reasonable doubt that the experimental manipulation necessary to obtain the needed samples or observations is not itself so stressful as to compromise the results?

Also, interpretation of the results of this kind of research is hampered by the fact that, by and large, there is not yet consensus as to the meaning of data on specific behavioral and hormonal changes in responses to stressors. What does it mean when an animal increases breathing rate by 250% in one environment compared with another? Does a 65% increase in plasma glucocorticoid concentration indicate the animal is under stress? If so, is the stress mild or severe? Scientists still do not understand how findings such as these relate to an animal's well-being, its health, and its productivity; consequently, we cannot rely on physiological or behavioral traits as valid indicators of the amount of stress an animal actually perceives, let alone how these might be related to the animal's health and productivity.

STRESS AND PRODUCTIVITY

Environmental stress generally alters animal performance (Curtis, 1981). The stress provokes the animal to react, and this reaction can influence the partition of resources among maintenance, reproductive, and productive functions in one or more of five ways:

1. The reaction may alter internal functions. Many bodily functions participate in productive processes as well as in reactions to stress. Survival responses may thus unintentionally affect productive preformance. For example, increased adrenal glucocorticoid secretion in response to stress can impair growth.

2. The reaction may divert nutrients. When an animal resonds to stress, it in effect diverts nutrients to use in higher-priority maintenance processes. Adaptive reactions are implemented even at the expense of productivity.
3. The reaction may reduce productivity directly. The animal's response sometimes partly comprises intentional reductions in productive processes. This generally frees some nutrients for maintenance uses. For example, an animal might reduce its productive rate in a hot environment in an attempt to re-establish heat balance with its surroundings.
4. The reaction may increase variability. Individual animals within a species differ from each other in functions, behaviors, and structures by what have been called "individuality differentials." Individual animals therefore differ in their responses to the same environmental stressor. In other words, two animals in the same group might successfully cope with the same stressful situation by calling different mechanisms into play. Then, if the complements of mechanisms used by the two individuals differ in the energy expenditure required to achieve them, the amount of energy diverted from productive processes will be different for the two animals. The result of this is that the amount of variation in individual performance in a group of animals tends to be related directly with the environmental adversities to which the animals are subjected.
5. The reaction may impair disease resistance. Because the animal's reaction to stress can impair disease resistance, that reaction influences the frequency and severity of disease. Of course, infection itself is a stress, so once established it in turn can influence the animal's productive performance. The mechanisms involved in the relations between environmental stress and resistance against infectious disease are just now being elucidated.

Kelley (1980) identified eight stressors: heat, cold, crowding, regrouping, weaning, limit-feeding, noise, and movement restraint. He documented the fact that all of these have been accorded a central role in stress-induced alterations of resistance against infection.

Having developed a framework for analyzing relations between adverse environments and animal productivity, it would be unrealistic to leave the impression that the link between stresses and productive processes are clear and simple. Consider two examples.

Lactating dairy cows held in a natural subtropical summer environment and provided no shelter are obviously under severe stress at mid-afternoon. They have markedly higher body temperatures and respiratory rates than their herdmates under the shade. Yet there might be no significant difference in fat-corrected milk yield between the two groups of cows (Johnson et al., 1966).

Socially and physically deprived animals often grow faster than do their counterparts in more enriched environments (Fiala et al.) So there is a risk in assuming that an animal stressed by a specific environmental complex is necessarily unfit for productive use in that environment. While one often might be justified in presuming that strain against stress reduces animal productivity, the animal can still be putting out an acceptable amount of product per unit of resource input.

Robert McDowell (1972) refers to "physiological adaptability" to environment (measured by physiological traits such as breathing rate) that is associated with survival responses, and to "performance adaptability" to environment (measured by productive-performance traits such as growth rate). These two often bear little positive relation to each other.

Thus, it is not sufficient for an animal producer to be concerned only with physiological and behavioral indices of environmental adaptability. Producers are more interested in the size of decrement, if any, in production associated with an animal's living in a particular environment. And to learn the quantitative effects of a given environment on animal performance, the productive traits themselves must be measured. After all, knowing a hen's breathing rate tells one little or nothing about her rate of lay.

There has been unfortunate ambiguity on this point among researchers and producers alike. An animal exhibiting marked strain has generally been assumed to be having markedly depressed performance. This is not necessarily so. Indeed, visible strain signifies that the animal is attempting to compensate for an environmental impingement. These attempts might succeed, and they might interfere with production only slightly or not at all.

The marvelous homeokinetic phenomena they possess make for resilient beasts and birds on our farms and permit profitable performance in a wide range of circumstances. The response flexibility that animals demontrate in the face of myriad stressors seem more remarkable than those instances when defensive reactions are inadequate and the environmental complex drastically reduces health or performance.

REFERENCES

Adler, H. C. 1976. Ethology in animal production. Livestock Prod. Sci. 3:303.

Archer, J. 1979. Animals Under Stress. Edward Arnold, London.

Bernard, C. 1957. An Introduction to the Study of Experimental Medicine. Dover, New York.

Bowman, J. C. 1977. Animals for Man. Edward Arnold, London.

Cabanac, M. 1971. Physiological role of pleasure. Science 173:1103.

Cannon, W. B. 1932. The Wisdom of the Body. Norton, New York.

Craig, J. V. 1981. Domestic Animal Behavior. Prentice-Hall, Englewood cliffs.

Curtis, S. E. 1981. Environmental Management in Animal Agriculture. Animal Environment Services, Mahomet, Illinois.

Dill, D. B. (Ed.) 1964. Handbook of Physiology. Section 4: Adaptation to the Environment. American Physiological Society, Washington.

Duncan, I. J. H. 1981. Animal rights-animal welfare: a scientist's assessment. Poul. Sci. 60:489.

Fiala, B., F. M. Snow, and W. T. Greenough. 1977. "Impoverished" rates weigh more than "enriched" rats because they eat more. Devel. Phychobiol. 10:537.

Fraser, D., J. S. D. Ritchie, and A. F. Fraser. 1975. The term "stress" in a veterinary context. Brit. Vet. J. 131:653.

Gordon, M. S. 1972. Animal Physiology: Principles and Adaptations. (Second ed.) Macmillan, New York.

Hafez, E. S. E. (Ed.) 1968. Adaptation of Domestic Animals. Lea and Febiger, Philadelphia.

Hafez, E. S. E. (Ed.) 1975. The Behavior of Domestic Animals (Third ed.) Williams and Wilkins, Baltimore.

Hale, E. B. 1969. Domestication and the evolution of behavior. In: E. S. E. Hafez (Ed.). The Behavior of Domestic Animals. William and Wilkins, Baltimore.

Johnson, H. D. (Ed.) 1976a. Progress in Animal Biometerology, Volume 1, Part I. Swets and Zeitlinger, Amsterdam.

Johnson, H. D. (Ed.) 1976b. Progress in Animal Biometerology, Volume 1, Part II. Swets and Zeitlinger, Amsterdam.

Johnston, J. E., J. Rainey, C. Breidenstein, and A. J. Gidry. 1966. Effects of ration fiber level on feed intake and milk production of dairy cattle under hot conditions. Proc. Fourth Int. Biometerological Cong., New Brunswick.

Kelley, K. W. 1980. Stress and immune function: A bibliographic review. Ann. Vet. Res. 11:445.

Mason, J. W. 1975. Emotion as reflected in patterns of endocrine integration. In: L. Levi (Ed.) Emotions--Their Parameters and Measurement. Raven, new York.

McDowell, R. E. 1972. Improvement of Livestock Production in Warm Climates. Freeman, San Francisco.

Perry, G. 1973. Can the physiologist measure stress? New Scientist 60 (18 October):175.

Russek, M. M. Fantino, and M. Cabanac. 1979. Effect of environmental temperature on pleasure ratings of odor and testes. Physiol. Behav. 22:251.

Selye, H. 1952. The Story of the Adaptation Syndrome. Acata, Montreal.

Stephens, D. B. 1980. Stress and its measurement in domestic animals: a review of behavioral and physiological studies under field and laboratory situations. Adv. Vet. Sci. Comp. Med. 24:179.

Visintainer, M. A., J. R. Volpicelli, and M. E. P. Seligman. 1982. Tumor rejection in rats after inescapable or escapable shock. Science 216:437.

27

ANIMAL WELFARE: AN INTERNATIONAL PERSPECTIVE

Stanley E. Curtis,
Harold D. Guither

Are today's intensive animal-production systems basically inhumane? This question is central to the issue of farm-animal welfare that has been developing in the U.S. for several years. Opinions vary across a broad spectrum. According to one extreme view, animals have the same mental experiences as humans ("anthropomorphism"), and thus they ought to be treated as humans. Ethical vegetarians believe we have no right to slaughter livestock or poultry for human consumption. A more widely held, more moderate position--and one where many animal welfarists and agriculturists find common ground--is that we should respect the animals we use and should not subject them to distress as a consequence of normal production pratices.

Most states have laws protecting domestic animals from neglect and abuse. But there is still debate about a third form of alleged cruelty to animals: depriving animals of opportunities to express certain supposedly necessary behaviors. Especially controversial are the practices of keeping laying hens in cages, gestating sows in stalls, and veal calves in crates. There are those who point out that some behaviors that animals express frequently in natural surroundings are not observed often in certain artificial environments (and vice versa), and they claim that this reflects undue stress on the animals in these unnatural surroundings. However, others say such differences in behavioral responses to the environments are to be expected because the array of behavioral triggers varies from one environment to another.

The general public's attention was first drawn to the welfare of food animals by the 1964 publication in England of the book Animal Machines by Ruth Harrison. As a result of the public interest this book generated, the British government appointed a committee that prepared a report on intensive animal-production systems. The report questioned the humaneness of several common husbandry practices. Since then, the debate over farm-animal welfare has spread all over northern Western Europe, the U.S. and Canada, as well as Australia and New Zealand. Conflict has arisen because livestock and poultry producers--and indeed producers of feed grains too--have perceived their economic interests as being threatened by some of the policies and regulations proposed by animal-welfare groups, whose views are based on ethical judgments rather than scientific evidence or economic feasibilities.

The extent to which, and the ways in which, a society uses animals for companionship, recreation, power, or food are ethical decisions, and they are heavily influenced by social and religious traditions. Like other public policy matters, these decisions should be made by our political system and not by any one sector with strong opinions or interests. But such public decisions can be made wisely only after all of the facts and consequences of proposed policies and regulations have been fully explored and analyzed.

Scientific research can contribute to discussions of animal welfare by producing scientific evidence on animals' relationships with their environments. Current gaps in knowledge lie largely in the areas of perception and stress, and these are where many investigations are now being focused.

Perception is the immediate discriminatory response of an animal to energy-activating sense organs. What do we know quantitatively, and especially relative to the human experience, about conscious perceptions of comfort and discomfort, pleasure and displeasure, pain or the absence thereof, by farm animals? Little or nothing, if purely anthropomorphic musings are ignored.

We should also recognize that the design of accommodations for humans, about whom much more is known in this respect, is still hampered by the paucity of quantitative data available and by the practical impossibility of meeting an organism's needs so precisely over time that it will never experience discomfort. Added to this is the complicating fact that individuals' perceptions of comfort differ so greatly. One architect has suggested that, as a practical matter, even facilities for humans cannot be designed to achieve some "comfort" zone--rather a "lack of discomfort" zone is the best that can be hoped for.

We know more about stress and its consequences in farm animals than about perception. An animal is under stress when it must make extreme functional, structural, or behavioral adjustments to cope with adverse aspects of its environment. Thus, an environmental complex is stressful only if it makes extreme demands on an animal.

Interpretation of stress parameters and indices is the real challenge as we continue to generate more knowledge and endeavor to use more completely what is already known for the purpose of increasing the fit between agricultural animals and their environments.

It is the unusual moment when any animal--in the wild or on a farm--is not responding to several stressors at once. Stress is the rule, not the exception, and nature has endowed animals with marvelous arrays of reactions to these impingements.

It is now generally recognized that the amount of stress an animal is under depends not only on the intensity and duration of a noxious agent, but on the animal's ability to modify its perceptions and the effects of the stressor, as well. There is increasing evidence that an animal's emotional feelings depend to some extent on the predictability and controllability of its environment. When an animal's expectations are being fulfilled, or it is able to control its surroundings, it feels more comfortable, even if it is responding to stressors. But, according to this theory, the animal feels uncomfortable or even distressed when its environment is unpredictable or uncontrollable.

Furthermore, the concept of alliesthesia holds that central perception of stress intensity depends on the context within which it occurs. For example, a human whose body temperature is below normal usually finds cold stimuli unpleasant and hot pleasant. Extrapolating to agricultural animals, perhaps stress of one sort actually primes an animal to receive pleasure from a stressful aspect of its environment; the cool of the night might prepare the animal to take comfort from the heat of a summer afternoon.

Scientists still do not fully understand how findings such as these relate to an animal's welfare, health, and productivity. But it is accepted that we cannot rely on physiological or behavioral traits alone as indicators of the amount of stress an animal actually perceives, let alone how these might be related to the animal's welfare, health, or productivity.

Another set of indicators must be taken into account, too. Environmental stress generally alters animal performance and also elicits physiological and behavioral responses. The stress provokes the animal to react in some way, and this reaction can influence the partition of resources among maintenance, reproductive, and productive functions in one or more of five ways: the reaction might 1) alter internal functions involved in economically important processes as well as reactions to stressors, 2) divert nutrients from productive or reproductive processes to maintenance processes, 3) reduce productivity directly, 4) increase individual variability in performance, or 5) impair disease resistance.

Still, it would be unrealistic to leave the impression that the links between stresses and productive and reproductive processes are clear and simple. Consider three examples: 1) lactating cows can be under severe heat stress each afternoon, but so long as adequate feed is available they might not suffer any reduction in milk yield; 2) animals kept in relatively barren environments where specific social or physical stimuli are absent sometimes grow faster than do their counterparts in more enriched surroundings; and 3) there is not always a clear correlation between signs of physical and social trauma in individual hens and their respective individual egg yields. The situation is a complicated one which needs further scientific investigation before the fit between food animals and their environments can be optimized in terms of welfare, health, and performance of the animals.

As a result of public pressure, funds have been allocated in many countries for more research on intensive production systems for laying hens, veal calves, and swine and on the responses of these animals to various stressors. But sometimes the public pressure for governmental action has been so strong that political decisions have been made before scientific evidence was available to support such decisions. This is despite the fact that, in most countries, governmental officials have been sympathetic to the economic realities farmers face in terms of animal-welfare regulations' increasing production costs. Legislators have been especially reluctant to enact regulations that would put their nation's farmers at a competitive disadvantage with producers in other countries.

Animal-welfare laws and regulatios already have been established as part of public policy in several European nations. Through government-appointed committees or commissions, views of animal welfarists, scientists, and producers have been brought together. This is the first phase of bringing the public's awareness of the animal-welfare issue to the point that they can participate intelligently in the policymaking process. When such a group issues its report, the views and proposals it contains provide the bases for more media attention and broader public awareness. For example, the _Report of the Technical Committee to Enquire into the Welfare of Animals Kept under Intensive Livestock Husbandry Systems_ in the U.K (1965), the National Council for Agricultural Research Committee of Experts' _Report on Animal Husbandry and Welfare_ in The Netherlands (1975), the Council of Europe's _European Convention for the Protection of Animals Kept for Farming Purposes_ (1976), and the House of Commons Agriculture Committee's report, _Animal Welfare in Poultry, Pig, and Veal Calf Production_ in the U.K. (1981) have provided reference bases for public discussion of animal-welfare issues. What the public might not realize is that these reports may or may not represent a majority of public opinion. They are simply the reference base upon which further discussion and debate will be carried out before final policy decisions are made.

Once these decisions have been made, agencies of government are then established to oversee the new laws and regulations. In some European countries, the regulations have dealt with practices progressive producers would have applied in their livestock and poultry operations anyway. But, of course, with a rgulation in place they no longer have a legal choice in the matter.

In the U.S., a joint resolution introduced into the Congress by Congressman Ronald Mottl of Ohio in July 1981 called for the creation of a 16-member committee to study animal-production practices in this country. Those concerned with current animal-production practices viewed this legislative approach as a means to further promote discussion of animal welfare and broaden public awareness of the issue.

Producer groups generally felt that no such discussion or study was needed. They believed they were using the most advanced production methods that scientific research and practical experience offered for the most efficient and profitable production of food. Although no hearings on this resolution were scheduled in 1982, the idea of a commission to study animal-welfare issues in the U.S. probably will continue as a goal of some animal-welfare organizations for a long time.

The experiences with animal-welfare policies and practices in Europe suggest that policymakers in the U.S. weigh scientific evidence carefully and have the benefits of broad public discussion on the issues--including ramifications impinging on the food distribution and pricing system--before attempting to set up legislation that would regulate the ways livestock and poultry are produced.

Part 7

FEEDS AND NUTRITION

28

NUTRIENT REQUIREMENTS AND SAMPLE RATIONS FOR HORSES

Doyle G. Meadows

Nutrition as a science is comparatively new. Research is revealing that horses can and should be fed according to their nutrient needs, rather than according to fables and trade secrets, as has been the case for many years. Figures on nutrient requirements that appear in this article are taken from "Nutrient Requirements of Horses," a publication of the National Research Council, National Academy of Science. These data are largely experimentally derived and represent the average for horses. Values for nutrient requirements, growth rates, and feed intake may be low for larger, more rapidly growing horses, and may be high for smaller, slower developing horses. However, these figures are "in the ball park" and the individual horseman may make minor adjustments to fit his horse.

Nutrient requirements are generally determined based on the following categories: maintenance, gestation, lactation, growth, and work. The horse is then fed to meet the sum total of these requirements. For example, a working horse would be fed to meet his maintenance and work requirement (horse simply maintaining zero weight change - not working, not in gestation, etc.) plus enough additional nutrients to fulfill the work requirements. Other examples could be maintenance plus gestation for pregnant mares, or maintenance plus lactation requirements for nursing mares.

Horse rations are balanced to meet the protein and energy needs of horses. These are the most costly nutrients and are of major concern. Vitamins and minerals are also important but are generally supplemented to the ration as needed after protein and energy needs are met.

Energy requirements are generally referred to in megacalories (Mcal), which is a quantitative term used to measure energy concentration in feeds. Protein requirements are generally expressed in terms of percentage crude protein required in the ration. The rations enclosed are "ball park" diets for the different classes of horses.

PROTEIN AND ENERGY

Maintenance

Protein. A 1,100 lb horse requires a minimum of slightly under an 8% crude protein ration (combination of hay and grain). This represents about 1.4 lb of protein per day. These values are indicative of a mature horse at maintenance that is idle, doing little or no work, and not in production. The protein requirement for this class of horse can be met very easily.

Protein Requirement for Maintenance (1,100 Lb Horse)

	Crude protein (lb)	Crude protein (%)*
Mature horse, maintenance	1.4	7.7

*Expressed on 90% dry matter basis.

Energy. A 1,100 lb horse requires about 16.4 Mcal of digestible energy per day to maintain body weight. A typical horse ration would contain about 1.0 Mcal of digestible energy per pound of feed (roughage and grain), therefore this horse would require 16.4 lb of this feed per day simply to maintain his present body weight. The maintenance requirements will vary according to the weight of the horse (i.e., heavier horse would require more energy).

Energy Requirement for Maintenance (1,100 Lb Horse)

	Digestible energy (Mcal)	Daily feed (lb)
Mature horse, maintenance	16.4	16.4

Sample Ration for a Mature Nonworking Horse

	Concentrate (lb)	Roughage (lb)
Legume or excellent quality grass hay	0	16-18*
Average quality grass hay	2-3	13 -16

*High quality pasture will substitute. To meet requirements without "hay belly" appearance, feed 10-11 lb grass hay and 5-6 lb (8% to 9% CP) concentrate.

Gestation

Protein. Feeding of the pregnant mare is an area in which many horsemen can improve their existing feeding program. Requirement for 1,100 lb mares in late gestation (last 3 mo) is increased to at least a 10% crude protein ration. This would be 10% crude protein content in the total ration; therefore, the crude protein content of concentrate portion of the ration would need to be higher to offset low protein hay. It is important to feed gestating mares a high-quality ration to achieve proper fetal development and to maintain the weight of the mare. The pregnant mare does not require much more protein than does an open mare, but the protein should be of high quality to meet her nutrient requirement.

Protein Requirement for Gestation (1,100 Lb Mare)

	Crude protein (lb)	Crude protein (%)*
Mare, last 90 days of gestation	1.7	10.0

*Expressed on 90% dry matter basis.

Energy. Mares in the last 90 days of gestation have an energy requirement above maintenance because of the fetal growth. Mares gaining at the rate of 1.2 lb/day would require 18.4 Mcal of energy per day or 2 Mcal of energy per day maintenance. The energy content of feed could be raised from 1.0 Mcal to 1.1 Mcal to meet the additional energy requirement.

Energy Requirement for Gestation (1,100 Lb Mare)

	Digestible energy (Mcal)	Daily feed (lb)
Mare, last 90 days of gestation	18.4	16.2

Sample Rations for Gestating Broodmares

Early to midgestation. During the first 8 mo of gestation, the fetus does not grow very rapidly and the pregnant mare's requirements are not increased. Feeding to maintenance requirements as listed above will meet the early to midgestation mare's requirements.

Late gestation. During the last 3 mo of gestation, the conceptus increases in size at approximately an average of 1 lb/day; therefore, more nutrient requirements are greatly increased. The pregnant 1,100 lb mare in late gestation should receive 18 lb (remember 130 lb of conceptus in abdominal cavity) of total daily feed containing 10% CP, .45% calcium, and .3% phosphorus in the total ration. Note that energy, protein, calcium, and phosphorus requirements increase in approximately the same quantity of feed, which suggests supplemental feeding of concentrate is necessary. Mares should be gaining .25 to .50 lb/day in addition to fetus weight increase. Requirements could be met as follows:

	Concentrate (lb)	Roughage (lb)
Legume or excellent quality grass hay	4-5	13-14
Average quality grass hay	6-7 (13-14% CP)	11-12

	Late gestation ration (Feed w/legume hay)	Late gestation ration (Feed w/grass hay)
Oats	45%	42%
Corn	42%	34%
SBM	--	10%
Wheat bran	10%	10%
Molasses	3%	3%
Limestone	--	1%

Lactation

Protein. Properly feeding the lactating mare is challenging and requires a large amount of high-quality feedstuffs to promote maximum milk production and foal growth. The lactating mare (1,100 lb) requires at least a 12.5% crude protein ration (combination grain and hay). The milking mare needs about 3.0 lb of protein, and to achieve this level of protein it would be necessary to feed approximately 22 lb of high-quality feedstuffs to each mare per day. This would require the mare to be fed a high concentrate ration (approximately 14 lb) and the remaining amount of ration to be provided in the form of long roughage. These values would be typical of an 1,100 lb mare in early lactation and these requirements would decrease as lactation progresses.

Protein Requirement for Lactation (1,100 Lb Mare)

	Crude protein (lb)	Crude protein (%)*
Lactating mare, first 3 months	3.0	12.5
Lactating mare, 3 months to weanling	2.4	11.0

*Expressed on 90% dry matter basis.

Energy. After foaling and throughout the milk production phase, the mare's energy requirement has dramatically increased. It takes 28.3 Mcal of energy per day to meet her requirement in the first 3 mo of lactation (1,100 lb mare). The energy requirement for lactation has almost doubled that of the 1,100 lb mare at maintenance. Therefore, the lactating mare has to be fed more of a higher energy ration to meet the additional energy requirement. The energy requirement for lactation declines in the latter stages of lactation due to decreased milk production.

Energy Requirement for Lactation (1,100 Lb Mare)

	Digestible energy (Mcal)	Daily feed (lb)
Lactating mare, first 3 months	28.3	22.2
Lactating mare, 3 months to weanling	24.3	20.6

Sample Rations for Lactating Broodmares

	Concentrate (lb)	Roughage (lb)
Legume or excellent quality grass hay	11-12(10-12% CP)	13-14
Average quality grass hay	13-14(14-16% CP)	11-12

Sample Rations

	Lactation ration (Feed w/legume hay %)	Lactation ration (Feed w/grass hay %)
Corn	46.0%	40.0%
Oats	44.5%	35.0%
SBM	5.0%	15.0%
Wheat bran	5.0%	5.0%
Molasses	--	3.0%
Limestone	--	1.5%
Dicalcium phosphate	.5%	.5%

Note: It is estimated that mares grazing a small-grain pasture can consume 2.5% of their body weight daily in dry matter. A 1,100 lb mare might consume 28 lb of dry matter. This will easily meet her protein requirement; however, an energy concentrate (low % CP) might be necessary to meet her energy requirements.

Growth

Protein. It is important to provide adequate quality and quantity of protein in the ration to promote rapid gains that are highly desired in the horse industry. Protein requirements for growth vary according to the age and (or) development of young horses. To ensure proper growth, it is necessary to provide an 18% crude protein ration to early-weaned or creep-fed foals. Weanlings (6 mo) and yearlings (12 mo) require a 14% (approximately) crude protein ration whereas the crude protein level is decreased to about 10% as the horses reach 2 years of age.

It is critical that high quality sources of protein be used in the complete ration. This means using feedstuffs that supply an adequate and well-balanced amino acid content. Amino acids are the building blocks of protein, and high-quality feedstuffs have excellent amino acid balance. Examples of high-quality protein sources would be soybean meal, linseed meal, and fish meal.

Protein Requirement for Growth

	Crude protein (lb)	Crude protein (%)*
Nursing foal 3 months old	1.7	16.0
Requirements above milk	.9	16.0
Weanling, 6 months old	1.7	14.5
Yearling, 12 months old	1.7	14.5
Long yearling, 18 months old	1.6	10.0
2-year-old	1.4	9.0

*Expressed on 90% dry matter basis.

Energy. Optimum growth and development of young, growing horses are a major concern to all horsemen. Foals at 3 months of age should be gaining at least 2.64 lb per day. In order for the foals to grow at this rate, they must receive an additional source of nutrients to supplement milk from the mare. This is generally provided in the form of creep feed. The foal's energy requirement is about 14 Mcal per day of which about half is provided in the milk, leaving 7 Mcal per day to be supplied by creep feed. As horses grow older, gain per day will decrease, but there will be a need for increased amounts of feed because the horse has more total body weight. Horses should reach their mature height and weight in 2.5 to 3 years of age. Many horsemen are growing horses at much faster rates than these outlined.

Energy Requirement for Growth

	Digestible energy (Mcal)	Daily feed (lb)
Nursing foal, 3 months old	13.7	9.2
Requirements above milk	6.9	4.9
Weanling, 6 months old	15.6	11.0
Yearling, 12 months old	16.8	13.2
Long yearling, 18 months old	17.0	14.3
2-year-old	16.5	14.5

Sample Rations for Young Growing Horses

	Creep ration (%)
Cracked corn	42.5
SBM	31.0
Rolled oats	23.0
Brewer's yeast	.5
Molasses	3.0
Dicalcium phosphate	1.0
Limestone	1.0
TM-salt	1.0

Sample Rations for Weanlings

	Concentrate (lb)	Roughage (lb)
Legume or excellent quality grass hay	7-8 (13-14% CP)	4-5
Average quality grass hay	8-9 (15-16% CP)	3-4

Sample Rations

	Weanling ration (Feed w/legume hay %)	Weanling ration (Feed w/grass hay %)
Corn	45.0	39.0
Oats	39.0	30.0
SBM	12.0	25.0(50%)
Molasses	3.0	3.0
Limestone	--	2.0
Dicalcium phosphate	1.0	1.0

Sample Rations for Yearlings* (700 Lb)

	Concentrate (lb)	Roughage (lb)
Legume or excellent quality grass hay	7 (10-12% CP)	8
Average quality grass hay (10% CP)	8 (14% CP)	7

Sample Rations

	Yearling ration (Feed w/legume hay %)	Yearling ration (Feed w/grass hay %)
Corn	46.0	42.0
Oats	42.5	35.0
SBM (50% CP)	8.0	15.0
Molasses	3.0	3.0
Limestone	--	1.0
Dicalcium phosphate	.5	.5

*These rations could be used throughout the yearling year. However, the protein level could be reduced somewhat in the last months of the yearling year.

Work

Energy. There is little research available to actually quantify the energy requirement for work of horses. One can calculate that a 1,100 lb horse, working at a moderate level for 1 to 3 hours per day, may need 28 to 30 Mcal of digestible energy per day, which represents some 12 to 14 Mcal per day above his maintenance needs. While it takes about 16.4 lb of feed to maintain a horse, it will take some 25 lb of feed to maintain him if he is going to be worked at a moderate level. Obviously, this varies with the amount of work done, and feeding programs are adjusted accordingly. The best way to measure the adequacy of the energy supply in mature working horses is monitoring body weight and condition. As the level of work increases, energy concentration of the ration should also be increased. Nutrient requirements of the horse in intense work could be met on a daily ration of 10 lb of good-quality grass hay and 13 to 15 lb of an energy concentrate (8% CP).

Minerals

Most mature horses can be maintained on pasture alone if the pasture is of good quality and the horses have an available mineral supplement. Although several minerals are needed in limited amounts, calcium, phosphorus, and salt are the minerals that are generally of primary concern. Trace mineral requirements are normally met with well-balanced rations or when horses have access to a trace-mineralized salt.

Calcium and phosphorus. Calcium and phosphorus are essential to sound bone development and other body functions. Calcium and phosphorus should be fed at ratio of 1.5 calcium to 1 phosphorus (1.5:1) although slight variations can exist with no adverse effects. All rations should contain more calcium than phosphorus. There is daily loss of these minerals and, therefore, they have to be supplied in the ration.

As would be expected, foals and young growing horses require more calcium and phosphorus than mature horses. Calcium requirements range from .85% for 3-month-old foals to .45% for 2-year-old horses. Phosphorus requirement for the 3-month-old foals is .60% and .35% for 2-year-old horses. Calcium and phosphorus requirements for the mature, idle horse (1,100 lb) are .30% calcium and .20% phosphorus (1.5:1). Both pregnant mares (last 90 days) and lactating mares require about .50% calcium and .35% phosphorus. Steamed bone meal, dicalcium phosphate, ground limestone and defluorinated phosphate are some common sources of calcium and phosphorus.

Calcium and Phosphorus Requirements of Horses

	Calcium (%)	Phosphorus (%)
Mature horses, maintenance	.30	.20
Mares, gestation and lactation	.50	.35
Nursing foals	.85	.60
Weanlings	.70	.50
Yearlings	.55	.40
Long yearlings	.45	.35
2-year-old	.45	.35
Mature working horses	.30	.20

Salt. The exact salt requirement of horses has not been determined, but if salt is fed at the rate of 0.5% to 1.0% of the diet or provided free choice, a deficiency is not likely to occur. Prolonged exercise and elevated environmental temperature will increase the need for sodium and chloride (salt is a chemical mixture of sodium and chloride); the quantity is thought to be about 60 g per day (.13 lb).

Vitamin requirements are closely associated with age, stage of production, and stress imposed upon the horse. Whether to add vitamins to the diet depends primarily on

1) type and quality of the diet (green, high quality forages are excellent sources for most vitamins), 2) extent of synthesis and absorption of vitamins in the digestive tract, 3) and in case of vitamin D, access to sunlight.

Vitamin A. Vitamin A is the fat-soluble vitamin of major concern in vitamin nutrition. Mature horses grazing good-quality forage will have an ample supply of vitamin A. Also, vitamin A can be stored in the liver and adequate vitamin A levels will exist in the blood for 3 to 6 months, even when consuming submarginal levels of vitamin A. Vitamin A requirement can be met by carotene, a plant precursor of vitamin A, or by synthetic vitamin A supplements. Stalled horses should be fed diets containing vitamin A.

A 1,100 lb mature, idle horse needs approximately 12,500 IU of vitamin A per day as compared to about 25,000 IU per day for gestating, lactating, or heavily stressed horses. As a horse reaches maturity, the vitamin A requirement increases.

Vitamins D, E, and K. Vitamin D requirements have not been adequately assessed but horses normally obtain sufficient vitamin D from sun-cured forages or from exposure to sunlight. The requirements for vitamin E have not been established. It is assumed that vitamin K is synthesized in adequate amounts by the intestinal microflora of the horse. No deficiencies have been reported.

B-Complex vitamins. The B-complex vitamins (i.e., thiamin, riboflavin, niacin, vitamin B_{12}) are usually supplied in adequate amounts in good quality forage. The amount supplied in forages, coupled with synthesis and absorption in the large intestine of the horse, is sufficient for most horses. Although exact levels are not known, severely stressed horses may have an additional B-complex vitamin requirement above maintenance.

Vitamins

Vitamin A	
Mature horse	12,500 IU/day
Gestating, lactating and heavily stressed horses	25,000 IU/day
Vitamins D, E, and K	
B-Complex vitamins	

Mineral and vitamin research is limited and definite requirements should be established in the future. However, at the present time, most complete-mixed horse rations do not warrant the use of expensive feed additives and vitamin-mineral premixes. When a requirement is not met, feed the horse only to meet that specific requirement. If the complete-mixed rations meet all necessary nutrient requirements, no increased performance will be obtained by top-dressing with a variety of supplements or premixes.

(Acknowledgement is extended to the Horse Section [Dr.Gary Potter, Dr. Doug Householder, and Mr. B. F. Yeates], Texas A&M University for providing material for this paper.)

29

FEEDING MARES FOR MAXIMUM REPRODUCTION EFFICIENCY

Melvin Bradley

The nutrients horses require are the same as those required by other warm-blooded animals: energy, protein, minerals, vitamins, and water. However, the energy needs of horses vary because of work and lactation. Energy is often expressed in total digestible nutrients (TDN). Idle horses may do well on 8 to 10 lb of TDN, but this requirement doubles when the horses are working hard and increases by 50% during lactation. In general, grains are three-fourths TDN, good hays are one-half, and poor hays are one-third TDN.

Protein is used for growth and lactation. Early weaned foals require 16% to 18% protein diets for muscular development, while adults can suffice on 8%. Good legume hays supply sufficient protein for all growing horses, except weanlings, while some fescue hays and winter pasture may furnish less than 4%.

There are two kinds of minerals: major minerals and trace minerals. The major minerals are calcium, phosphorus, and salt. Calcium and phosphorus are used to grow teeth and bones, thus, their percentage in the weanling's diet is 0.7% and 0.6% respectively, compared to 0.27% and 0.18% in the diet of a mature horse that has finished growing bones and teeth.

Potassium is usually considered a major mineral in cattle feeding, but horses' needs are usually met in the feeds supplied. Trace minerals are magnesium, sulfur, iron, zinc, manganese, copper, cobalt, iodine, selenium, and fluorine. These can be taken care of, for the most part, by feeding a trace-mineralized salt. It is unwise to feed three or four sources of trace minerals, because some of them are toxic, especially iodine.

There are two broad categories of vitamins: fat soluble and water soluble. Fat soluble vitamins are A, E, D, and K. They dissolve and store in fat and in the liver. Therefore, these vitamins can be fed intermittently. Generally speaking, the horse that has been on good pasture will have a 60 to 90 day supply of fat soluble vitamins stored in its body. On the other hand, the B vitamins are water soluble, dissolve in water, and must be available of vitamin A could result in enough uterine infection to make settling difficult. Mares that consume carotine, a precursor to vitamin A from hay and grass, manufacture all they need. However, most horse breeders begin to feed brood mares approximately 20,000 units per head per day during late winter when grass is dormant. To meet the different nutritional needs of mares, you should know their nutrient requirements (table 1). Note the dramatic change in nutrient needs of a dry mare as compared to one in lactation. These requirements create a management problem when the producer tries to put all mares together--that is, to group open mares, those in late pregnancy, and those in lactation. The open mares obviously become too fat, while lactating mares do not get enough feed to sustain adequate milk production. Fortunately, foals come in early spring when pastures are at peak production.

TABLE 1: NUTRIENT CONCENTRATION FOR MARES*

	TDN lb	C.P. %	Ca %	P %
Mature	8.25	7.7	0.27	0.18
Late pregnancy	9.25	10.0	0.45	0.30
Lactating	14.25	12.5	0.45	0.30

*Source: 1976 NRC

In formulating rations it is important to understand kind and amount of nutrients supplied by common horse feeds. These can be compared in table 2. Note the high TDN in grains, the range in their protein level from 9% to 12% and the extreme deficiency of calcium in the grains horses eat. Also, these grains have substantial amounts of phosphorus, with wheat bran being extremely high. The implications are that when heavy grain feeding is combined with nonlegume hay, a bad calcium-phosphorus imbalance occurs--unless calcium is added. One percent feeder-grade limestone can be added to the grain ration, or the ration can be fed with legume hay, which is very high in calcium. Alfalfa often will range almost 2% calcium compared to 0.3% for timothy. Forages are just average in phosphorus supplying 0.2% to 0.35%.

Is alfalfa damaging to the kidneys of your horse? If the hay is moldy, it can be. Grooms make this claim because they have noticed that stable horses urinate often when fed high amounts of alfalfa. This is because extra protein in alfalfa is not needed by an adult horse and it is broken down into nitrogen. Nitrogen is a salt and salts are voided through the urine. Since urine is constant at about 3% salt, horses must drink much more water to void the excess

TABLE 2: AVERAGE ANALYSIS OF SOME COMMON HORSE FEEDS

	TDN %	C.P. %		Ca %		P %
Grain						
Barley	70	12		.05		.32
Corn, NO2	80	9		.02		.26
Molasses, liquid	54	3		.0		.0
Oats	66	12		.09		.32
Wheat bran	60	16		.10		1.30
Hay						
Alfalfa						
Early cut	58	16		1.4		.2
Average	49	11		1.2		.2
Late cut	48	10		1.0		.2
Timothy						
Good	50	7		.3		.2
Poor	48	6		.3		.2
Mixed hay	50	9		.6		.2
Mineral						
Defluorinated rock phosphate				30.0		13.00
Dicalcium phosphate				22.0		19.00
Limestone				33.0		0.00
Commercial						
No. 1				15.0		14.00
No. 2				12.0		8.00
No. 3				10.0		5.00
No. 4				20.0		5.00
No. 5				6.0		2.00
Protein Supplement						
Cottonseed meal	65	41		.26		1.00
Linseed meal	62	35		.38		.78
Soybean meal	76	44		.30		.64

*Source: UMC Forage Lab

nitrogen when they are consuming large amounts of alfalfa. There is little reason to feed an adult mare more than 4 to 6 lb of alfalfa as this supplies her vitamins, minerals, and most of her protein when supplemented with a nonlegume hay.

Note the mineral supplements shown in table 2. Dicalcium phosphate and limestone are most often used in commercial horse rations; they are high in calcium, and dicalcium phosphate is 19% phosphorus. In general, foal rations with a base of two-thirds oats and one-third corn, use 10 to 15 lb of dicalcium phosphate and 20 to 25 lb of limestone per ton, if nonlegumes are the source of hay. A much better ration would be to feed the best quality legumes to these foals and reduce the calcium in the ration to about 10 lb.

Of the protein supplements, soybean meal is the best choice because it has the best balance of amino acids, a higher protein level, and is cheaper.

Years ago, linseed meal was the choice of horsemen. This was because of its laxative effect and the bloom of the hair coat. More recently linseedmeal has been extracted chemically from the bean, and the fat that bloomed the hair coat is now removed in processing. Because of its high price, poor balance of amino acids, and loss of linoleic acid, (the fatty acid that bloomed the hair coat), there is not much reason to use it in the diet of horses. Two tablespoonsful of corn oil per day will bloom the hair coat of a show horse. It can be purchased at most grocery stores.

FEEDING BARREN MARES

The biggest problem in nutrition of barren mares is that of regulating their condition to reduce fat, to tone muscles, and to provide exercise. If they are ridden, so much the better; however, most bands of mares have a dominance hierarchy that results in the most dominant mares getting most of the feed. Condition can be improved if overweight mares can be subdivided into like groups with reduced feed. Regulation of condition should begin around the first of the year, when their feed is more expensive than in the spring. It takes about 2 or 3 months to get a fat mare down to the correct condition without producing trauma that will cause wood chewing. It may be necessary to reduce feed intake through use of a late-cut hay. Since vitamins A, D, and E are stored, they can be fed in salt or supplied twice weekly in a small amount of grain.

Mares that are not fat should be flushed. Research shows that if adequately flushed a month prior to breeding, mares in medium condition have a better chance of conception. The energy and general diet levels are increased 10% to 15% one month before breeding. Usually, this is automatic if the mares are bred in the spring when on a diet with good grass; otherwise, grain feed may be indicated.

FEEDING PREGNANT MARES

Feed the pregnant mare during the first two-thirds of pregnancy just as you would an adult horse. No additional nutrients are needed, because there is little growth of the fetus. As winter approaches, regulate the ration to maintain correct condition. Supply minerals and water at least twice daily if it is no colder than 45°F. If the mare drinks through chopped ice on a pond, she may break through, fall, or not get enough water to adequately move the mass of hay through her intestinal tract. Impaction may result and an operation be required and the foal, or even the mare may be lost.

Be sure mares are not forced to eat moldy hay. Mares are more susceptible to colic from moldy hay than any other large domestic farm animal. Large round bales tend to be moldy in the core. When mares eat from each end of moldy bales, because of the leached outer surface, they are running the risk of abortion.

COLIC CONTROL

Be sure the mare is on a good worm program because strongyles vulgarus triggers more colic than any other single cause of colic in horses. Treatment should be aimed at this devastating parasite. There are a number of new horse wormers in tube, feed, or paste that can be given in pregnancy. Do not neglect this phase of your management program.

The last third of pregnancy requires improved nutrition. An 80 to 90 lb foal develops most in the last 60 days of pregnancy. Also, more nutrients are needed to prepare the mare for greater milk production. Pay particular attention to more energy; be sure that the mare is receiving 10% to 12% protein and that she has access to salt, calcium, phosphorus, and vitamins. The easiest way may be to feed legume hay and a small amount of grain, dependent upon her condition. Mares that foal late on good pasture automatically receive increased levels of nutrients. if mares are in confinement when they foal, they become much less tolerant of each other and tend to injure each other in kicking, fighting, or chasing. Timid mares must be isolated to prevent injury. This also allows the feeder to regulate feed according to condition.

Since fescue pasture constitutes a large part of our pasture in the U.S., and since so much trouble with foaling mares has been experienced on it, some special precautions may be necessary. It is not uncommon to see mares that are on winter fescue be fed a reasonable level of other feeds and still abort 2 to 4 weeks prior to foaling. If this occurs in your mare herd and the placenta is extremely thick, it probably is a result of some toxin or some lack of nutrient in fescue pastures. It is important to get mares off fescue immediately; when one aborts all are likely to follow suit. If they are removed to another pasture or to a dry lot where feeds other than fescue hay are supplied, they may recover in a week or 2 weeks and foal normally. If left on the pasture, most foals are likely to be lost. Another daily. A healthy adult horse that is not parasitized or traumatized by intestinal surgery will receive all of the B vitamins it needs from microbial synthesis in its intestinal tract. However, a confined horse receiving poor quality hay (that has been rained on or is more than a year old) may be deficient in vitamin A. Commercial rations usually supply 4,000-5,000 IU of vitamin A per pound of mixed grain. Many horse feeders routinely feed a mixture of vitamin A, D, and E to supply 20,000 units of vitamin A, 4,000 units of D and 20 to 40 units of E per head per day. These premixes can be purchased at a very minimal fee at many feed stores.

REPRODUCTIVE NUTRIENTS

The three reproductive nutrients are energy, phosphorus, and vitamin A. An abundance of energy causes fat or obese mares. It is most important in your breeding program to maintain the dry mare in an athletic condition. if she is obese, her reproductive performance will be decreased and she will be much more difficult to get settled. Also, fat pregnant mares have more trouble foaling than do those in working condition. It is much more difficult to take weight off the mare than to put it on. To reduce weight, decrease the palatability and nutrient density of the diet. Cut out grain and feed a nonlegume hay that is not palatable enough to result in heavy consumption. Legumes, such as alfalfa, will maintain and actually create obesity. If feed is severely restricted, wood chewing results. Feeds grown on poor soil may have less than 0.18% phosphorus; horses eat such feeds reluctantly. It is important to offer horses a mineral mixture containing calcium and phosphorus throughout their lives. If they are consuming rough feeds, such as nonlegumes, pasture, and/or hay in winter, the mineral should have a calcium-phosphorus ratio of about 1:1 or, as an example, 15% calcium and 14% phosphorus. Grain is a good source of phosphorus, thus for horses eating several pounds of grain per day, the ratio of the calcium-phosphorus block should be two parts calcium to one part phosphorus. Horses should not depend on calcium and phosphorus for their salt requirement but should be fed salt separately.

It is crucial that the calcium-phosphorus ratio be maintained correctly. If phosphorus exceeds calcium in the diet, young horses will be lame and stand on their toes as a result of shortening tendons. This lameness will rotate from one foot to another and result in permanent damage. When the ratio of phosphorus exceeds that of calcium by approximately 2:1 for long periods of time, older horses can be afflicted with big head.

Vitamin A has been called the anti-infection vitamin. The exterior of the skin and the interior of the uterus and the digestive organs are maintained by vitamin A feeding that nourishes epithelial cells; therefore, a short supply symptom is a lack of milk when mares foal normally or have weak foals. For this reason, udders of mares that are feeding on fescue should be watched carefully prior to foaling. If the udders are not developing, grain feeding is indicated. If the mare has a foal while feeding on fescue and has little or no milk, the foal is likely to die for lack of colostrum. The milk flow must be restored quickly. Some breeders have restored this milk flow by feeding 10 lb of grain divided into three different daily feedings to get milk production going again. This is not without the risk of foundering the mare; however, the foal will die without sufficient milk flow.

FEEDING LACTATING MARES

The mares nutrient needs increase at least 50% when she foals. Her peak lactation is about the 4th or 5th week. This means she should be fed well or grazed on the best pastures. Mares in dry lot that foal early should be given the best of hay and probably up to 10 lb of lactation ration. The ration should meet the nutrient requirement of the mare. Continued feeding of poor quality nonlegume hay is not a satisfactory management practice for your mare herd. Table 3 shows a suggested mare grain ration for the last third of pregnancy and the first 3 months of lactation.

TABLE 3: MARE GRAIN RATION DURING LAST THIRD OF PREGNANCY

Ingredients				Lb
Corn, cracked				850
Oats, crimpled				850
SMB				100
Molasses, liquid				150
Dicalcium				3
Limestone				22
Salt				10
Vitamins, A, D, E				2
Total				1,987
	C.P.	Ca	P	
Analysis, %	11.4	0.45	.30	

In summary, nutrition for a breeding herd simply involves supplying mares with the essential nutrients at the appropriate time period in their production cycles. Maintain dry mares in athletic condition, but avoid a shortage of any nutrient at any time. The condition of pregnant mares can be regulated during the winter to reduce cost, but they must be well fed the last third of pregnancy. Lactation is the time to offer the best and the most feed. After a month or two, the mare's milk flow will decline and her feed can be reduced at the same time the foal begins use of a creep feeder which results in a well-grown foal ready to wean with minimum of trauma.

30
FEEDING GROWING HORSES FOR SOUNDNESS

Melvin Bradley

Pregnant mares and young growing horses are often subjected to poor mineral feeding practices. Some are fed excessive carbohydrates and protein, resulting in heavy bodies with weak legs because:

- Minerals are not plentiful in the standard feeds that young horses consume.
- Mineral requirements are high for young growing horses.
- Calcium and phosphorus ratios are often incorrect.
- Minerals are often excessively and/or improperly supplemented by the horse feeders.

The effects of poor mineral feeding are not readily apparent but can result in irreversible damage to bones that appears as lameness in later life. Finally, the trend toward large size and heavy feeding increases the risk of lameness. The young horse that approaches 14 hands in height and 900 lb in weight as a yearling cannot tolerate error in its mineral feeding program. Leg bones must grow more than an inch per month during the first year. With perfect nutrient balance in the ration, the ligaments and tendons of some young horses will not grow enough in length to accommodate rapid rate of leg-bone growth. The result is straight, erect pasterns, lameness that shifts from one leg to another, and temporary and/or permanent unsoundness.

THE FEEDER'S CHALLENGE

The horse nutritionists' (feeders or owners) challenge is to analyze the feeding program, recognize areas of deficiencies or excesses, formulate or select satisfactory rations for different ages and stages of growth and work, and to monitor the horses' feed with routine chemical analysis.

Most horse owners are complacent about their feeding programs. If the animals appear to be in good condition, or even fat, concern for nutrient balance or parasite infestation may lag. A knowledgeable feeder begins by familiarizing himself with the nutritional requirements of each age group of horses. Following are percentages of energy (TDN), protein (CP), calcium (Ca), and phosphorus (P) by age groups for a horse's ration.

Caution is indicated when interpreting figures in table 1. National Research Council (NRC) Standards are minimum amounts based on research of "average" horses under controlled management conditions fed "average" feeds. In reality, there may be a "peck order," different ages in group-fed horses, feed wastage, and great variation in quality of feeds offered. Forages vary widely in quality and commercial horse grain mixtures tend to vary more than most customers realize.

One is impressed with the high CP, Ca, and P requirements of creep-fed and early-weaned foals (table 1) compared to their needs one year later. This is because rapid growth requires high levels of protein, calcium, and phosphorus to produce muscle and bone.

CREEP FEEDING FOALS

It is much less expensive to feed the foal directly than to heavy feed the mare all through lactation. The mare's milk flow will begin to decline, usually at the 5th week. For this reason, creep feeding is recommended. Foals usually begin eating grain from their mother's trough when they are 7 to 12 days of age. This is a good time to build a small creep or area for the foal that is near to, but excludes the mare. Some excellent legume hay can be placed in this creep to entice the foal to enter. In the trough place a commercial foal-creep ration. In early feeding the foal needs 0.8% calcium and 0.6% phosphorus. Do not feed adult horses grain rations to foals because they are too low in protein, calcium, and phosphorous. Since small quantities are used, many companies do not stock them. If you have a large number of foals and are unable to find a creep ration, the one offered in table 2 will suffice.

Feed all the foal will eat but clean out the stale feed and feed it to the mare. As the foal grows, he eventually will be eating 1 lb for creep feed per 100 lb body weight and will be in very good condition when weaning time arrives.

FEEDING PRECAUTIONS

- Feed grain ration free choice with good legume hay to foals for 2 weeks of age to weaning, or to early weaned foals from 3 to 8 months of age.
- Do not feed weaned (or older) foals on this ration because it is too high in protein and calcium unless fed with non-legume hay. At a year

TABLE 1. DAILY NUTRIENT NEEDS—1,100 LB MATURE WEIGHT (AS-FED BASIS)*

Class	Total feed lb	TDN lb	CP		Ca		P	
			%	lb	%	gm	%	gm
Mature horses at maintenance	18.00	9.10	8.60	1.54	0.30	26	0.20	16
Mares, last 90 days of gest.	18.00	10.20	11.10	1.85	0.50	38	0.40	34
Lact. mare, 1st 3 mo	25.00	15.80	14.00	3.30	0.50	56	0.40	34
Lact. mare, 4 mo to weaning	23.00	13.60	12.25	2.70	0.45	46	0.30	30
Creep feed (supplemental)	—	—	18.00	—	0.90	37	0.60	22
Foal (3 mo)	12.00	8.35	18.00	1.85	0.90	37	0.60	22
Weanling (6 mo)	14.00	9.90	16.00	1.95	0.70	38	0.50	28
Yearling (12 mo)	16.00	9.50	13.30	1.85	0.60	34	0.40	24
Long yearling (18 mo)	17.00	9.70	11.10	1.75	0.45	31	0.33	21
2-yr-old (lt. training)	18.00	9.20	10.00	1.55	0.45	28	0.33	19
Mature working horses:								
Lt. work	20.00	13.30	8.60	1.55	0.30	26	0.20	16
Moderate work	30.00	20.00	8.60	1.55	0.30	26	0.20	16
Intense work	40.00	24.50	8.60	1.55	0.30	26	0.20	16

*Adapted from 1973 and 1978 "NRC Nutrient Requirements for Horses."

of age (or sooner) replace this ration with the ration in table 3.

- Be sure preparation of the ration does not result in dust or "fines."

TABLE 2. FOAL CREEP RATION TO PROVIDE 18% CP, .88% Ca, AND .60% P

Ingredients	1/2 Ton	1 Ton
Oats, crimpled or crushed	440.00	880.00
Corn, coarsely cracked	220.00	440.00
Soybean meal, 44%	240.00	480.00
Molasses, liquid	70.00	140.00
Dicalcium phosphate	15.00	30.00
Limestone	10.00	20.00
Salt, trace mineral	5.00	10.00
Vitamins, A,E,D, to supply 4,000 IU/lb	1.00*	2.00*
Total, lbs	1,001.00	2,002.00

*A premix containing 4,000,000 IU of vitamin A, 1,000,000 IU of vitamin D, and 1,000 IU of vitamin E is desirable

TABLE 3. INADEQUATE WEANLING DIET (6 MONTHS) OF TIMOTHY AND OATS

	TDN, (lb)	CP, %	Ca, %	P, %
Requirements[1]	9.90	16.00	0.70	0.50
Nutrients in feed[2]				
Oats, 7 lbs		12.00	0.09	0.32
Timothy, 7 lbs		7%	0.30	0.20
Total		19.00	0.30	0.52
Average, %		9.5	0.29	0.26

[1]From Table 1
[2]See Feeding Mares for Maximum Efficiency, Table 2, this book

FEEDING WEANLINGS

For weanlings a good feeding program consists of: (1) buying a commercial ration from a reputable company that supplies the percentage nutrients listed in table 1, (2) feeding equal parts of it with all the free-choice legume hay (alfalfa, red clover, lespedeza, and others) the weanling will eat, (3) offering a free-choice calcium-phosphorus mineral with a 2:1 ratio and (4) feeding a separate source of trace mineralized salt.

Some trace minerals are toxic and a since there are wide variations in grains and forages in different areas of the country, local professionals, such as county extension agents or state horse specialists, should be consulted when feeding them. Trace mineralized salt is usually sufficient without risk of toxicity or deficiency.

To a lesser degree, some concern for overfeeding of fat soluble vitamins (A,E,D,K) is justified because vitamin D is toxic in rather low levels. The kind of forage or hay necessary for satisfactory growing-horse performance will supply the animal's vitamin needs. When fat soluble vitamins are purchased for feeding, the ratio is important. International Units (IU) of vitamin A should exceed D by 5 to 10 times, and D usually exceeds levels of E by 80 to 100 times. Feeding is simplified by offering 10,000 to 20,000 IUs of vitamin A and the others will be in correct proportions.

Water soluble B vitamins are usually synthesized in the intestines of the healthy horse in sufficient quantity to meet its needs. Beware of the temptation to "spoon in," "pour on," or "force-feed" something with "magic properties" unless you know what the feed contains and the levels of the ingredients in the ration.

The standard weanling diet of oats and timothy hay in equal parts falls far short of the weanling's need.

This ration falls far short of the nutrient needs of the weanling for satisfactory body growth and bone development. One reason many foals survive such a diet is because its protein deficiency (41%) produces such slow growth that under-mineralized leg bones are able to carry the foal's lightweight body. Adding a protein supplement to the oat diet without adding Ca and P would worsen a bad situation by increasing feed intake and growth, resulting in a heavy body. Symptoms of swelling of the ankles, shifting lameness and erect pasterns can be expected to appear with the higher carbohydrate intake.

Levels of 0.29 Ca and 0.26 P are far too low (requirements are 0.70 and 0.50) and the ratio is too narrow for even poor growth. While free-choice feeding of Ca and P is recommended, foals will not consume enough to correct such a large deficiency. Instead of a desired Ca:P ratio of 1.5:1 we have 1.1:1, which is dangerously close to being too narrow. Conversely, a ratio of over 3:1 is probably too long. If either mineral is in excess, insoluble salts that are not well absorbed are formed in the gut.

When large numbers of foals are involved, the grain ration shown in table 3 may be economical. Be sure the feed company that mixes it has sufficient machinery to prepare it without excess "fines" and that the formulator understands it will be chemically checked for appropriate nutrient levels (See the accompanying paper on Quality and Quantity Control Horse Feeding, in this book).

The amounts of grain that weanlings can safely consume are about 1% of body weight or 1 lb per hundred weight. As long as the amount of grain by weight does not exceed that of good legume hay, few digestive or nutritional problems surface. One exception is noted with some individual foals when fed excellent quality alfalfa hay that may supply about as much energy as grain. If symptoms described above should occur, 3 lb hay can replace 2 lb grain and the condition will likely improve.

PLEASE NOTE:

- Feed this grain ration to weanlings with good legume or at least half-legume hay in the amount of 1 to 1 1/2 lb of grain per 100 lb body weight. Feed hay free-choice.
- Do not stuff weanlings with 15-20 lb of any grain feed.
- If you "cut" this ration by feeding half oats or half corn with it, the level of calcium will be too low, unless excellent alfalfa hay is fed free-choice.
- Change to the ration in table 4 by 14-16 months of age for better growth and economy.

TABLE 4. FOAL CREEP RATION TO PROVIDE 16.31% CP, 0.75% Ca, AND 0.55% P

Ingredients	1/2 Ton	1 Ton
Oats, crimpled or crushed	440.00	880.00
Corn, coarsely cracked	270.00	540.00
Soybean meal, 44%	190.00	380.00
Molasses, liquid	75.00	150.00
Dicalcium phosphate	10.00	20.00
Limestone	10.00	20.00
Salt, trace mineral	5.00	10.00
Vitamins, A,E,D, to supply 4,000 IU/lb	1.00*	2.00*
Total, lb	1,001.00	2,002.00

*A premix containing 4,000,000 IU of vitamin A, 1,000,000 IU of vitamin D, and 1,000 IU of vitamin E is desirable

FEEDING YEARLINGS AND TWO-YEAR-OLDS

If the creep and weanling feeding programs have been adequate, yearlings and two-year-olds will be well grown, in athletic condition but not fat, and will have good feet and bones that should wear for many years. Such animals can utilize much grass and less grain, with more non-legume hays in the diet until they begin to work.

Both, yearlings and two-year-olds, should be separated from competition from adult horses and each age should be grouped together. Good pasture with shade, minerals, water and fly protection is recommended.

If grain feeding is needed, a commercial ration with nutrient levels recommended in table 1 or the custom ration in table 4 will suffice.

When serious training starts, grain must be fed. It should take 2 to 3 weeks for the young horse to reach levels of 10 or 12 lb of grain daily with free-choice quality hay. Feed half the grain ration after rest at night, with one-fourth fed at two other feedings (table 5).

Sweet feeds (liquid molasses) will develop a taste that results in more intake for "picky" eaters. Keep troughs clean and supply abundant water.

Stress and colic are reduced by regular feeding times each day with quiet handling and quality feed. An effective worming program is of vital importance with growing horses.

TABLE 5. YEARLING, TWO-YEAR-OLD, LATE PREGNANCY AND LACTATING MARE RATION TO PROVIDE 14.3% CP, 0.61% Ca, AND 0.43% P

Ingredients	1/2 Ton	1 Ton
Oats, crimpled or crushed	440.00	880.00
Corn, coarsely cracked	340.00	680.00
Soybean meal, 44%	130.00	260.00
Molasses, liquid	70.00	140.00
Dicalcium phosphate	5.00	10.00
Limestone	10.00	20.00
Salt, trace mineral	5.00	10.00
Vitamins, A,E,D, to supply 4,000 IU/lb	1.00*	2.00*
Total, lb	1,001.00	2,002.00

*A premix containing 4,000,000 IU of vitamin A, 1,000,000 IU of vitamin D, and 1,000 IU of vitamin E is desirable

PLEASE NOTE:

- Feed this ration at the beginning of the yearling year with good legume or at least half legume hay or good pasture. Regulate intake to control the desired degree of condition. Four to eight pounds daily should suffice.
- As growing horses approach 18 months of age, non-legume hay is sufficient with adequate grain to maintain condition.

- Feed mares in late pregnancy and early lactation 6 to 10 lb of grain as needed to regulate condition and sustain good milk production. If no pasture is available, feed good mixed hay free-choice.
- If mares are obese in late pregnancy, they need no grain but may be maintained on quality legume or mixed or non-legume hay.

SUMMARY

Adequate nutrients in young horse's diets don't just happen. It takes planning! After planning and formulating good rations on paper, things can go wrong in the process of preparation, handling or feeding. If you really want to know what your horse is consuming, monitor his feed by chemical analysis. If not, live dangerously and listen to and follow all of the advice people of less knowledge than you have to offer.

31

FEEDSTUFF EVALUATION AND NUTRIENT VALUE FOR HORSES

Doyle G. Meadows

Horsemen and other livestock producers have traditionally classified certain feeds as "horse feeds" and would not feed other available feedstuffs because the feeds did not carry a connotation for being "horse feeds." Many horsemen felt they had to feed oats and timothy hay, and generally paid high prices for these feedstuffs.

From a nutritional standpoint, such higher costs were unjustified. Today, horsemen are becoming more aware of alternate feed sources for horses--and that feeds such as oats, corn, grain sorghum, timothy hay, alfalfa hay, and soybean meal, are simply sources of basic nutrients for the horse. There is nothing magic about any of the feeds, they are merely a way for the horse to meet his nutrient requirements.

Feedstuffs are divided into three categories: 1) roughages--pasture and hay, 2) energy concentrates--grains, and 3) protein concentrates or supplement premixes--feeds such as soybean meal. It is very important to note that nutrient content of plants is greatly affected by such things as stage of maturity, season of year, climate, variety, and length of storage.

ROUGHAGES

Roughages or hays are divided into grasses or legumes. Legumes are hays (pastures) such as alfalfa, clover, and lespedeza, while grass hays are Bermuda, timothy, orchard grass, meadow and prairie. Most legume or grass hays can be utilized as roughages for horses, if managed and fed properly. However, horsemen need to be aware of the nutrient content of these feedstuffs to develop good feeding programs.

The energy content of legumes (alfalfa and clover) is generally no higher than 1.0 Mcal of digestible energy per lb. The protein content of typical quality legume hay is about 15% crude protein, with a range of 12% to 19% protein on a dry basis. Legumes also contain high levels of calcium. Good quality legume hays are excellent feeds for horses. They do not cause kidney damage but may aggravate existing kidney problems due to their high nitrogen content. Horses should be gradually accustomed to rich legume hay.

Grass hays range in energy content from about .6 to .9 Mcal of digestible energy per lb. Protein content, generally low to moderate, will range from 6% to 11% crude protein. Grass hays are variable in nutrient content as they are readily affected by maturity, fertilization rates, and available moisture. Neither grasses nor legumes are good sources of phosphorus.

Legumes are higher in protein than the grasses and tend to have slightly higher energy values. Roughages that are green, leafy, dust free and aromatic will provide at least part of the daily vitamin requirement. This type roughage also contains carotene, which can be converted to vitamin A. If roughages are old, bleached, weathered, dark, and dusty, they will not contain sufficient vitamin content.

Good quality roughages may be used as the sole source of feed for mature, idle horses provided they have access to a mineral supplement. Roughages are not to be considered good sources of energy for hard-working or highly productive horses.

PROTEIN FEEDS AND SUPPLEMENTS

Protein concentrates, or high-protein feeds, are made up primarily of the oil-seed meals, by-products from the distilling and brewing industry, and certain animal by-products. Examples are soybean meal, cottonseed meal, linseed (flaxseed) meal, fish meal, and milk protein. Fish meal and milk protein are usually costly. Not only are these feeds high in protein, they are also fairly good sources of energy.

The value of a protein supplement not only depends on its protein content but also its protein quality (amino acid balance). Soybean meal is the highest quality protein concentrate in terms of meeting amino acid deficiencies of the cereal grains. Soybean meal, fish meal, and milk protein have protein quality superior to that of cottonseed meal or linseed meal. Most commercially mixed feeds and mixed supplements contain a mixture of protein sources (animal and plant) to ensure an adequate supply of all amino acids.

In addition to these feedstuffs, one can consider commercial vitamin and mineral premixes that may be added to rations when sufficient stress or increased production warrants a need. These vitamin or mineral concentrates should be fed according to directions. The levels to feed vary according to the particular product, but emphasis should be placed on not overfeeding these products. Rations for horses that are in heavy training, rations for lactating mares, and rations for young, growing foals may be fortified to good advantage with concentrated premixes to supply both

Table 1. Nutrient Content of Selected Legumes

Hay	Digest. Energy Mcal/lb	Protein %	Ca %	P %	Vitamin A* IU/lb
LEGUMES					
Alfalfa (mid-bloom)	0.92	15.00	1.21	0.19	5448
Alfalfa (mature)	0.87	13.50	0.64	0.14	2584
Lespedeza (pre-bloom)	0.91	16.02	1.02	0.23	____
Lespedeza (full bloom)	0.84	12.06	0.93	0.20	____
Clover	0.89	15.21	1.27	0.16	5726

* Values for Vitamin A activity in hays as given here represents maximums if hay is of excellent quality and fresh. Hay that is over six months and of average to poor quality has little or no Vitamin A activity.

Table 2. Nutrient Content of Selected Grasses

Hay	Digest. Energy Mcal/lb	Protein %	Ca %	P %	Vitamin A* IU/lb
GRASSES					
Bermuda (Coastal)	0.79	8.5	0.41	0.16	5972
Sorghum X Sudan grass	0.78	7.8	0.35	0.15	____
Timothy	0.79	7.4	0.37	0.16	1766
Orchard grass	0.75	8.0	0.36	0.29	4842
Meadow	0.66	8.1	0.51	0.15	____
Prairie (Midwest)	0.92	6.7	0.41	0.15	____

* Values for Vitamin A activity in hays are given here represents maximums if hay is of excellent quality and fresh. Hay that is over six months and of average to poor quality has little or no Vitamin A activity.

Table 3. NUTRIENT CONTENT OF SELECTED GRAINS (AS FED BASIS)

Grain	Digest. Energy Mcal/lb.	Protein %	Ca %	P %	Vitamin A Equivalent IU/lb.
Oats 36 lb/bu.	1.35	12.0	0.09	0.32	____
27 lb/bu.	1.19	12.0	0.09	0.32	____
Corn	1.58	9.0	0.02	0.31	1362
Barley	1.46	12.4	0.08	0.42	____
Sorghum Grains	1.36	10.0	0.03	0.29	____

Table 4. NUTRIENT CONTENT OF SELECTED HIGH PROTEIN FEEDS

Feed	Digest. Energy Mcal/lb	Protein %	Ca %	P %	Vitamin A Equivalent IU/lb.
Soybean Meal	1.45	44	0.32	0.67	___
Cottonseed Meal	1.37	41	0.17	1.31	___
Linseed Meal	1.38	35	0.4	0.82	___
Pelleted Supplements	Nutrient content will be specified on the feed tag. Most will fall in the range of 22 to 35 percent protein.				

water- and fat-soluble vitamins in addition to major and minor minerals, particularly when quality of feed is in doubt.

Many good horsemen feed some commercially mixed concentrate feed. Tags on these feeds specify protein, fiber, and fat content as well as a list of ingredients. The fiber content of these feeds is related to the energy content--lower fiber feeds contain higher energy levels. The following table shows a relationship between fiber content of a mixed feed and expected digestible energy content.

(Acknowledgement is extended to the Horse Section [Dr. Gary Potter, Dr. Doug Householder and Mr. B. F. Yeates] Texas A&M University for providing factual material for this paper.)

Table 5. RELATIONSHIP OF CRUDE FIBER TO EXPECTED DIGESTIBLE ENERGY IN MIXED CONCENTRATE FEEDS

Crude Fiber %	Digestible Energy Mcal/lb
2.0	1.62
4.0	1.55
6.0	1.45
8.0	1.35
10.0	1.25
12.0	1.15

32

PROTEIN SUPPLEMENTS FOR LACTATING MARES AND EFFECTS ON FOAL GROWTH

Doyle G. Meadows

INTRODUCTION

A major concern of every horse owner, whether a large horse-breeding farm or an individual with one pleasure horse, is proper feeding techniques and correct use of feedstuffs. Everyone is concerned with the nutritional well-being of their horses. However, many horse owners have been misled by unsupported feed recommendations and have fallen victims to fads in lieu of scientific concepts on horse nutrition and basic feeding principles.

Today, a large unanswered question exists among horse nutritionists regarding the ability of the nonruminant herbivore to utilize nonprotein nitrogen (NPN) or poor quality protein supplements. Some evidence indicates that NPN (urea) can be used for maintenance but the extent to which a horse can use NPN for production purposes is not clear. If urea could be substituted for the expensive high-quality soybean meal (SBM) in equine rations and productivity remained the same, the horseman could maintain the present level of production while decreasing total feed costs. The level of nitrogen and (or) amino acid balance of rations fed to mature horses to support production requirements above maintenance has not been studied sufficiently in detail in the equine. However, the research discussed here helps to answer some of the questions concerning the qualitative and quantitative protein requirements of the lactating mare.

PROCEDURE

Two trials using 27 (trial I) and 28 (trial II) mares of predominantly quarter horse breeding were conducted to determine the quantitative and qualitative protein requirement of the lactating mare. Mares were blocked by expected foaling date and mating selection and assigned to dietary treatments containing different levels and sources of dietary nitrogen. Treatments were evaluated by foal growth, nitrogen balance, milk yield, and milk composition. Rations in trial I were formulated and fed to contain a slight nitrogen deficiency in the control and a significant excess in the supplemented diets. Concentrate diets were formulated to contain different levels and sources of nitrogen as follows: control (corn and oats) 10% CP; control + soybean meal (SBM), 15% CP; control + urea, 15% CP. Average quality (9.96% CP) coastal Bermuda grass hay was fed with each diet at 10 lb per mare daily.

Trial II rations were formulated and fed to create a severe nitrogen deficiency in the control while the supplemented diets would just meet protein requirements of the lactating mare. Dietary treatments for trial II were: control (corn and oats) 10% CP; control + SBM, 17% CP; control + SBM/urea, 17% CP; and control + urea, 17% CP. A low quality (4.01% CP) coastal Bermuda grass hay was fed at the rate of 10 lb per mare. Concentrate diets were fed to meet energy requirements of lactating mares.

Mares were placed on treatment immediately following parturition, were fed individually (separated from foal) twice daily and maintained (together with foal) at all other times in dry lots with free access to water. All mares were weighed weekly and fed to maintain constant weight for an 84-day experimental period. Foal growth measurements (weight, height, and heart girth) were taken on days 0, 7, 14, 21, 28, 56, and 84. Foals did not have access to creep feed.

Milk production was determined by the weigh-suckle-weigh technique on approximately day 45 of lactation. In trial I, foals were allowed to nurse every 4 hours for 72 consecutive hours. An average of the 3-day production estimates was used as the milk production estimate for each mare. Foals in trial II were allowed to nurse every 3 hours during a 12-hour day and 12-hour night milking schedule. Daily milk production values were a combination of day and night milk yield values.

RESULTS, DISCUSSION, AND CONCLUSIONS

Utilization of urea or other non-protein nitrogen sources has not been well-established in the horse. Cattle, on the other hand, through fermentative digestion, can improve protein quality of low-quality feedstuffs provided an adequate N supply is available in the rumen. In this manner, ruminants are able to utilize NPN supplies in the diet. Unlike cattle, the nonruminant herbivore does not have a rumen to break down feedstuffs and alter protein quality of the diet prior to the site of absorption. The horse is able to alter dietary protein composition by the addition of N to a C-skeleton through fermentative digestion in the lower tract. However, this occurs beyond the site of major absorption. The extent of microbial protein synthesis in the cecum and large intestine of the horse is not clear. Researchers differ in their views concerning the effective site of digestion and absorption of protein in the horse.

There are two different schools of thought that concern N metabolism in the mature horse. Most reports suggest that the majority of protein digestion and absorption of amino acids takes place in the small intestine and that horses are susceptible to amino acid deficiencies like other nonruminants. Therefore, horses should be fed high-quality rations containing preformed protein. Other researchers contend that protein digestion and absorption does not take place solely in the small intestine and that the horse has some ability to synthesize and digest microbial protein in the lower tract, thus being indifferent to dietary protein quality. This theory suggests that addition of a NPN source could be useful in formulating horse rations.

Milk yield was determined in trial I (table 1) and trial II (table 2) using the weigh-suckle-weigh technique. Mares fed the SBM diet tended to produce more milk; however, differences in measured milk yield were small and not statistically significant (P>.05). It appears from these data that milk yield was underestimated. Since there were small observed effects of dietary treatments on milk yield, it is likely that the weigh-suckle-weigh technique reflected the total volume of milk produced only from mares producing less milk. Mares fed diets containing SBM tended to produce milk containing higher levels of protein and fat (table 3).

Nitrogen balance data obtained in trial I suggests that supplemental nitrogen fed in the form of urea was utilized poorly by the lactating mare. Also, these results indicate that SBM was of much higher biological value and was utilized to a greater extent in the mare than was urea. Therefore, the lactating mare is sensitive to differences in dietary protein quality.

Foals from mares fed SBM diets grew faster (weight and heart girth) than foals from mares fed the control diet in trial I (P<.25). It appears from these growth data that 183 g N/day fed to the mare in early lactation is insufficient to promote adequate foal growth in terms of body weight and heart girth gains. However, in the latter stages of lactation, 183 g N/day appears to be adequate to promote acceptable foal gain. This represents a 10% CP ration fed to meet the mare's energy requirement. The magnitude of growth differences in trial I was suppressed due to the quantity of preformed protein in the control diet. Mares fed the control diet actually received sufficient nitrogen in the latter stages of the trial, thus little response was seen to nitrogen supplementation. Mares fed the urea diet showed little response in foal growth over that of foals from mares on the control diet (table 4).

In trial II (table 5), preweaning growth rate (weight and heart girth gains) was significantly influenced (P<.05) by the addition of SBM as a supplemental nitrogen source to the mares' diet as compared to the other three dietary treatments. Mares fed diets containing urea did not show any response in additional foal performance over foals from mares fed the control diet. Neither protein quality nor

TABLE 1. MILK PRODUCTION BY MARES IN TRIAL I*

	Treatment		
	Control	SBM	Urea
Range, lb	18.63-24.31	19.14-24.79	17.49-26.33
Milk production, lb	20.44	22.31	21.23
% of control	--	109	104
Mare weight, lb	1124	1161	1126
Milk production as % of mare weight	1.82	1.79	1.89

* Represents value for 3-day collection period.

TABLE 2. MILK PRODUCTION BY MARES IN TRIAL II

	Treatment			
	Control	SBM	SBM/Urea	Urea
Range, lb	19.47-26.97	23.47-29.45	20.48-27.96	21.47-29.46
Total daily milk production, lb	23.47	26.25	24.68	24.97
Day, lb	12.28	13.49	12.50	12.21
Night, lb	11.20	12.76	12.19	12.76
Total daily milk production, % of control	—	112	105	106
Mare weight, lb	1135	1133	1141	1139
Milk production as % of mare wt.	2.08	2.36	2.16	2.19

TABLE 3. COMPOSITION OF MARE'S MILK IN TRIAL II (%)

	Treatment			
	Control	SBM	SBM/Urea	Urea
Milk fat	1.29	1.66	1.74	1.66
Milk protein	1.99[b]	2.11[ab]	2.16[a]	1.98[b]
Milk solids	10.36	10.37	10.41	10.22
Protein content of milk solids	20.61	21.88	22.49	20.24

[ab] Means in the same row not sharing same superscript are significantly different (P<.05).

quantity in the mares' diet affected height gain of foals, indicating that adequate long bone growth was not affected by dietary treatments used in the study (tables 4 and 5).

TABLE 4. AVERAGE DAILY WEIGHT GAIN, TOTAL HEART GIRTH GAIN, AND TOTAL HEIGHT GAIN OF FOALS IN TRIAL I

	Treatment		
	Control	SBM	Urea
Avg daily wt gain, lb	1.87	2.07	1.97
% of control	--	111.00	104.00
Total heart girth gain (in.)	11.69	12.83	11.50
% of control	--	111.00	99.00
Total height gain (in.)	7.75	7.05	7.24
% of control	--	91.00	96.00

TABLE 5. AVERAGE DAILY WEIGHT GAIN, TOTAL HEART GIRTH AND TOTAL HEIGHT GAIN OF FOALS IN TRIAL II

	Treatment			
	Control	SBM	SBM/Urea	Urea
Avg daily wt gain, lb	1.74[b]	2.31[a]	1.94[b]	1.72[b]
% of control	—	133.00	111.00	99.00
Total heart girth gain (in.)	10.51[c]	13.54[a]	11.89[b]	9.92[c]
% of control	—	128.00	113.00	94.00
Total height gain (in.)	6.54[b]	7.24[ab]	7.71[a]	7.28[ab]
% of control	—	111.00	118.00	111.00

[a,b,c] Means in the same row not sharing same superscript are significantly different ($P<.05$).

It can also be noted that 247 g N/day (13.5% CP ration) fed to mares did not produce faster growing foals than mares fed 224 g N/day (12.5% CP ration) of the same quality protein. These data indicate that 224 g N/day is sufficient for the mare in early lactation to maximize foal growth. Further, 183 g N/day appears adequate in latter stages of lactation.

This study indicates there was little if any improved performance of foals when urea was fed as the sole supplemental nitrogen source to the lactating mare. Although urea may have been utilized for microbial protein synthesis in the hind gut of mares, it does not appear that this protein was absorbed and used for protein synthesis at the cellular level. Therefore, it may be concluded that urea cannot be utilized to a major extent by the lactating mare and that she was sensitive to protein quality in the diet.

Statistically significant treatment effects were observed throughout the growth trial yet no significant differences were observed in the milk yield data. This indicates that the weigh-suckle-weigh technique to determine milk production was less sensitive in determining the value of a diet than was foal-growth measurements. Foal growth was a better indicator of milk production than direct measurements based on the weigh-suckle-weigh technique and was more useful in measuring effects of diet composition on milk production in the mare.

Nitrogen utilization in the horse has been tested using different techniques--one of which involved feeding urea as a supplemental N source in combination with high- and low-quality feedstuffs. Results from feeding almost identical diets leave conflicting interpretation regarding mechanisms of nitrogen utilization in the horse. Some researchers reported mature horses were unable to make effective use of urea while others reported no change in nitrogen balance when up to 75% of the dietary protein was replaced by urea. It is accepted that horses are able to utilize urea for maintenance when added to a low-protein basal diet (less than 6%). It is also suggested that the young growing horse is unable to utilize urea for growth.

The major objective of this study was to determine if urea could be used for productive purposes in the lactating mare. This objective was pursued by measuring foal growth, milk production, milk composition, and determining nitrogen balance of lactating mares fed different levels and sources of dietary nitrogen. Two separate trials were conducted to evaluate dietary treatments.

Results from this study support two concepts regarding N utilization in lactating mares. First, mares fed diets containing SBM as the supplemental N source produced the fastest growing foals as indicated by body weight gain and heart girth circumference. Second, the addition of urea to a corn-oat diet fed to lactating mares resulted in little, if any advantage in foal performance over a low-protein, unsupplemented corn-oat diet. Although mares may be able to utilize urea to promote microbial protein synthesis in the hind gut, it appears that little absorption of essential amino acids from microbial protein takes place. It has also been suggested the end product of proteolysis in the large intestine in principally ammonia rather than free amino acids and the results of this study support that view.

Milk yield of mares was increased when increasing levels of SBM were added to the control diet. However, mares fed urea as the sole source of supplemental nitrogen showed little response in milk yield over mares fed control diets. Also, mares fed diets containing SBM tended to produce higher quality milk as indicated by higher milk protein and fat levels. This indicates that mares fed a high quality preformed protein supplement produced more total volume of higher quality milk than those fed urea supplements.

Nitrogen balance data revealed that mares were able to utilize a much higher percentage of N from a preformed protein than from urea. Although adult horses are able to utilize urea for maintenance, these data indicate that lactating mares need a high-quality preformed protein diet to support adequate milk synthesis.

When evaluating diets for lactating mares, the qualitative as well as quantitative aspects of protein in the diet must be considered. It appears that both level and source of dietary supplemental nitrogen affects milk production in the mare and subsequent foal growth and development.

REFERENCE

Meadows, D. G., G. D. Potter, W. B. Thomas, J. Hesby, and J. G. Anderson, Texas Agricultural Experiment Station, College Station.

33 QUALITY AND QUANTITY CONTROL IN HORSE FEEDING

Melvin Bradley

INTRODUCTION

Most information about horse feeding is passed by word of mouth or by advertising a product. Many recommendations conflict, and few will stand the test of a chemical analysis. While horse research lags behind that of food animals, enough scientific knowledge is available when followed to ensure good feeding practices. The feeder must rely on the National Research Council for nutrient requirements for each age and stage of production (just as all other animal species do, including humans). The feeder must feed a balanced ration and monitor his feeding program with chemical analysis to be assured his horses are eating appropriate amounts of essential nutrients.

With increased pressure for early growth and use, young horses cannot tolerate poor rations even as well as in the past. Slow-maturing breeds, given time to grow and mature, can tolerate nutrient deficiencies to a degree, but it should be the goal of all horse owners to know that their horses are receiving appropriate kinds and amounts of nutrients. Until feeds are routinely analyzed, one can never be sure.

There are few topics today as controversial as "what to feed your horse," and no subject with more volunteers offering conflicting advice about it. One wonders what the late Harry S. Truman, who said of his economic advisors, "If you laid them end-to-end, they will point in every direction," would say about those who so freely advise others about horse feeding. Credibility can be established only by a chemical check of your horse feeding program.

HORSE FEEDING PROGRAM

Horse owners can be reasonably sure of correct feeding by a routine chemical analysis of feeds consumed by their horses. Feed analysis is a routine practice of every horse nutrition consultant, because he is legally liable for his recommendations and must have evidence to support his recommendations. If the basic nutrients of energy, protein, calcium, phosphorus, salt, vitamins, and water are offered according to the horse's needs, about 90% of the practical feeding errors in most stables are eliminated. Contrary to popular opinion and sales pitches, there are no magic or secret potions offered in small amounts that spare or replace basic nutrients or that give your horse a performance "edge" over his competition after the basic nutrients are supplied. With modern analytical methods, no ingredient secrets can long be kept by one company from a competitor. This is not to imply that all feeds are the same. Far from it!

- Know the requirements of your horses (table 1).
- Select appropriate grain to feed with the hay available.
- Chemically analyze this grain to be sure you are buying what your horse needs.

Chemical Analysis

Chemical analysis is a method of checking the accuracy of ration formulation and comparing amounts and ratios of one nutrient to another. Such analysis confirms or conflicts with the feeder's ideas about what he thinks he is offering his horses.

Analysis helps to:

- identify inadequate rations before damage occurs.
- monitor accuracy of your feed mixing service.
- identify quality feeds.
- stimulate better skills and understanding of feed formulation.

The horse feeder who selects commercial feed from a reputable company and feeds it according to directions with appropriate legume hay or nonlegume but excellent quality hay, has minimum need for chemical analysis. However, few horses are fed this way. The feeder is likely to make changes in his feeding program that results in excesses, imbalances, or inadequacies that can be readily identified by analysis. Listed below are a few common feeding errors.

- Good oats and nonlegume hay fed to weanlings. Analysis shows large shortages of protein, calcium, and phosphorus.
- Adult horse grain feeds fed to young growing horses. Adult horse grains are approximately 11% protein, 0.4% calcium and 0.3% phosphorus--all too low for growing horses. Use grains formulated for each age group.
- Bad calcium-phosphorus ratios. This is a very common condition of heavy grain feeding.
- Cutting a commercial grain mixture with another grain such as adding half oats or corn. Analysis reveals this practice reduces calcium by about half and leaves phosphorus almost unchanged.

- The feeding of home-grown grains without adding calcium (limestone, etc.). Oats or corn has very little calcium and rather good levels of phosphorus. Unless supplemented or fed with legume hay, the calcium level and mineral ratio are disastrous.

The trend is toward large size and early use of horses. Those genetically capable of fast growth have narrow tolerances for ration errors, especially minerals that produce bone growth. When performance is demanded from two--year-olds, the need for correct feeding practices becomes evident.

Sampling Technique for Hays

The data you get from the laboratory can be no better than the sample you submit. Be sure the sample is representative of the feed from which it is drawn.

Baled hay is best sampled with a Pennsylvania State forage sampler available from that institution and various equipment companies. It consists of a tube with a shank for a half-inch drill that bores into the end of bales and withdraws inch-long samples from approximately 16 inches within the bale. Randomly select 15 to 20 bales for your quart-size sample and pack it tightly into a plastic bag for mailing to the laboratory. You are interested in percentages of protein, TDN, calcium, phosphorus, acid detergent fiber, and moisture.

Small sections from 15 to 20 bales can be taken and cut into 3-inch lengths for chemical analysis without the use of a forage sampler. Be careful not to lose leaves when using this method. Remember, you are seeking a representative sample of your hay. For this reason only hay of like types is used in a single sample. Separately sample different cuttings (first, second, etc.) of the same hay and hays from different fields.

Table 1 shows results of an appropriately sampled lot of good third-cutting alfalfa hay. Consider the "as fed" column when evaluating feeds for horses. The moisture figure of 13.28% is good, leaving 86.72% of its weight as dry matter. The crude protein percentage of 18.08 is excellent, although some alfalfa will exceed 20%. Acid detergent fiber (ADF) gives a more accurate measure of net energy and reflects stage of plant maturity when cut better than the old method of crude fiber analysis. The level of 20.20% ADF is good for hay which may range up to 50% in very poor hay. The estimate of 60.73% total digestible nutrients (TDN) is excellent and exceeds that of poor quality oats. TDN levels of poor hay may approach 30%. The 1.4% calcium and 0.20% phosphorus levels are standard.

When evaluating results of your hay compare it to figures given for feeds in the article "Feeding Growing Horses for Soundness" in this book.

Sampling Techniques for Grains

A very real problem with horse grains is too many fine particles ("fines") that are not eaten and the settling out (into the fines) of important nutrients such as calcium, phosphorus and vitamins. Even with well-mixed dry rations, this can be a problem. Adding 7.5 to 10% liquid molasses helps to keep these nutrients in suspension and reduces dust.

Representative areas from 5 or 6 sacks will suffice for your quart sample if appropriately taken. Lay the sack on its side on a table and remove most of the sack before taking large dips with a tablespoon from top to bottom and side to side. Fines tend to filter to the bottom and may be missed when only the top is sampled.

Bulk grains are best sampled with a grain sampler. This is a pipe within a pipe that upon being twisted permits grain to enter small "windows" from top to bottom of the pipe. It is excellent for both bulk and sack sampling.

"Grab" samples can be made from unloading augers or downspouts by taking small amounts until the process of unloading is complete.

Beware of taking grain only from the top of sacks, bins, or loaded trucks for your sample because you may miss the fines.

Table 2 shows the analysis of a ration formulated for adult horses to yield 12% protein, 70% TDN, 0.5% calcium and 0.35% phosphorus. These levels exceed National Research Council (NRC) requirements for adult working horses. However, research indicates digestibility declines with age, and many of these horses were over 20 years of age.

This analysis sheet is a pleasure to the nutritionist, feed company, and horse owner because it analyzes almost exactly what was formulated, which means the feed was mixed, sampled, and chemically analyzed correctly.

The moisture level (even with 7 % molasses) is good and the protein, calcium, and phosphorus almost perfect.

Acid detergent fiber is usually predictable and seldom taken in a grain sample. However, if feed company personnel are aware that an occasional ADF will be taken, it prevents alfalfa hay or other high fiber ingredients from being added to the mixture. An ADF of under 10% indicates the use of specified feeds.

TABLE 1

RECEIVED 11/17/81

SAMPLE NO. 81-17-K-4
SAMPLE I.D. ALFALFA HAY

	As Fed Basis	Dry Matter Basis	As Fed Basis	Dry Matter Basis	As Fed Basis	Dry Matter Basis
MOISTURE......o/o	13.28	0.00				
DRY MATTER....o/o	86.72	100.00				
CRUDE PROTEIN.o/o	18.08	20.84				
AVAIL PROTEIN.o/o						
ADF-NITROGEN..o/o						
PEPSIN A PROT.o/o						
A.D. FIBER....o/o	20.20	23.29				
N.D. FIBER....o/o						
T.D.N. ....o/o	60.73	70.02				
NE LACT...MCAL/LB	.66	.76				
N E GAIN..MCAL/LB	.37	.43				
N E MAINT.MCAL/LB	.60	.69				
CRUDE FAT.....o/o						
PH............o/o						
ASH...........o/o						
NITROGEN......o/o	2.89	3.33				
CALCIUM.......o/o	1.40	1.62				
PHOSPHORUS....o/o	.20	.23				
MAGNESIUM.....o/o						
POTASSIUM.....o/o						
SODIUM........o/o						
SULFUR........o/o						
IRON..........PPM						
MOLYBDENUM....PPM						
COPPER........PPM						
MANGANESE.....PPM						
ZINC..........PPM						
ALUMINUM......PPM						
NITRATES......o/o						

NAME..............CALLAWAY STABLES
ADDRESS...........

SUBMITTED BY......DR. BRADLEY
ADDRESS...........

TABLE 2

RECEIVED 6/7/82

SAMPLE NO 82-7-F-14
SAMPLE I.D. COMPLETE RATION

	As Fed Basis	Dry Matter Basis	As Fed Basis	Dry Matter Basis	As Fed Basis	Dry Matter Basis
MOISTURE..... o/o	11.96	0.00				
DRY MATTER... o/o	88.04	100.00				
CRUDE PROTEIN o/o	12.12	13.76				
ADJ CRUDE PROTo/o						
AVAIL PROTEIN o/o						
ADF-NITROGEN. o/o						
PEPSIN A PROT o/o						
A.D. FIBER... o/o	8.63	9.80				
N.D. FIBER... o/o						
TDN...........o/o	73.62	83.62				
NE LACT...MCAL/LB	.69	.78				
N E GAIN..MCAL/LB	.49	.55				
N E MAINT.MCAL/LB	.73	.83				
CRUDE FAT.... o/o						
PH........... o/o						
ASH.......... o/o						
NITROGEN..... o/o	1.94	2.20				
CALCIUM ..... o/o	.54	.61				
PHOSPHORUS... o/o	.36	.41				
MAGNESIUM.... o/o						
POTASSIUM.... o/o						
SODIUM....... o/o						
SULFUR........o/o						
IRON......... PPM						
MOLYBDENUM... PPM						
COPPER....... PPM						
MANGANESE.... PPM						
ZINC......... PPM						
ALUMINUM.....PPM						
NITRATES......o/o		NEGATIVE				

NAME............. STEPHENS COLLEGE
ADDRESS.......... COLUMBIA, MO

SUBMITTED BY..... DR. M. BRADLEY
ADDRESS.......... UMC-COLUMBIA

34

FEEDING MANAGEMENT OF THE HORSE

Doyle G. Meadows

Horses as individualists. Optimum feeding of horses is an art. Horses are individualists, as compared to other livestock species; therefore, different horses may need to be managed and fed differently. The good horseman must learn what is normal for a given horse and, during daily observation, quickly detect any abnormalities in feeding behavior. Inadequacies of the feeds being utilized also can be detected and necessary adjustments made.

Feed storage. Concentrates and hays must be stored properly to maintain their quality. Boxes, bins, drums, large cans, etc., should have tight-fitting lids so that feed can be kept dry and free of contaminants. Pallets work well for keeping baled hay off damp floors.

Feeding by weight. Feed horses by weight, not volume. The coffee can or the bucket with the painted marks inside are popular feed-measuring items. Horses do not require a certain volume of nutrients but a certain weight proportional to their body weight and status. Feeding solely by volume is dangerous because feeds weigh different amounts per unit volume. For example, 1 gallon (3 lb coffee can) of oats weighs 1 to 3 lb. One gallon (3 lb coffee can) of corn weighs 4 to 5 lb.

Weigh feeds using standardized measuring items and get a feel for hay weights (remember, types of hay vary in weight depending upon baling method). You can feed by volume, but you should weigh any new feeds and adjust volume measures accordingly. Feeds then do not have to be weighed at each feeding.

Roughage. All rations for horses should contain some roughage. Without roughage in the ration, horses have a tendency to chew things, particularly wood, and are susceptible to colic and founder.

Horses on pasture. Roughages can be supplied with well-managed and well-fertilized pastures. In addition to supplying the roughage requirement of the ration, pastures also supply an excellent source of energy, protein, vitamins, and minerals and are a good place for exercise. Poor pastures do not supply adequate roughage or nutrients and should not be relied upon to any large extent.

Stalled horses. Stalled or penned horses should receive .5 to 1.0 lb of hay per 100 lb of body weight. A 1,000 lb horse should receive 5 to 10 lb of high-quality hay per day (horses grouped in pens and kept on limited-roughage diets may chew on other horses' manes and tails and may practice coprophagy--eating of feces). Some of the roughage can be fed in complete pelleted feeds; however, some long roughage is usually needed.

Feeding regularity. Regularity in feeding time is essential. Have a set feeding time every day, including weekends and holidays. It is important to have approximately equal time intervals between daily feedings: for example, feed at 6:30 a.m. and 6:30 p.m. when feeding twice per day. Horses are less likely to "go off feed" if fed on a regular schedule. This is particularly important in young growing horses where maximum feed intake is desired.

Uneaten feed. Check the feed box and hay rack for refusals before the next feeding. Refusal of feed and (or) hay suggests the horse 1) was overfed, 2) was fed the correct amount of feed but something was wrong with the feed or hay, or 3) is sick. Refusal of hay and overeating of concentrate for a period of time can lead to serious digestive disturbances.

Sanitation. Sanitation around the feeding and watering area is mandatory. Uneaten (possibly spoiled) feed or hay should be removed from feed boxes and hay mangers, and water troughs should be scrubbed routinely. Feeding on the ground is a major factor in parasite infestation due to feed exposure to fecal material containing parasite eggs.

Hay feeding methods. There are several methods for feeding hay. Some horsemen have reasoned that horses should eat from the ground, because horses in the wild ate that way. However, possible intake of dirt, sand, parasite eggs, and wastage are problems with this method. Corner mangers on the ground in stalls alleviate some of these problems. Racks, nets, or mangers mounted at chest level are popular; however, nets are time-consuming to fill. Mangers should not be too high to prevent the horse from eating in a natural position.

Frequent feeding. Horses should be fed as frequently as possible to maximize efficiency. Mature idle horses or horses used infrequently can be fed once daily. The thumb rule is that you can feed one time per day if total grain intake is less than .5% of body weight. Growing horses, milking mares and performance and working horses require high feed intakes and should be fed twice daily. Feeding more than two times per day is practiced under some management regimes to encourage horses to consume more feed (e.g., race horses and halter horses).

Total daily feed. The total daily feed should generally be divided with equal proportions offered at each of the feeding times. When hay and concentrate are fed, proportion both at each feeding rather than feeding, for example, grain in morning and hay at night. Hard working horses should be fed the majority of their hay at night or when there is ample eating time.

Ration changes. Avoid sudden, abrupt changes in rations. Changes in the physical characteristics of a ration (from pelleted to meal or vice versa; cubes to roughage or vice versa) or in the ration ingredients (whole oats to sorghum or corn grain), or from one commercial feed to another can cause horses to go off feed and have diarrhea that often leads to serious digestive disturbances. Take several days and preferably a week to change horses over to a new ration.

Changes in feed and exercise. "Letting down" and "feeding up" should be done gradually. Horses brought from sales and (or) brought off show or racing circuits are accustomed to large quantities of feed and a strenuous exercise regimen. Conversely, horses in the rough going into a fitting program for show or racing may not be accustomed to the larger amounts of feed and exercise. Increase or decrease both feed quantity and exercise slowly to prevent harmful effects.

Controlling eating habits. Aggressive horses should be discouraged from eating fast. Some horses are aggressive eaters and they gulp mouthfuls or "bolt" their feed. This situation occurs when overly aggressive horses are fed in deep, narrow troughs. To slow down the over-aggressive horse 1) feed him in a larger, shallower box where the feed cannot build up and 2) place small rocks, bricks, salt blocks, or other objects (that the horse cannot swallow) in the trough. Some horses root their feed out of troughs. Rings mounted on the top of troughs and lipped troughs prevent rooting and feed wastage. Slowing down the eating pattern of the aggressive horse will enhance digestion and will help prevent other digestive disturbances.

Provide atmosphere to encourage timid horses to eat. Little can be done to force a timid horse to eat; however, he should be fed where he is not bothered and (or) afraid to eat. Solid partitions or partially solid partitions (just at feed box between stalls) are good as they prevent horses from "fussing" across the fence at feeding time. This is particularly true if the horses are fed at different times.

Some horses also exhibit anxiety when noise and activity from people and (or) other horses are present at feeding time; therefore, barn activity should be minimized at feeding times.

Individual feeding. Horses should be fed individually for maximum growth or performance. Individual feeding is ideal for young horses being prepared for show and (or) performance. Performance horses in training (e.g., race horses) are usually stalled and fed individually. Individual feeding would be ideal for all classes of horses; however, it is impractical in some instances (e.g., broodmares and foals).

Group feeding. If horses must be group fed, they should be fed by class. Horses require different types of rations and quantities of nutrients to meet their requirements at certain ages and stages of growth, production, or use. These classes of horses are idle mature horses, working mature horses, growing horses, gestating mares (last 90 days gestation), and lactating mares. Where management allows, horses should be divided into these classes and not all run and fed together.

Group feeding management of older horses needs to take into account "dominance hierarchies." Horses in a group, particularly older horses, will establish a "pecking order" or their dominance hierarchy, as is seen in other species. This suggests that the more aggressive horses consume larger quantities of feed at the expense of the more timid horses that do not get enough to eat. Management systems should be utilized to encourage individual animals eating within the group. For example, provide sufficient individual feed tubs (that cannot be tipped over), spaced far apart (50 ft), so that every horse can eat at the same time. Use run-in sheds with adequate room where there is no direct competition for feed.

Group feeding of foals will work if done properly. Foals should be started on a creep ration as soon as they will eat (less than 1 month old). Creep feeding discourages foals from eating with their mothers. The ration should be high in protein and energy and fortified with vitamins and minerals. The creep ration should be available to the foals at any time they choose to eat. Pelleted creep rations are preferred because every bite the foal consumes is a balanced mouthful (he can selectively eat or pick at a meal ration). Fresh creep ration should be placed in the creep on a daily basis to assure that no foal consumes spoiled feed. The creep should be roomy and located where the broodmares frequently gather (e.g., feeding area, watering area). Weaning is much less stressful to foals who have learned to eat creep feed.

Feeding time. Work easily and quietly around horses when feeding and (or) handling them. This is particularly important for young horses. Feeding time is a convenient time to observe horses and look for any injuries or other possible abnormalities that should be checked further. Novice horsemen should be encouraged to go into the stall at feeding and look around--not merely throw the feed in from the front of the stall. "An ounce of prevention here is worth a pound of cure."

Watering. Working horses or any hot horse should be watered with care. Control the water intake of the hot horse immediately following hard work. While cooling off, the horse may be given a few sips of water, but he should be cool (respiration rate back to normal) before getting a small drink. When the horse is completely cooled and relaxed, he may have full access to water and feed.

Clean fresh water should be available to all horses at all times except the hot horse. Sources of water in order of preference are purified water from commercial water plants, good wells, running streams, tanks, or ponds. Automatic waterers are used for stalled horses by many horsemen; however, some horsemen water in buckets or other containers so that they will know if, for some reason, the horse is not drinking. Very cold or very hot water discourages water intake; a range of 45°F to 65°F seems desirable. Water consumption is highly correlated with dry matter intake. The normal idle, mature horse should drink 4 to 8 gallons of water per day. Milking mares and horses in training will require more water due to milk production and sweating, respectively.

Exercise. Provide exercise on a regular and frequent schedule. Seldom do horses on pasture or with access to free exercise develop abnormal feeding habits, digestive disturbances, or other upsets. Stalled horses fed large quantities of feed and given limited exercise and infrequent exercise are sure to develop irregular eating habits, undesirable behavior patterns (pawing, kicking, wood chewing, etc.), and (or) digestive disorders. Daily exercise is desirable for stalled horses whether it be riding, lounging, free exercise in turnout trap, mechanical walker, ponying, swimming, treadmill, etc.

Salt intake. Horses should have access to salt to meet their requirements. Virtually all commercially prepared feeds contain salt (approximately .5% total feed) or TM-salt (.5% to 1% total feed). This level of salt generally will meet the salt requirements of all horses. Salt needs may vary between horses, particularly in periods of increased sweating; therefore, supplemental salt may be offered in block or loose form to horses consuming commercial rations. Because of boredom, stalled horses sometimes consume excessive amounts of salt that leads to excessive water intake and excessive urine production. Horses on pasture, receiving no commercially prepared feed, should always have access to free choice TM-salt. It is difficult for horses to lick their requirement of salt from a block; however, in humid areas, blocks crumble easily and salt can be consumed faster. Salt in the pasture should be fed from a clean container, preferably protected from wind and weather.

Overfeeding. Do not overfeed horses. Some horses suffer from obesity malnutrition and others from deficiency malnutrition. Generally horses are overfed because owners do not understand nutrient requirements and overfeed as a safeguard against underfeeding. Some people feed horses in the same way that they themselves like to eat. The extremely fat horse that is not receiving adequate exercise is predisposed to colic or founder, which could render the horse useless for the remainder of his life (Arab proverb: "Two of the horse's greatest enemies...fat and rest.")

When feeding conditioners or supplements, feed exactly as recommended on the label. It is human nature to think that the recommended quantity of one dipper full couldn't possibly do a 1,000 lb horse that much good. Therefore, the horse gets two dippers full each feeding. Supplements and conditioners are designed to be fed exactly as indicated and overfeeding could and often does cause toxicity problems.

Wood chewing. Preventative measures can discourage horses from chewing wood. Chewing wood and eating bark from trees is often observed in horses. Horses probably chew wood and tree bark for reasons other than nutrient deficiency: for example, boredom, lack of exercise, and lack of adequate long roughage. Splinters, toothwear, and colic are possible results for the horse. The bark of some trees (locust) and leaves of others (wild cherry, apricot, peach, almond) may be toxic to horses. Wood chewing is an acquired habit and an entire herd or barnfull may pick it up from one horse.

Wood chewing should be prevented at the outset. If horses have access to wood, these suggestions can help:

- Design stalls or pens so that horses can't get at the wood.
- Use hardwood lumber (oak) and (or) treated lumber (creosote or penta).
- Paint lumber with repellants.
- Use metal flashing over boards, etc.
- Wrap trees in pastures with old fence wire. (Older trees can be painted with repellants.)

Horse droppings. Routinely and frequently check the horses' droppings. The characteristics of droppings may vary some from horse to horse; however, any unusual change in quantity, consistency, odor, color, or composition indicates a possible disorder. For example, the normal adult horse defecates 35 to 50 lbs of feces per day. Stoppage of or limited feces production indicates upcoming colic conditions.

Poisonous plants. Plant poisoning is not a common problem in horses. Most poisonous plants are unpalatable and have a low concentration of the toxic principle. Generally, poisoning is precipitated only when horses are starved into eating poisonous plants or when lush clippings are fed to horses in confinement. Plants that can be poisonous to horses are: bracken fern, castor bean, fiddleneck, golden weed, horsetail, Japanese yew, jimsonweed, locoweed, oleander, prince's-plume, rattleweed, Russian knapwees, tansy ragwort, whitehead, wild cherry, wild onion, wild tobacco, woody aster, yellow-star thistle.

(Acknowledgment is extended to the Horse Section [Dr. Gary Potter, Dr. Doug Householder, and Mr. B. F. Yeates] Texas A&M University for providing material for this paper.)

Part 8

PASTURE, FORAGE, AND RANGE

35
UNDERSTANDING RANGE CONDITION FOR PROFITABLE RANCHING

Martin H. Gonzalez

INTRODUCTION

In many parts of the world, improvement of the livestock industry is believed to be dependent on two primary factors related to animals: genetics and disease control. Another nonanimal consideration seems to be rain! It is common to listen to ranchers talking about a new bull they have or plan to import, about the last rain--or complaining about the lack of it (even when the normal rainy season is still months ahead--or about prices of livestock on the market and the latest government regulations. However, less often we find a group of ranchers here or in Mexico, discussing range conditions, the forage production on the ranch, or range improvements.

Perhaps because we are overgrazing and because our pasture and forage production is poor, we manage to put the blame on the rain, the market, or the government, we do not like to recognize that a great part of our attention should be devoted to the basic things on the ranch: soil, grass, and water.

In the U.S. and Canada, many ranchers might be familiar with the range condition concept and its implications, but in other countries the situation is different. Even when the common sense, the personal ability, and the experience of a producer is outstanding, he will require a good and basic knowledge of the condition of the range because it is the key to profitable ranching. And this is true, particularly, today, when the livestock industry needs to reduce the cost of production--and can by using desirable range plants--still one of the cheapest sources of forage.

SOIL DEVELOPMENT, PLANT SUCCESSION, AND RANGE MANAGEMENT

Range management as an art and as a science has its roots in ecological principles. A range user has to deal with the basic components of the ecosystem around him and his operation, and should understand how they function and how they interact. Profitable ranching requires an understanding of basic ecology.

Figure 1 will help us to understand how the basic ecological processes relate to range management, exemplified by a rangelands ecosystem.

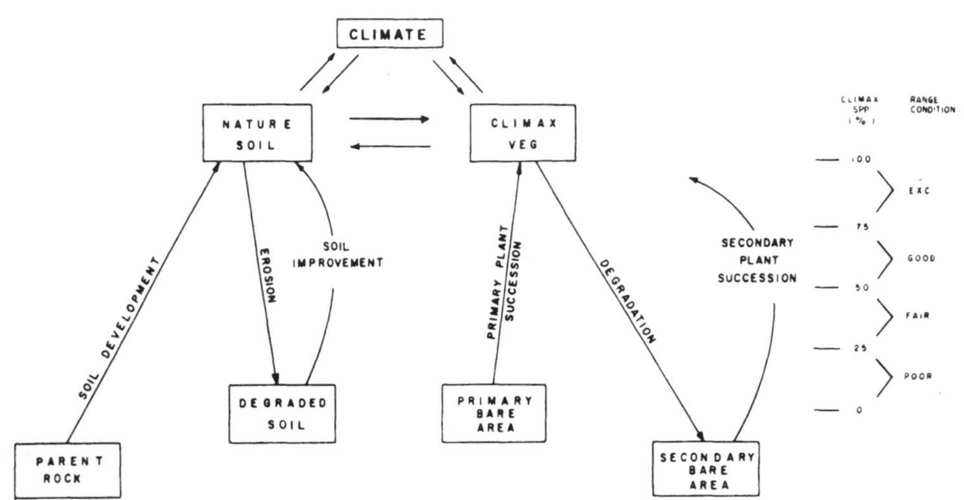

Figure 1. Relationship between soil, vegetation, climate, and range condition

Soil has developed for thousands of years from parent material. After a long process under various climatic conditions, it developed freely and reached maturity in its most productive phase. During that process, plant life had a parallel development, beginning from a primary bare area until reaching its climax stage (or maximum degree of development). This series of natural vegetative changes toward the climax stage are known as the primary plant succession. At this point, a fully developed soil and climax vegetation were in equilibrium with the climate for that particular area.

However, mainly due to man's actions, the soil has often been eroded, caused by degradation of the vegetative cover, and the pristine or climax condition has been destroyed. That which took centuries to build was destroyed by man in a much shorter period: by plowing of rangelands, by excessive fire, by timber exploitation, by over-grazing, by erosion, etc. It was then that the mature soil was converted into a degraded soil, and the climax vegetation was converted into a secondary bare area very far from its productive potential.

But man had not always been abusive of his resources. After he realized the serious damage caused to soil and vegetation and, of course, to the total productivity of an area, he started a series of soil improvement practices, together with some actions to accelerate the rehabilitation of the vegetative cover (secondary plant succession). These man-activated or "artificial" improvement practices (usually very expensive) were particularly needed in those areas where disturbances had been so severe that a spontaneous recovery was impossible.

And it is here, during different stages of this secondary plant succession, that most of our ranches are found today. The degree of disturbance they once experienced (or still are) and the degree of advancement within this succession, relate directly to range condition, according to the percentage of climax (desirable) forage species in the botanical composition found in the different pastures. It is impossible, under practical grazing conditions, to expect a range to remain in its primitive, climax state, and very few pastures can be found in excellent condition. There are many well-managed ranches in "good" condition, but most of the land, particularly in Latinamerican native ranges, would be characterized as in "fair" and "poor" condition. Many of these lands have the potential--under some grazing restrictions and good management--to improve and to recover productivity. However, on the lands, some agronomic improvement practices (soil conservation, brush control, range reseeding, etc.) must be used to reach their potential. We must pay this high price for the misunderstanding and mismanagement of our range resources.

QUANTITATIVE ECOLOGY AND RANGE MANAGEMENT

The Range Condition Concept

Range condition reflects the health of a range or pasture. It is based on the relation between the present vegetation and the vegetation that, potentially, a given site should have.

Dyksterhuis (1948) classified range plants in three categories according to their response to use by animals, and determined range condition based on the quantity of each of these species when sampling the range. These species are decreasers, increasers, and invaders. Figure 2 shows these relationships that every rancher should keep permanently in his mind.

Extensive field work has been done to classify plant species as decreasers, increasers, or invaders. Based on this work, range condition guides have been developed for every different site within any given vegetative type.

These include most of the plants that fall within each category and the percentage of each that is permitted in the botanical composition--based on what the climax vegetation should be. This permitted percentage, particularly for the increaser species, has been determined by experience and comparisons of the same species in different sites and under different grazing pressures.

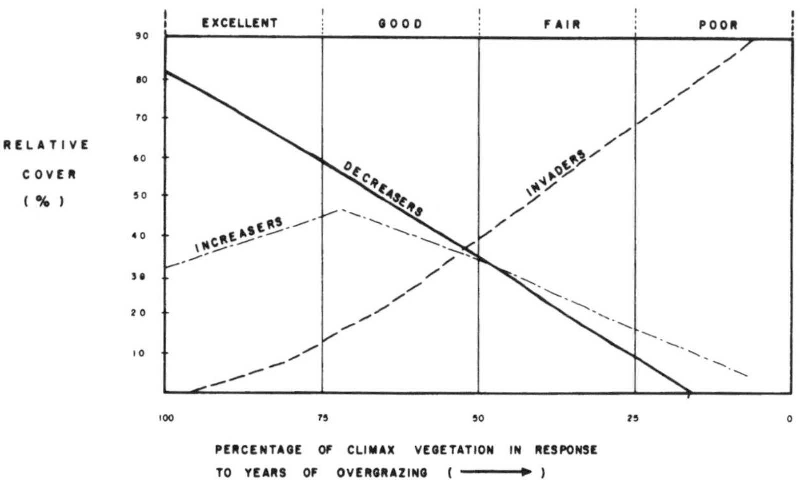

Figure 2. Quantitative basis for determining range condition (Dyksterhuis, 1948)

The decreasers are usually components of the climax community and are perennial, highly productive forage plants that are palatable and desirable. They are permitted in the percentages found in the botanical composition. The invaders are not permitted at all in the composition, so the percentage they accomplish for is substracted from the total. The increasers, most of which belong to the climax community, are permitted in varying percentages according to the range site and the associated species. The permitted percentage of each increaser species for the different sites has been calculated according to their response to grazing. This percentage has a maximum that will not interfere with the most desirable species--and a maximum that will not benefit the expansion of invaders. These changes in composition take place only when overgrazing occur over long periods, thus causing the deterioration of the range.

Dyksterhuis established a quantitative basis to determine range condition as follows:

Excellent Condition describes the botanical composition including more than 76% of climax or desirable forage plants. In other words, the vegetation is a mixture of decreasers and allowable increasers. There is almost no erosion and plant density is high. Almost no invaders are present. Litter is abundant.

Good Condition describes the biological composition that contains desirable species. The proportion of decreasers is less than that found in excellent-condition composition and few invaders may be evident. Litter is not as abundant but does occur.

Fair Condition describes the composition that has only 26% to 50% of the plants of climax or desirable species. In this condition, the signs of deterioration are more noticeable, particularly in marginal areas. Invaders appear in the same proportion as do decreases and increases; erosion has begun to be evident and density is generally low.

Poor Condition describes a badly degraded range with less than 25% of the plants in the composition of desirable or climax species. Forage production is very low. Erosion is a problem and invaders constitute a majority in the botanical composition. Almost no litter is evident. Spontaneous revegetation may not be economical under this condition and rehabilitation can be achieved only through range improvement practices.

We must realize that these conditions cannot be compared among different sites. Range site separation must be done before determining condition since their characteristics and potential are different.

Forage Production and Range Condition

The condition of the range and the amount of desirable forage produced are directly related for all types of range vegetation. However, from a production standpoint, a disclimax condition (just below the climax) may present a better diet for the grazing animals since some of the "increasers" in a given site could be more palatable or have a higher nutritive value than do some "decreasers" or climax species.

Figure 3 shows comparisons in the productivity of five selected rangeland types in northern Mexico, each for different condition: excellent, good, fair, and poor (Gonzalez, 1969).

It is observed that, even when there are substantial differences in kg/ha of usable forages between good an excellent conditions, there are greater differences between fair and good conditions and still greater differences between the poor and the fair conditions. This is true for all the vegetative types represented. Such large differences can serve as an index in estimating the effort, the expense and the time that may be needed to bring back the poorer conditions to a more productive stage.

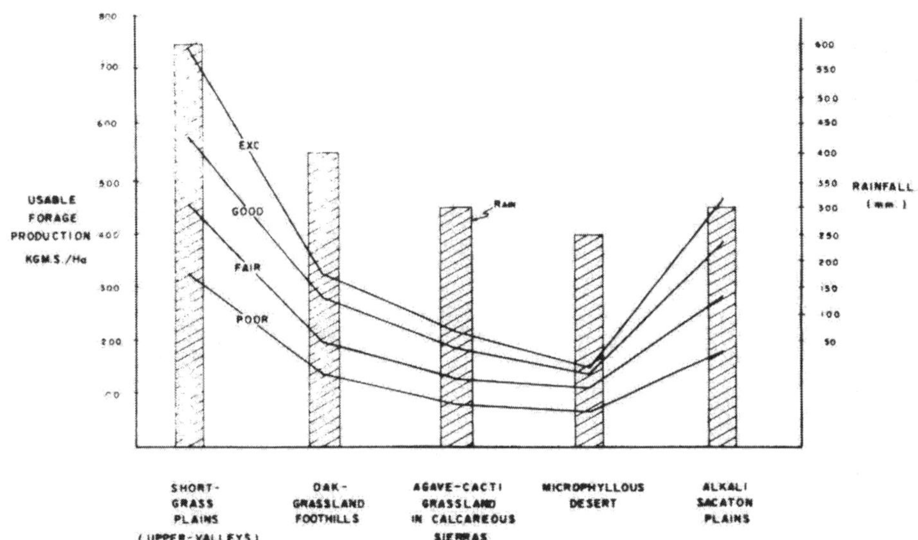

Figure 3. Forage production in five vegetative types under different range conditions in northern Mexico (Gonzalez, 1969)

RECOGNIZING RANGE CONDITIONS ON A RANCH

There are now sufficient range-condition guides for most of the rangeland areas of North America. For example, the Soil Conservation Service in the U.S. and COTECOCA-SARH in Mexico represent the agencies that developed such guides and that are using them extensively, along with many other government and educational institutions. It is not difficult to learn how to use these guides, but a basic knowledge of the terrain and the vegetation is needed.

Hoffman and Ragsdale (1974) from Texas A&M recommend for following steps in using the range-condition guides:

- Determine the range site.
- List the plants found on the site.
- Estimate the percentage of each plant that could be in the composition.

- Refer to the guide to determine the percentage of each plant that could be in the composition in the vegetational areas and site for which you are judging.

Carrying Capacity

Each guide includes the carrying capacity that is estimated for the different conditions in every site because range condition directly reflects the forage production in that particular site. A good condition means a stable, well-managed range; fair and poor conditions are indicators of heavy or severe use. On the other hand, if a site is in excellent condition, perhaps it is under-utilized. These guides provide an efficient, easy way to determine the carrying capacity (AU/ha) that the ranch should have--as compared with that under current management.

After the condition of the range has been determined, some adjustments may be necessary to keep the ranching operation flexible and to give the range the opportunity to recover. Adjustments may be necessary on:

- Livestock numbers (usually reducing them if fair and poor conditions are dominant).
- The present grazing system.
- The most economically convenient improvement practices to accelerate rehabilitation (brush control, range reseeding, etc.)--if condition is poor and hope for natural recovery is low.

REFERENCES

Dyksterhuis, E. J. 1948. Guide to condition and management of ranges based on quantitative ecology. Amer. Soc. Agron. App. Sec. Mimeo. p 25 (Abstr).

Gonzalez, M. H. 1969. Coeficientes de agostadero para el estado de Chihuahua. Memoria COTECOCA-SARH. Mexico.

Hoffman, G. O. and B. J. Ragsdale. 1974. How good is your range? Tex. Agr. Ext. Service Bull. Texas A&M Univ., College Station, Texas.

36
RANGE IMPROVEMENT PRACTICES AND COMPARATIVE ECONOMICS

Martin H. Gonzalez

Grazing lands of the world's rangelands have suffered abusive management that has, in turn, caused reduced productivity. This low productivity often is attributed to critical drought periods; however, the direct effect of a drought is to aggravate the poor management of the land.

The world demand for livestock products is increasing but the production from rangelands is decreasing because: faulty management has reduced the condition of the land and consequently the productivity; other demands for the land (agriculture, industrial, urban, highways) are deminishing the number of hectares available.

World evidence confirms the desertification of our ranges and the destruction of millions of highly productive hectares that have become unproductive and denuded.

A study conducted in Mexico (CFAN-CID, 1969), which included a survey in nine central and northern states and covered around 100 million hectares, indicated that overgrazing was evident in 85% of the land; light or advanced erosion was a problem in 87.5% of the overgrazed area, and 49.7% of the land was infested by undesirable plants, mostly brush. Overgrazing is a problem in almost all countries where degradation of rangelands has been so severe that economically it is impossible to expect a natural recover. Before the rangeland reaches this point, various range improvement practices, basically agronomic, are needed to facilitate and accelerate the secondary plant succession for returning those lands to a productive stage.

In this paper, emphasis will be given to the principles governing the most common improvement practices. More specific information for a particular problem may be obtained from neighboring experimental stations, the extension and university people, or the county agent.

MAIN RANGE IMPROVEMENT PRACTICES

Of the four elements applied in the rehibilitation of grazing lands (climate [rainfall], soil, vegetation, and grazing management), climate is the only one that man <u>cannot</u>

manipulate. However, manipulation of water from rainfall once it hits the ground is one of the most important aspects of range improvement. On the other hand, with correct planning and management of the other three elements, man may rehabilitate deteriorating lands. Some of the most common (and needed) improvement practices on rangeland are the following:

Water Conservation

Efficient use of rainfall enhances all the other improvement practices. When properly managed, even excess water can be beneficial. The principle in water conservation is to supply the forage species with moisture in such a way and at the right time so that the effects last the longest.

Water conservation can be managed in two ways: (1) by retaining each drop of rain where it falls and (2) by diverting excessive surface runoff to the highest-producing sites on the ranch. These two steps can be accomplished by either simple or complicated structures. However, the best way to conserve water for plant use is to have a good vegetative (grass) cover so that rain infiltrates into the soil and runoff and erosion are avoided.

Figure 1 shows how different grass covers on the rangelands of Chihuahua affect infiltration of water (Martinez, 1959). This figure averages data for tests on different short-grass ranges where bluegrama (Boutelouagracilis) was dominant. It was found that soils in an exclosure where they had been protected from grazing for 7 years had an infiltration rate 118% higher than bare ground, 63% higher than an overgrazed area, and 25% higher than a moderately grazed area. In turn, the soil that was moderately grazed absorbed 30% more water than the overgrazed site and 74% more than the bare area.

Table 1 shows the results of similar tests comparing different types and density of vegetative cover to conserve rainfall on different soils. On short-grass plains, sandy loam soil under excellent condition absorbed 450 mm more rain in a period of 105 minutes than a bare area; 270 mm more than range in poor condition, and 170 mm more than the soil with a cover in good to fair condition. The results were the same for the other two vegetative types--the oak-bunchgrass in the stony foothills and on the alkali flats with their heavy clay and deep soils (Sanchez, 1972).

Among the commonly used water conservation practices on ranches are contour furrows, range pitting (small, medium, or large pits), subsoiling, and low-retention fences. The conservation practice(s) to use depends on soil type, topography, slope, plant cover, costs, and storm intensity and frequency.

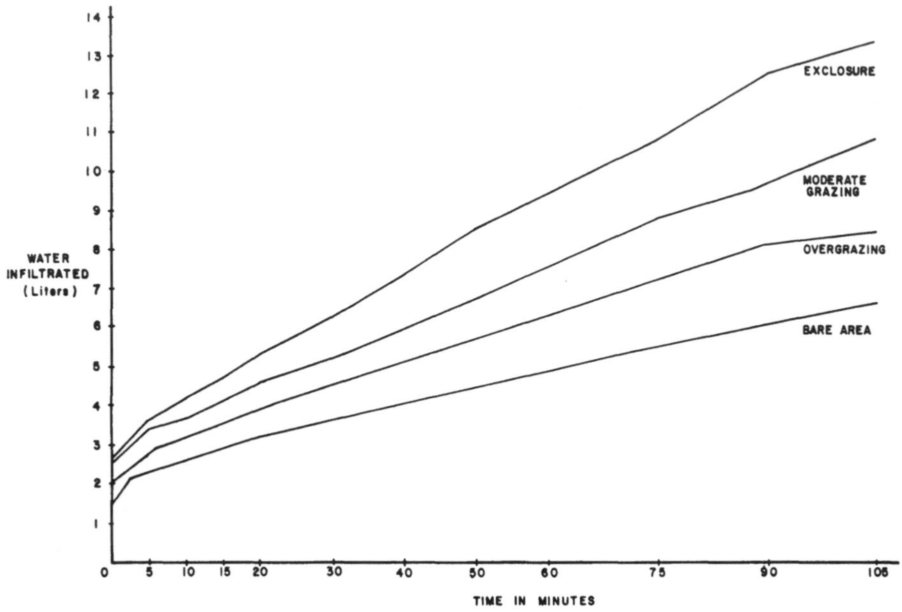

Figure 1. Water infiltration in short grass range under different covers. Chihuahua, 1964.

TABLE 1. INFLUENCE OF RANGE CONDITION ON WATER INFILTRATION[1] IN THREE VEGETATIVE TYPES AT LA CAMPANA, CHIHUAHUA

Range condition	Shortgrass plains	Oak bunchgrass foothills	Alkali flats bottom lands	$\overline{X}$
	mm	mm	mm	mm
Excellent	650	450	150	417
Good-fair	480	375	120	325
Poor	380	230	81	230
Bare area	200	190	73	154

[1] Mm of rain or equivalent absorbed in 105 minutes.

Soil Conservation

The most effective practice for soil conservation on rangeland is the same as for water conservation--a permanent cover of desirable plants. All soil conservation practices have a double purpose: to conserve water and to avoid erosion. Contour furrows, gully control by using different structures, terracing, and wind breakers are all effective soil conservation practices. As for water control, the practice to be selected will depend on the terrain. Heavy, sophisticated equipment for soil conservation is easily replaced by using some of the natural materials found on the ranch or in the pastures. For example, a gully can be controlled by filling an area with rocks, yucca plants, palmilla (sacahuiste) plants, old trees, or the undesirable vegetation nearby. Gullies can serve as dump sites for those materials. In a very few years the plant material, rocks, and soil eroding from upstream will compact, cause a dam to form, and erosion downstream will be reduced.

Control of Brush and Other Undesirable Plants

Unfortunately, most of the world's rangelands in use are infested, by varying degrees, with undesirable vegetation of which the woody species are the most common. This invasion is the direct result of the degradation caused by the different activities of man while using his grazing resources. Control of undesirable brush can be done by different methods: mechanical, chemical, and biological.

Mechanical control. Mechanical control is usually done with heavy or with light equipment. Heavy equipment includes dozing, chaining, disking, and heavy shredders or crushers. Light equipment includes the use of smaller tractors and some manual tools for shredding and disking. the equipment to use depends on the type of vegetation and its density, conditions of the terrain, availability of equipment, and costs. When using mechanical brush control, try to move or remove as little top soil as possible.

Research and rangeland experience all over the world demonstrate the competition between undesirable brush and grasses. The woody species of brush requires nearly four times more water to produce one kg of aerial growth and provides much lower forage quality than some of the best perennial grasses.

On Rancho Experimental La Campana, Chihuahua, Mexico, mechanical control of shrubs in a short-grass range invaded by *Acacia*, *Mimosa*, *Eysenhardtia*, and *Brickellia* resulted in a 102% range increase in forage production (Gomez and Gonzalez, 1978). Chemical control of "chaparrilo" (Eysenhardtia spinosa) produced 428 kg DM/ha more than the untreated areas, which had only 171.1 kg DM/ha. This represents 250% increment in forage production (Gomez and Gonzalez, 1976).

Chemical control. The chemical control of undesirable plants has developed intensively during the last two decades. The chemical products (herbicides) on the market for almost any brush-control program can be applied either by aerial spraying or ground spraying. Ground applications can be foliar, on the trunk or stumps, or directly on the ground in the form of pellets.

Herbicides work in different ways:

- contact herbicides are those that kill the plant or parts of the plant directly exposed to the chemical. This type is used on annual weeds. Examples: diesel oil, diquat, and paraquat.
- Translocated herbicides (also called hormonal or systemic)--are used in low concentrations. The toxic substance is carried through different parts of the plant by the plant's own liquids. Examples: 2,4-D, 2,4,5-T, silvex, dicamba, and picloram.
- Selective herbicides are the ones that kill a particular species or group of species without damaging others. If heavy dosages are used these may act as nonselective. This type of herbicide is the most widely used for range improvement because it does not affect grass or grass-like plants (monocots). Examples: 2,4,5-T picloram, TCA, and MCPA.
- Nonselective herbicides kill or damage all plants where applied. Example: AMS, amitrol, PCP, diesel oil, kerosene.
- Soil sterilants are generally in the form of pellets and are applied directly to the soil. When the pellets are diluted by rain, they are absorbed by the roots. Examples: Atrazine, bromocil, dicamba, diuron, and fenuron. The effect may be permanent or temporary, selective or nonselective. A detailed treatise on herbicidal plant control including products, methods, advantages, problems, etc., is found Range Development and Improvements (Vallentine, 1971).

Phenological stage of the plants, time of year, and meteorological conditions are three of the important factors to consider in brush-control programs.

Biological control. Another, but complicated, way to control undesirable plants and shrubs is through biological control. This can be achieved by insects or domestic or wild animals. There are insects that specifically live on certain plants, feeding from them, and eventually damaging them. Examples are the mesquite twig girdler (Oncideres rhodosticta) and two species of leaf-feeding beetles (Chrysolina gemellata and C. hyperic*y*) that attack the klmath weed (Hypericum perforatum).

Although biological control of shrubs and other undesirable plants by insects is complicated, the use of domestic animals is not. Studies in Central Chihuahua (Fierro, et al., 1979) used goats over a period of 3 years to defoliate five different woody species. Results indicated the goats' preferred shrubs to grasses. In moderate vs intensive browsing and in shredded and not shredded areas, intensive defoliation by goats caused a 36% mortality of "Chaparrilo" plants (Eysenhardtia spinosa); moderate defoliation killed 14% of the plants. A similar trend was observed for Acacia and Mimosa.

An additional benefit of using goats for biological control of noxious shrubs was their milk production--an average 30 ml per goat per day for 4 mo a year. The production of perennial grasses increased from 94 kg/ha to 360 kg/ha in the first year (1977) and to 950 kg/ha 3 years later with moderate browsing.

The main factor to consider if planning to control shrubs with sheep or goats is the type of vegetation. On many ranges invaded by shrubs and other plants that cattle will not eat, grazing systems combining small and large herbivores have been established. Even in some areas where toxic plants for cattle existed, changing to goats (or perhaps sheep) allowed those areas to be utilized.

Fire. The cheapest and most effective brush control tool is fire. However, there are certain rules that have to be followed: (1) the amount of fuel (grasses and associated vegetation) must be high enough to carry the fire at the desired intensity; (2) wind direction, velocity, and soil moisture content determine the season and time of day to use fire. Extreme care must be taken to avoid letting the fire run wild and damage adjacent areas. Vallentine (1971) in his book, Range Development and Improvements, describes in detail the procedures of burning for brush control and its consequences.

Numerous controlled burning experiments have been done to control undesirable vegetation. Glendening and Paulsen (1955) obtained 52% kill of young mesquites having basal diameters of 0.5" or less; however, only 8% to 18% of the taller trees were destroyed by fire. In other areas, only 9% were destroyed by burning (Reynolds and Bohning, 1956), but White (1969) reported a wild fire that killed 20% of the trees with moderate and severe burns. Cable (1965) reported that variable results were obtained from different fuel covers: 4500 kg/ha of fuel (ground cover) killed 25% of the trees when burned; areas with 2200 kg/ha of fuel killed only 8% of the trees. Other species damaged by fire were Ocotillo (Fouqeria splendens) with 40% to 67% killed; Sotol (Dasylirion texanum) had 97% dead plants, and creosote bush (Larrea tridentata) was reported with a "very high" percentage killed.

June fires killed 44% of the choya cactus (Opuntia fulgida) and 28% of the pricklypear (O. engelmanii) (Reynolds and Bohning, 1956).

Range Reseeding

Reseeding of denuded areas is an improvement practice that obtains fast results. However, many risks are involved in this type of operation, and extreme care must be exercised to establish and produce good forage.

Range reseeding has proven successful in many areas in the U.S., Canada, Mexico, and some Latin American countries. In arid and semiarid lands it is combined usually with some water catchment devices or structures in order to assist in the establishment of the new plants. Commonly associated with land clearing and brush control, reseeding is a must in many parts of the world.

The following are some points to help in understanding, planning, and managing a range reseeding operation.

What is reseeding? Reseeding refers to the artificial, not natural, revegetation of the range. It includes planting native or introduced species with one or several of the following objectives: 1) to improve plant cover/density, 2) to improve forage production, 3) to improve forage quality, 4) to improve efficiency of water use, and 5) to improve the diet of grazing animals.

Why reseeding? Reseeding is initiated for two reasons:

- To rehabilitate unproductive areas and put them into production again. Nongrazed areas, because of lack of forage, are a waste of money.
- To accelerate secondary plant succession in areas where natural revegetation is too slow or almost impossible.

When is reseeding necessary and justifiable? Reseeding is needed and justified:

- when percentage of desirable vegetation in the botanical composition is very low (below 18-20%) and plant density is rare.
- when the range site to reseed has the agronomic potential to respond to the treatment (topography, soil quality, etc.).
- when climatic and meteorological conditions are favorable and meet the minimum safe requirements: amount and distribution of rainfall, and frost-free period.
- when the proper seed is available.

How to reseed. There are some basic rules that one has to follow to minimize risks when trying to improve a range by revegetation:

- Choose the best adapted species, native or introduced, based on experimental evidence or on regional experience.
- Select a high quality seed with a high PLS (pure live seed) percentage; do not sacrifice quality for a low price.

- Seed at the beginning of the <u>formal</u> rainy season--when you are sure the rains will continue with regular frequency. Do not plant late in the summer because the plants will not have a chance to get established before the first frosts.
- Use the recommended seeding rates on a PLS basis.
- Try to use a grass seed driller that will assure the best distribution of seed on the ground. If no machinery is available and you have to broadcast, cover the seed lightly with branches from shrubs.
- Broadcast a mixture of seeds separately according to size, i.e., the small seeds (weeping live, clover, bluepanic) should be mixed with fine sand for a better distribution and planted separately from larger or fluffy seeds (side oats, crested wheat, buffel). Drillers usually have different types of boxes for different sizes of seeds and can be calibrated separately for the desired rate.
- Plant grass seeds no more than 1 to 1.5 cm deep in a firm seed bed. If possible, press the soil down lightly on the seeds to make contact between seed and soil. If this is not possible, and broadcast is used, cover lightly with a rake made of shrub branches.
- Do not fertilize at the time of planting since many weeds will take advantage of the fertilizer and will compete strongly with the planted species. It is better to wait until there is an established stand of the seeded species.
- Protect the seeded area from grazing at least during the first growing season. Depending on the density of the stand, the area could be grazed lightly in the second year. Remember: reseeding is expensive and risky, and proper management is necessary for a long-lasting stand.

<u>Range fertilization</u>. Recent increases in prices of fertilizers have limited their use on rangelands. However, areas receiving annual rainfall above 350 mm have responded significantly in terms of forage production, forage nutritive quality, and carrying capacity so that it might still pay to fertilize.

Nitrogen is the most commonly used fertilizer for semi-arid and temperate rangelands; however, nitrogen plus phosphorous have demonstrated to be better in some rangelands. For instance, experiments in Central Chihuahua, Mexico, using different levels of N and P and their combinations, indicated that either N or P alone produces more forage than the nontreated areas, but the application of N and P together was much better. By applying 80 kg of N and 50 kg

P_2O_5/ha, forage production increased 132% the first year; crude protein per hectare changed from 31 to 107 kg and phosphorus per hectare increased from 0.42 to 1.46 kg. In this same experiment, 73% of the cost of fertilizers was recovered during the first year (Gonzalez, 1972).

In spite of the "energy crisis" and the high price for fertilizer, it is important to fertilize rangeland because of the growing demand for land and because of the need to intensify as much as possible by getting the maximum out of every hectare of range. That is why fertilizers have an important role in areas where ecological conditions are favorable for their use.

Type of fertilizer, apportionment/ha, time of application, and method of application are some of the aspects to consider once you have had a soil analysis and determined the need to fertilize.

PRIORITIES IN SELECTING IMPROVEMENT PRACTICES

Every ranch has its own problems and own needs for range improvement. All of the range improvement practices discussed are expensive, but each must be considered carefully when bringing rangeland back into production.

The following are some guides to help in the selection of which improvement practice(s) to use.

- First, evaluate the present range condition to determine if an improvement is necessary. Usually ranges in poor or fair condition need some help, but ranches in better condition may need only a change in management.
- Second, define the problem. Brush infestation? Low plant density? Poor botanical composition? Erosion? . . . Establish your priorities. Treat the best sites first because they respond better and faster to treatment. Define which improvement practice(s) to use and, if several are needed, establish your priorities.
- Third, in selecting the practice to improve your ranch, costs are basic. There must be an evaluation of costs in terms of the potential to improve forage production and pay back the expenses. One must be alert to prices and cost fluctuations.
- Finally, once the practice has been selected, plan <u>how</u> to do it: 1) what to use, 2) when to do it, and 3) who is going to do it. Before starting the program define the type of equipment and materials, the time of the year, the number of hectares to treat, and the people responsible for doing the job.

In analyzing all the parameters, a combination of two or more practices (if needed!) may be the most economically productive.

For example, figure 2 is a classic for understanding the priorities for improvement in northern Mexico in relation to the interaction of rain and vegetation. Of the 340 mm average annual rainfall, 40 mm, almost 12%, fall in the winter time when it is not used readily by plants. Of the remaining 300 mm about 100 mm are lost through evapotranspiration and surface runoff; only 200 mm are available for the vegetation. According to studies on over 100 million hectares of rangelands in Central and Northern Mexico (CFAN-CID, 1968), botanical composition in range pastures includes more than 50% of undesirable plants; therefore, at least 125 mm of rain are used by shrubs, forbs, and weeds, and the final 75 mm for desirable forage species--the equivalent of a mere 22% of the total rain recorded.

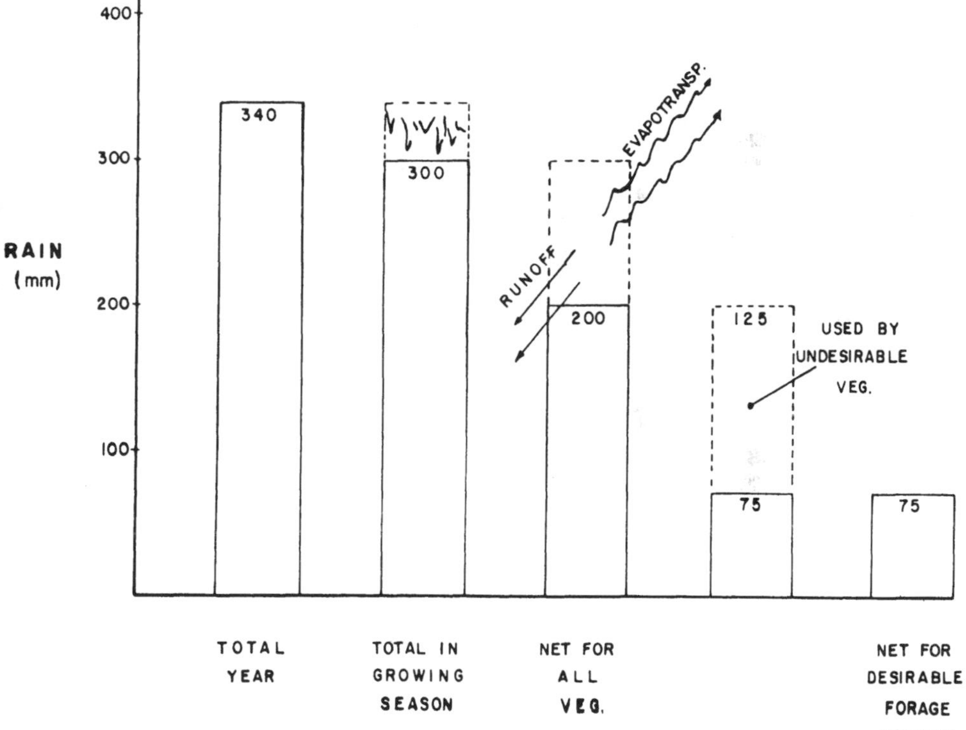

Figure 2. Used and wasted rainfall. Northern Mexico.

This simple rain-vegetation relation shows that range improvement practices should be directed towards: (1) obtaining a better conservation of water and (2) eliminating undesirable vegetation so the limited available rain is used by desirable plants.

ESTIMATED PRODUCTION INCREASE DUE TO IMPROVEMENT PRACTICES

Based on evidence accumulated by research at Rancho Experimental La Campana, in Chihuahua; CIPES, in Carbo, Sonora; and Vaquerias Experimental ranch, in Jalisco, as well as field data in different ranches in northern and central Mexico, the increase in range production due to several improvement practices is illustrated in table 2 (Gonzalez, 1977). This table includes averages of results of different vegetative types where improvement practices have been conducted. These indicate the benefits in a 4- to 6-year period (short-term) and the projected increases that can be obtained in 8 to 12 years (long-term).

TABLE 2. ACTUAL AND POTENTIAL INCREASE IN FORAGE PRODUCTION IN RANGELANDS OF CENTRAL AND NORTHERN MEXICO

		Potential	
	Present	Short-term (actual)	Long-term (projected)
Forage Prod. kg DM/ha	210	420	580
% increase due		100	176
Grazing systems and intensities		40	46
Soil and water conservation		15	40
Brush control		30	60
Range reseeding		15	30
Carrying capacity Ha/AU	20.6	10.3	7.5

Source: M. Gonzalez (1977).

In the short-term period, the forage production of the range increased 100% from an average of 210 kg DM/ha to 420 kg DM/ha. This increment was due to the adjustment of the carrying capacity of the ranches and the modification of their grazing systems (40%); to soil and water conservation (15%); to brush control (30%); and range reseeding (15%). These improvements increased the yearly carrying capacity of the range from 20.6 ha/AU to 10.3 ha/AU.

The projected increase in production for the long-term period was based on the continuation of adequate management and improved range conditions. For an 8- to 12-year period, it was estimated that forage production would reach 580 kg DM/ha or 176% above the present. Carrying capacity would be 7.5 ha--almost triple of the present one.

The range improvement will be reflected by increased livestock production, both in quantity and quality, because of better forage in the pastures.

REFERENCES

Cable, D. R. 1965. Damage to mesquite, lehmann lovegrass and blackgrama by a hot June fire. J. Range Mgt. 18:326.

Fierro, L. C., F. Gomez and M. H. Gonzalez. 1979. Control biologico de arbustivas indeseables utilizando ganado caprino. Bol. Pastizales, Vol. X No. 3 Rancho Exp. La Campana INIP-SARH, Mexico.

Glendening G. and H. A. Paulsen. 1955. Reproduction and establishment of velvet mesquite and relation to invasion of semidesert grasslands. USDA Tech. Bull. 1127.

Gomez, F. and M. H. Gonzalez. 1976. Evaluacion de cinco mezclas de herbicidas en elcombate de chaparrillo, Eycenhardtia spinosa. Bol. Pastizales Vol III, No. 2 Rancho Exp. La Campana INIP-SARH. Mexico.

Gomez, F. and M. H. Gonzalez. 1978. Efecto del chapeo mecanico en el incremento de un pastizal invadido por arbustivas Bol. Pastizales, Vol. IX, No. 5 Rancho Exp. La Campana INIP-SARH, Mexico.

Gonzalez, M. H. 1972. Aumentos en la produccion de carne con la fertilizacion de un pastizal. Bol. Pastizales Vol. III, No. 1 Rancho Exp. La Campana, INIP-SARH, Mexico.

Gonzalez, M. H. 1976. El potencial de las tierras no cultivables en la produccion de alimentos. Ingenieria Agronomica, Vol. II, No. 2, Col. Ing. Agr. de Mex. AC. Mex.

Martinez, J. 1959. Pruebas de infiltracion en un pastizal mediano-abierto. Tesis. Esc. de Ganaderia, Univ. de Chihuahua, Mexico.

Reynolds, H. G. and J. Bohning. 1956. Effects of burning on a desert grass shrub range in southern Arizona. Ecology 37.

Sanchez, A. 1972. Efecto de la condicion de pastizal en la infiltracion del agua de lluvia Bol. Pastizales, Vol. III, No. 3. Rancho Exp. La Campana INIP-SARH, Mexico.

Vallentine, J. F. 1977. Range Development and Improvements. Brigham Young Univ. Press. Provo, Utah.

White, L. D. 1969. Effects of a wild fire on several desert grass land shrub species. J. Range Mgt. 22.

37
WHAT TYPE OF GRAZING SYSTEM FOR MY RANCH . . . ?

Martin H. Gonzalez

INTRODUCTION

Ranchers have developed a growing interest in grazing systems during the last decade. The introduction of the new, "intensive" systems has received more attention from producers and researchers than the traditional, continuous, and extensive rotation systems we knew.

A never-ending controversy could evolve from a group of ranchers discussing the pros and cons of each one's favorite system. All of the systems have been used under different conditions and have their particular advocates but, at the same time, everybody knows that a specific system cannot be universally implemented. Each ranch has its peculiar objectives, ecological features, market situations, financial capabilities, and the personal preference of its owner. As Penfield (1982) wrote, "The most important element in a grazing system involves a commitment by you to make it work." And at the same time the rancher must be committed to working with nature not against it--a principle to be kept in mind when selecting a grazing system.

Both native rangelands and cultivated forage communities and their biological and economic ecosystems lead ultimately to the production of livestock products used by man. Most lands under livestock or wildlife production have unbalanced ecosystems in which one or more of the major constituents (plant, animal, or man) has a chronic or seasonal lack of nutrients, water, or other environmental requirement (Whythe, 1978). It is the responsibility of the soil and crop scientists to help overcome these deficiencies. But it is the rancher, the manager of the range, who has to decide whether the proposals of the scientists and the new "discoveries" are economically acceptable; the rancher, after all, is part of the ecosystem in which he lives with his domestic livestock.

TYPES OF GRAZING SYSTEMS

The number and variation of grazing systems is almost infinite. Every ranch has its peculiar characteristics--

even adjoining ranches with similar ecological conditions differ in objectives, management, and financial capabilities. For this presentation we will consider some of the most common grazing systems, knowing that within each of them the variations are numerous. Any rancher can identify with one of these models and design his own system.

Continuous Grazing

Continuous grazing is the constant use of forage on a given area, either throughout the year or during most of the growing period. This type of grazing does not always result in range decline (overgrazing). Light, continuous use results in range improvement, but returns per hectare are lower than those obtained from a four-pasture, deferred-rotation grazing (Merrill, 1980).

Under continuous grazing, stocking rates are set with only minor adjustments through the year. The number of animals does not vary greatly except during long drought periods (Kothmann, 1980). Under continuous grazing, stocking rate is the only variable the manager can adjust; this allows little flexibility in responding to drought seasons (Vaughan-Evans, 1978).

Rotation System

A rotation system moves animals from one pasture to another according to a fixed schedule. Within this system are the extensive (or non-intensive) and the intensive systems. In the extensive system, several pastures are grazed and one is resting for a certain number of months. The rest period is rotated so that the same pasture has the rest period during different months each year. In the intensive system, all the animals graze one pasture (for a short period) while all the other pastures rest.

Deferred rotation. Under deferred rotation, each pasture is deferred during each season (Ambolt, 1973). The design includes a fixed number of pastures for each herd of livestock. The goal is to improve the vegetation condition of the resting unit, but the length and intensity of grazing on the remaining units should not be so long as to cause range deterioration (Merrill, 1980). Carrying capacity is determined by the total land area in the system and should be set conservatively. Deferment periods usually vary from three to six months, but may be longer (Kothmann, 1980).

The purpose of such a system is two-fold: to improve the vigor and production of forage, which in turn increases the desirable forage species, the carrying capacity, and the animal response to the improved forage. The result is higher returns per unit area. Figure 1 shows the Merrill 4-pasture deferment-rotation grazing as a good example of this system.

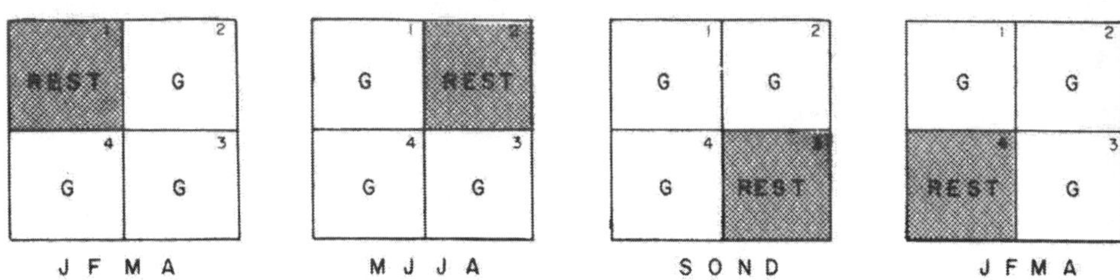

Figure 1. Merrill's four-pasture, deferred-rotation grazing system

<u>Rest rotation</u>. In the rest-rotation system, a pasture is not grazed for a full year and the rest period rotates among pastures. Deferments are provided for seed production and for seedling establishment. The number of pastures in the system may be from two to five (Hormay, 1970). The carrying capacity is based on that portion of the range that is grazed each year. Generally a much smaller percentage of the total area is available to grazing during the growing season (Kothmann, 1980).

These two basic nonintensive rotation systems sometimes are combined to take advantage of both the rest and the deferment on a varying number of pastures.

A disadvantage for some ranches is that pastures should be reasonably uniform in both size and carrying capacity. This often requires additional fencing and watering (figure 2).

<u>Intensive systems</u>. The two most common types of intensive grazing systems are the high intensity-low frequency, and the short-duration grazing. Both of these systems are based on grazing all the animals in one pasture while the other pastures rest. The difference between the two systems is the length of time given to the grazing and rest periods. The advantages of the long rest period are: it provides maximum vegetation improvement, eliminates seasonal rotations, and concentrates the livestock in one area which makes handling and working easier. The disadvantages of the intensive systems are: an increase in the need for water storage and daily output on individual pastures, generally lower livestock gains per head, and a possible parasite problem if sheep and goats are present (Merrill, 1980).

High intensity-low frequency (HILF)--When designing an HILF system, the forage species present are evaluated and a system that favors the plants we want to produce is designed. The time plants need for recovery from grazing determines the rest periods needed in the system (figure 3).

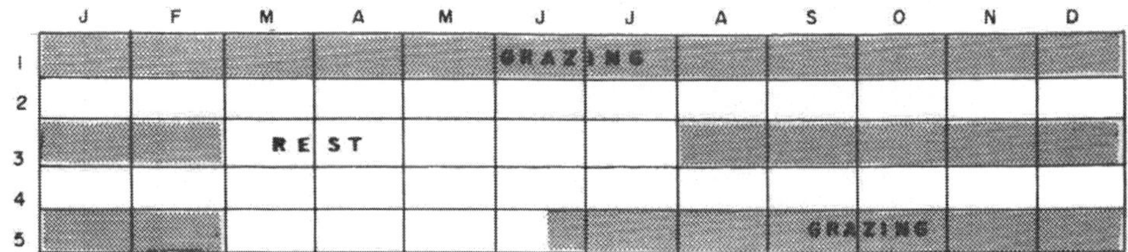

Figure 2. Rest rotation system with five pastures

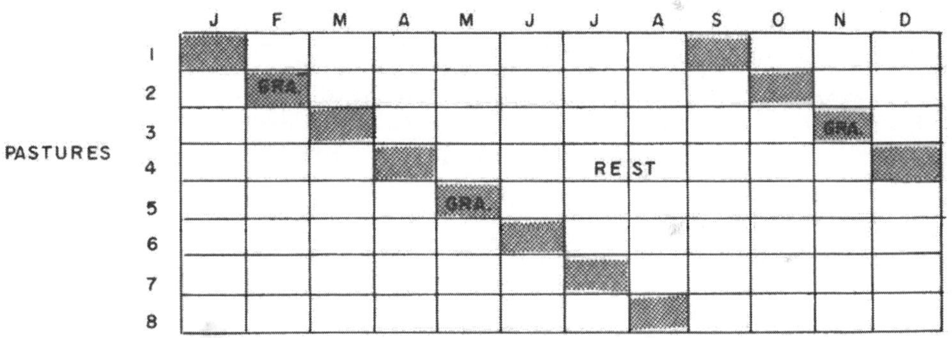

Figure 3. High-intensity--low-frequency grazing system

The HILF systems are based on the use of intensive grazing periods with relatively long rests. At least three pastures are required per herd. The grazing periods are generally more than two weeks, rest periods longer than 60 days, and grazing cycles over 90 days. Carrying capacity is based on the total land area in the system and moderation in length of grazing cycle to prevent an excessive decline in animal production.

Short duration--Like the HILF, this system also must have three or more pastures per herd, but the grazing and rest periods are shorter. Grazing periods should be less than 14 days (preferably 7 days or fewer and rest periods should vary from 30 to 60 days, but never exceeding 60 days (figure 4).

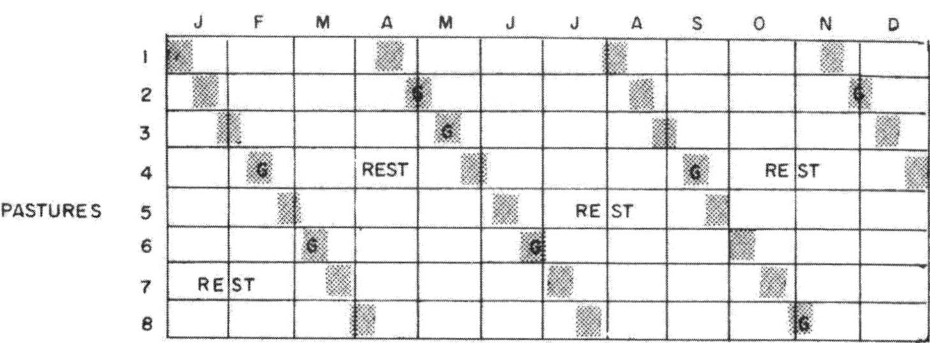

Figure 4. Short-duration grazing system

Grazing cycles are generally short enough to allow six or more full rotations per year (Savory, 1979). Carrying capacity is based on the total land area in the system. Because of the high degree of control over both frequency and intensity of defoliation (greater than under other systems), a higher degree of use in this system is possible without detriment to plant or animal production (Kothmann, 1980).

PARAMETERS FOR THE EVALUATION OF GRAZING SYSTEMS

Objectives

Apart from the specific objectives that a rancher might have, there are a number of general objectives for selecting his grazing system. Some of the most important are the following:

- To obtain a better livestock grazing distribution and, consequently, a more uniform utilization of the range or pasture.
- To maintain or improve forage density.
- To maintain or improve the botanical composition.
- To meet the nutritional needs of the grazing animals.

All of the objectives are affected by the major objective--to increase the overall efficiency of the operation.

Considerations

Thee are some basic premises that the rancher must consider when choosing his grazing system. The following deserve his attention:

- Viability of the system. Will all factors combined in its planning make it work?
- Cost of additional infrastructure (fencing, watering, pens, etc.) and labor involved.
- Minimizing the effect of adverse weather.
- Stocking rate.
- Pasture size.
- Location of water, salt, and supplements.
- The number of pastures in the system.
- The number of herds per system, if more than one system is to be used.
- The length of the grazing cycle and the rest period--the heart of the system.

Consequences

A well-planned and implemented grazing system will have some beneficial consequences. On the contrary, if the development of the system is not in agreement with the basic objectives and considerations, the results are likely to be poor. Some of the positive effects of a well-designed and managed system are:

Effects on the vegetation.

- The system will produce more forage.
- Seed production is increased.
- Botanical composition of pastures is improved.
- Chemical composition of forage plants (animal's diet) is improved.
- Over-all range resources are improved by the rapid increase in range condition.
- The amount of forage consumed by livestock due to nonselective grazing is increased.

Effects on the animal.

- Minimal or no stress on animals.
- Improved animal performance.
- Less handling of animals in general.
- Reduced supplemental feeding.
- Minimized labor costs.
- Higher individual weight gain.
- Better calving intervals.
- Increased stocking rate.
- Interrupted disease and parasite cycle.

A possible added benefit would be the increase in wildlife population and an improvement in their condition.

PLANNING THE SYSTEM FOR YOUR RANCH

Assuming that you have decided to develop your own grazing system or to adopt one of many models available, a logical sequence of action is recommended.

Inventory and study carefully your range resource.
- Vegetative types
- Range sites within vegetative types
- Your pastures:
  - a Number
  - b Size
  - c Forage production
  - d Season of use
  - e Present range condition

Reaffirm your objectives for:
- Type of livestock operation
  - a Cow-calf
  - b Feeder cattle
  - c Combination
- Type of grazing system
  - a Extensive rotation
    1. Deferred rotation
    2. Rest rotation
  - b Intensive rotation
    1. High intensity-low frequency
    2. Short-duration grazing

Design your grazing system according to:
- Your objectives
- Present number and size of pastures
- Projected number and size of new pastures
- Expenses involved in additional proposed infrastructure
  - a Fencing
  - b Water development
  - c Labor

Expand the management plan.
- For the range-pastures
  - a Grazing cycle
  - b Range improvement practices needed
- For the livestock
  - a Breeding and calving
  - b Nutrition program
    - 1 Range supplements for cows
    - 2 Possibilities of creep feeding
    - 3 For replacement heifers
  - c Sanitation program
  - d Marketing
- For monitoring
  - a Forage utilization
  - b Range condition and trend
  - c Animal production

1 Percentage calf crop
2 Weaning weights

d Market situations

We must emphasize that the best grazing system is the one best adapted to your particular interest and the conditions on your ranch. Even if economic resources and infrastructure are limited, a grazing system can be improved through better management. Work with what you have and follow the basics for the grazing system that best suits your ranching business.

REFERENCES

Hormay, A. 1970. Principles of rest-rotation grazing and multiple use land management. USDA, F. S. Training Test 4(2200).

Merrill, L. B. 1980. Considerations necessary in selecting and developing a grazing system. What are the alternatives? Proc. Grazing Mgt. Systems for S. W. Rangelands Symposium. The Range Impr. Task Force. New Mexico State Univ., Las Cruces, N. M.

Kothmann, M. M. 1980. Integrating livestock needs to the grazing system. Proc. Grazing Mgt. Systems for S. W. Rangelands Symposium. The Range Impr. Task Force. New Mexico State Univ., Las Cruces, N. M.

Penfield, S. 1982. Are you ready for a grazing system? Rangelands 4(1). Soc. Range Mgt.

Savory, A. 1979. Range management principles underlying short-duration grazing. Beef Cattle Sci. Handbook. Agr. Services Found., Clovis, Ca. 16:375.

Vaughan-Evans, R. H. 1978. Short-duration grazing improves veld conservation and farm income in the Oue group of I.C.A.'s Cenex reports. Mimeo. Rhodesia.

Wambolt, Carl L. 1973. Range grazing systems. Coop. Ext. Serv. Bull. 340. Montana State Univ.

38

DETERMINING CARRYING CAPACITY ON RANGELAND TO SET STOCKING RATES THAT WILL BE MOST PRODUCTIVE

John L. Merrill

Every rancher has to decide the stocking rate for his pasture--how many animals for how long. If he makes the wrong decision, the problem will be compounded by continuing indecision or unsound decisions.

Successful range management depends upon proper use which is defined as "stocking according to forage available for use--always leaving enough for production, reproduction, and soil protection." This is easier said than done. Other important proper-use factors are the species of livestock and wildlife to be grazed, the proper distribution of grazing, and the grazing method employed, but a key concern is setting and adjusting the stocking rate as determined by carrying capacity.

Experience and research have helped the grazing-land manager estimate and/or calculate the carrying capacity of either range or tame pasture for making better stocking-rate decisions and adjustments. By using several different methods simultaneously, the grazing-land manager's judgment and accuracy are improved.

Over a period of many years the USDA Soil Conservation Service (SCS) has developed range site descriptions and range condition guides for determining carrying capacity for almost every area of privately owned rangeland in the U.S. The safe-starting stocking rates derived by the SCS are fairly conservative so as to avoid recommendations that might cause trouble.

A range site is a distinctive kind of rangeland of relatively homogeneous climate, soils, and topography that will support a typical group of plants and level of production. Range condition refers to its current productivity relative to its natural capability to produce in climax condition. Range condition is classified by the percentage of climax plants present and usually is expressed as excellent, good, fair, and poor by 25% increments. For each classified condition there is a corresponding carrying capacity or safe-starting stocking rate range expressed in acres per Animal Unit Year Long (AUYL) or Animal Unit Month (AUM). The range allows for varying forage production from year to year within a given range condition.

Climax plants include decreasers, which are those plant species dominant in the original-climax vegetation that decrease with continued overuse. Increasers are secondary-climax plants that increase for a period of time as decreasers decline from overuse, but they in turn will also decrease from continued overuse. Invaders are those plants that are not found (or found in small numbers) in the climax situation that begin to cover bare ground as the climax plants decline and, therefore, are not counted in range-condition classification.

For many years SCS range-condition estimates were based on a visual estimate of percentage by ground cover (basal density and/or canopy cover) of climax plants with percentages of decreasers counted, but with no greater percentage of increasers counted than were estimated to be present in climax condition. This system was easy for both range technicians and ranchers to learn and use and provided a reasonably accurate track of changes in plant composition if recorded year after year (appendix A).

In an attempt to refine that system, SCS range-condition guides now assign a percentage of each plant or group of plants similar in production and reaction to grazing pressure. The percentage present is determined by estimating the pounds of air-dry forage per acre of these plants or groups divided by the total pounds per acre (appendix B).

This system is much more difficult and complicated to learn and use and confuses annual forage production (which fluctuates widely from year to year) with actual changes in plant composition (which are much slower to change) so that neither is tracked accurately and separately. For these reasons, the present system seems to be more complicated and time-consuming, less useful, no more accurate than the former--and the estimate of total forage production per acre is not used per se in estimating the carrying capacity.

For the past twenty years, the author has developed and used a third method, proven accurate in practice, that is fast and easy to learn and use in a year-long grazing area. It is based on estimating the total forage production per acre in the fall at or near the end of the growing season to determine the carrying capacity through the dormant season. The pounds of forage per acre may be estimated as an average per acre for the whole pasture. In either case the pounds per acre are multiplied by the number of pasture acres to determine the total forage available in the pasture.

To determine the amount of forage available for animal intake, the total forage present should be divided by four to allow for the amount already grazed during the growing season, the amount lost to weathering and trampling, and the amount that should be left for soil protection at the end of the grazing season before the following year's growth is initiated. To determine the amount of usable forage required per animal unit, multiply 30 lb/day (3% of body weight) by the number of days to be grazed (dormant season plus 30 to

45 days to allow for a late spring or a shorter period if livestock are to be rotated). If the pasture is being grazed yearlong, the winter stocking rate will approximately equal the yearlong stocking rate.

The eye of the technician or rancher can be "set" for visual estimates by clipping all the forage from a 21 in. radius circle. Weigh in grams, multiply by 10 to convert to pounds per acre; adjust for moisture content to determine total pounds of air-dry forage per acre. The moisture adjustment can be made by using SCS tables on forage type and stage of growth or by carefully heating the clippings (Avoid fire!) in the kitchen oven and adding 10% to get air-dry forage amount. Usually, only two or three clippings are necessary to condition the eye to recognize typical areas before proceeding with visual estimates. This procedure comes very naturally to those accustomed to estimating livestock weights (appendix C).

A fourth method for estimating carrying capacity involves using the total forage producer per year required to support an animal unit (but provide for forage loss and soil protection in areas of differing annual rainfall) and dividing by the actual total annual forage production per acre as determined by clipping or ocular estimate to determine acres per animal unit on a yearlong basis. This method was developed independently by Hershel Bell, H. L. Leithead, and a number of other workers based on their field observations and experience. The following table, published by the Texas Agricultural Extension Service, uses this method.

Forage requirement per animal unit of domestic livestock

Average annual rainfall	Annual forage requirement
More than 30 inches	30,000 pounds
20 to 30 inches	24,000 pounds
15 to 20 inches	20,000 pounds
Less than 15 inches	15,000 pounds

A fifth method of annually setting or adjusting stocking rate is used by Dick Whetsell on the Oklahoma Land and Cattle Company in the Osage area of northeastern Oklahoma--an area of 32 in. to 34 in. average annual rainfall. Since no one can predict future forage production accurately, he bases his adjustments on the previous year's forage production. In an average area where 8 acres are required per animal unit throughout the year, if the previous year has been above average in forage production, the allowance would be reduced by 1A to 7A. If another good year followed in succession, the allowance would be reduced to 6A but never below 6A. If that succeeding year had been average, the stocking rate would remain at 7A; if below, it would have returned to 8A. In years of reduced forage production, reductions would be made in 1A increments. To adjust this method for areas of less rainfall larger increments of adjustment would be needed. An area requiring 16 A/AU would require adjustment in 2A increments.

A sixth method uses the net-energy method to estimate carrying capacity by balancing animal energy needs with energy available in the amount of forage per acre with energy provided by supplements deducted to arrive at the number of acres required per animal (appendix D).

A seventh method, simple and widely used, is to make adjustments of stocking rate based on the degree of use and overall range trend. The present degree of use is the indicator of range trend, followed by plant vigor, production and reproduction, with litter, organic matter, ground cover, and livestock production following. A stable trend would indicate leaving livestock numbers as they are with concomitant additions of livestock if the trend is upward and decreased numbers if the trend is down.

Data for all of these methods can be obtained rather rapidly for the purpose of original inventory or subsequent monitoring by moving through each pasture and recording data in abbreviated form site by site and/or pasture by pasture on aerial photographs or other maps. Site can be noted by initials, as can classification of condition (E, G, F, or P), the amount of forage present in pounds per acre, current degree of use (L--light, M--medium, or H--heavy)and trend (+, 0, or - to represent upward, stable, or downward). An example map notation follows:

Fair condition		Moderate use
Deep upland site	DU/F/2500/M/+	
2500 pounds of forage		Upward trend

Calculations of an example by methods one and three might be done as follows:

North pasture - 328 acres   November 1 Observations

	AUYL by Forage		AUYL by Condition
28 ac.	S/F/1500/H/- =	42,000	@ 18 AU = 1.56
200 ac.	CL/G/2500/M/0 =	500,000	@ 10 AU = 20.0
100 ac.	BL/G/3000/M/+ =	300,000	@ 12 AU = 8.33
		842,000	29.89 = 30 AU
		÷ 4 = 210,500	
		÷ 5400 = 38.98 = 39 AU	

(30#/AU/DA x 180 DAS = 5400#/AU)

If your stocking rate the previous year had been 35 AU and this had been an average forage producing year with a total forage production of 3500 lb/A, a crosscheck by method four would indicate 38 AU and by methods five and seven would be

35 AU. With these calculations as background, you might confidently decide to stock on the basis of 37 AU in that pasture.

Before determining the number of domestic livestock to be grazed, be sure to deduct the animal unit equivalent of grazing wildlife, if numbers are significant. Species and classes of livestock should be chosen according to forage and water resources, topography, markets, facilities, and knowledge available. Stocking rate usually can be increased if rotation grazing and/or multiple species are used to increase efficiency.

Animal Unit Equivalents Guide

Kind and classes of animals	Animal-unit equivalent
Cow, dry	1.00
Cow, with calf	1.00
Bull, mature	1.25
Cattle, 1 year of age	.60
Cattle, 2 years of age	.80
Horse, mature	1.25
Sheep, mature	.20
Lamb, 1 year of age	.15
Goat, mature	.15
Kid, 1 year of age	.10
Deer, white-tailed, mature	.15
Deer, mule, mature	.20
Antelope, mature	.20
Bison, mature	1.00
Sheep, bighorn, mature	.20
Exotic species	(to be determined locally)

Stocking-rate calculations are exactly the same situation as calculating a feed ration. They are used to decide on a reasonable course of action to be followed by continuous close observation of results; in this case, the response of both forage and livestock are observed and the necessary adjustment made to achieve the desired results. No amount of calculation can take the place of good judgment, keen observation, and timely action, but used properly, they certainly can contribute to wiser decisions and less costly errors.

Since years are seldom alike in forage production, good ranchmen plan for adjustments in stocking rate to match forage available without wrecking breeding management, cash flow, and tax management. Understocking, overstocking, and even constant stocking at the same rate increases costs and reduces profitability. A constant, knowledgeable effort to stock properly will pay good dividends.

APPENDIX A

TECHNICAL GUIDE TO RANGE SITES AND CONDITION CLASSES (WORK UNIT--FORT WORTH)

KEY CLIMAX PLANTS	Plant Percent Allowable by Sites			INVADING PLANTS
	DU	RP	VS	
Little bluestem	-	-	-	Annuals
Big bluestem	-	-	-	Texas grama
Indiangrass	-	-		Tumblegrass
Switchgrass	-	-		Hairy tridens
Perennial wildrye	-			Scribner panicum
Sideoats grama	10	15	0	Fall witchgrass
Hairy dropseed		5	-	Halls panicum
Vine-mesquite	5	5		Sand dropseed
Meadow dropseed	5	15		Roemer senna
Silver bluestem	5	10	15	Gray goldaster
Texas wintergrass	10	5	5	Dyschoriste
Texas cupgrass	5	5		Buckwheat
Buffalograss	Inv.	5	10	Curly gumweed
White tridens	5	5		Mealycup sage
Seep muhly	Inv.	5		Nightshades
Tall hairy grama	Inv.	5	10	Milkweeds
Rough tridens		Inv.	15	Western ragweed
Perennial threeawn	Inv.	Inv.	10	Baldwin ironweed
Climax forbs	5	5	10	Texas stillingia
Woody canopy *	5	5	5	Prickly pear
				Mesquite
				Yucca
				Sumac
Maximum Total Allowable Increasers	20	30	35	

Range Condition	Safe Starting Stocking Rates - Ac./AUYL			% Climax Vegetation
Excellent	7	10	16	76-100
Good	9	12	20	51-75
Fair	14	16	28	26-50
Poor	20	23	35	0-25

LEGEND

Blank = Not significant
- = Decreaser: all allowed
5, etc. = Increaser: allowable percentage
Inv. = Invader

Range Sites and Major Soils:

DU= Deep Upland (GP) 2-San Saba clay; 2X-Denton, Krum clay; Lewisville clay loam.

RP= Rolling Prairie (GP) 18c-Denton-Tarrant complex; 18-Denton clay, shallow phase.

VS= Very Shallow (GP) 24c-Tarrant stony clay, unfractured substrata.

* Use open canopy method to determine percentage of all brush species. On savannah sites, determine range condition by the percentage composition of understory vegetation. To estimate available grazing, reduce the acreage in the site by the percent that climax brush exceeds the indicated allowable. On prairie sites, estimate percentage of invading brush species by loosely compacting the canopy to simulate total shading. The percentage figure thus obtained is "counted" against range condition.

APPENDIX B

Page 2
Clay Loam
Section II-E
November 13, 1972

RELATIVE PERCENTAGE

Grasses 90%		Woody 5%		Forbs 5%	
Little bluestem	35	Elm	5	Engelmanndaisy	5
Big bluestem	20	Hackberry		Maxmilian sunflower	
Indiangrass	15	Pecan		Yellow neptunia	
Switchgrass	5	Plum		Catclaw sensitive-briar	
Virginia and Canada wildrye		Liveoak	T	Prairie clovers	
Sideoats grama	10			Scurfpeas	
Texas wintergrass				Gaura	
Tall dropseed	5			Heath aster	
Vine mesquite				Trailing ratany	
Texas cupgrass				Blacksamson	
White tridens				Golden dalea	
Silver bluestem				Wildbeans	
Hairy grama				Tickclovers	
				Gayfeather	
				Prairie bluets	
				Bundleflower	

b. As retrogression occurs, big bluestem decreases rapidly followed by Indiangrass, little bluestem, and switchgrass. Sideoats grama, tall dropseed, and Texas wintergrass increase initially and then decrease as retrogression continues. Buffalograss, Texas grama, tumblegrass, red threeawn, western ragweed, Baldwin ironweed, queensdelight, mesquite, sumac, lotebush and common honeylocust invade the site.

c. Approximate total annual yield of this site in excellent condition ranges from 3000 pounds per acre in poor years to 6500 pounds per acre of air-dry vegetation in good years.

4. WILDLIFE NATIVE TO THE SITE: Dove and quail inhabit this site.

5. GUIDE TO INITIAL STOCKING RATE:

a. Condition Class

Condition Class	Climax Vegetation	Ac/AU/YL
Excellent	76-100	6-9
Good	51-75	8-12
Fair	26-50	10-16
Poor	0-25	14-20

b. Introduced Species

Species	Percent of the Area Established 100-76	75-51	50-26	25-0
King Range bluestem	8-10	10-14	14-18	18+
Common bermudagrass	7-10	10-13	13-18	18+
Kleingrass	7-10	10-13	13-18	18+

APPENDIX C

YIELD DETERMINATIONS

CIRCLES

RADIUS	UNIT OF WEIGHT	CALCULATION
3.725 ft. (3 ft. 8 3/4")	pounds	x 1000=lbs./ac.
2.945 ft. (2 ft.11 3/8")	ounces	x 100=lbs./ac.
1.75 ft. (1 ft. 9 in.)	grams	x 10=lbs./ac.

SQUARE

LENGTH OF SIDE	UNIT OF WEIGHT	CALCULATION
6.6 ft. (6 ft. 7 1/4")	pounds	x 1000=lbs./ac.
5.22 ft. (5 ft. 2 5/8")	ounces	x 100=lbs./ac.
3.1 ft. (3 ft. 1 1/4")	grams	x 10=lbs./ac.

Percentage of Air-Dry Matter in Harvested Plant Material at Various Stages of Growth

Grasses	Before heading; initial growth to boot stage	Headed out; boot stage to flowering	Seed ripe; leaf tips drying	Leaves dry; stems partly dry	Apparent dormancy
	Percent	Percent	Percent	Percent	Percent
Cool season.............	35	45	60	85	95
wheatgrasses					
perennial bromes					
bluegrasses					
prairie junegrass					
Warm-season					
Tall grasses..........	30	45	60	85	95
bluestems					
indiangrass					
switchgrass					
Mid grasses...........	40	55	65	90	95
side-oats grama					
tobosa					
galleta					
Short grasses.........	45	60	80	90	95
blue grama					
buffalograss					
short three-awns					

Trees	New leaf and twig growth until leaves are full size	Older and full-size green leaves	Green fruit	Dry fruit
	Percent	Percent	Percent	Percent
Evergreen coniferous.......	45	55	35	85
ponderosa pine, slash pine-longleaf pine				
Utah juniper				
rocky mountain juniper				
spruce				
Live oak..................	40	55	40	80
Deciduous.................	40	50	35	85
blackjack oak				
post oak				
hickory				

APPENDIX C (con't)

Percentage of Air-Dry Matter in Harvested Plant Material at Various Stages of Growth

Shrubs	New leaf and twig growth until leaves are full size	Older and full-size green leaves	Green fruit	Dry fruit
	Percent	Percent	Percent	Percent
Evergreen..................	55	65	35	85
big sagebrush				
bitterbrush				
ephedra				
algerita				
gallberry				
Deciduous..................	35	50	30	85
snowberry				
rabbitbrush				
snakeweed				
Gambel oak				
mesquite				
Yucca and yucca-like plants.....................	55	65	35	85
yucca				
sotol				
saw-palmetto				

Forbs	Initial growth to flowering	Flowering to seed maturity	Seed ripe; leaf tips dry	Leaves dry; stems drying	Dry
	Percent	Percent	Percent	Percent	Percent
Succulent..................	15	35	60	90	100
violet					
waterleaf					
buttercup					
bluebells					
onion, lilies					
Leafy......................	20	40	60	90	100
lupine					
lespedeza					
compassplant					
balsamroot					
tickclover					
Fibrous leaves or mat.......................	30	50	75	90	100
phlox					
mat eriogonum					
pussytoes					

Succulents	New growth pads and fruits	Older pads	Old growth in dry years
	Percent	Percent	Percent
pricklypear and barrel cactus......................	10	10	15+
cholla cactus................	20	25	30+

APPENDIX D

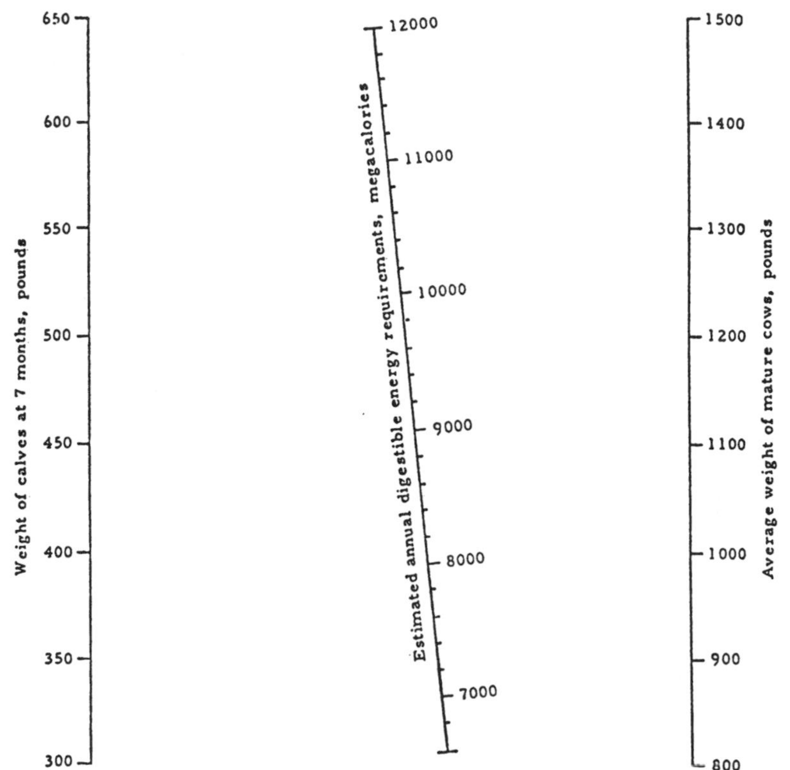

Figure 6. Estimating annual digestible energy requirements for a cow and her calf (2 miles travel).

			Example		Your herd 1000#	1300#
1*	Energy requirement for cow and calf (1,000 lb. cow—500 lb. calf)		8,817		8800	11300
2**	Minus energy in supplemental feed					
	(300 lb. CSC 1.3 megacalorie)	390				
	(150 lb. hay 0.9 megacalorie)	135		-390		
		525	−525		8410	10910
3	Energy needed from pasture forage		8,292			
4	Megacalorie per pound of forage		÷1.0			
5	Total pounds of air-dry forage required for each cow and calf		8,292		8410	10910
6	40 to 65% of the forage used by cow and calf (40 to 50% for western native pastures) (55 to 65% for eastern improved pastures)		÷.40			
7	Total pounds of air-dry forage required per cow		20,730		21025	27275
8	Pounds of air-dry forage produced per acre		1,300		2000	2000
9	Number of acres per cow and calf		.16		10.5	13.6ac

*Use your average cow weight and 7 month calf weight and determine the digestible energy requirements from figure 6.

**Use your planned supplemental feeding program to calculate this figure. If grain is fed, use 1.60 megacalories for this fraction.

Part 9

HEALTH, DISEASE, AND PARASITES

39

INFECTIOUS LIVESTOCK DISEASES: THEIR WORLDWIDE TOLL

Harry C. Mussman

By good fortune, plus nearly a century's worth of cooperation between stockmen and government, the United States has managed to wipe out or keep out the most serious of the world's livestock diseases.

Contagious bovine pleuropneumonia, introduced in the mid-1800s, was the first major disease we battled together and eradicated. Hog cholera was the last to be wiped out--just four years ago. In between these two efforts were nine outbreaks of foot-and-mouth disease, which has not been back since the last remains were buried in 1929.

These diseases and others still bring major losses and reduce livestock productivity in much of the world. This is particularly true in the developing nations of Africa, Asia, and Latin America. The popularly held notion that Third World livestock are infected with all of the most dreaded diseases is not true, but many of these countries do live with at least a few of them, often superimposed one on the other. Losses due to animal disease in some developing countries are estimated at 30% to 40% annually--twice as great as losses recorded by most industrialized nations.

Looking at meat and dairy productivity, developing nations produce one-fifth the beef and veal per animal that developed countries do, one-half the pork, one-half the eggs, and one-eighth the milk. North America and Europe are roughly four times more efficient in mutton, lamb, and goat meat production than is South America, and 25% more efficient than Africa.

The so-called Third World is hungry already, and is projected to host 90% of the global growth in human population expected by the year 2000. Obviously, greater and more reliable sources of protein must be found and developed, or many of these people will not survive. A diseased animal that may not survive can hardly be counted a productive resource. A virus or other disease agent can wipe out entire herds in a very short time--or, in a chronic state, can leave them debilitated, more expensive to feed, and capable of producing far less milk or meat.

Of all the factors involved with production, animal diseases have by far the most dramatic impact. Unfortunately, disease control is not improving rapidly enough in the Third World. And many of the most feared diseases--because of the tremendous economic toll they can take--are appearing in locations where they never existed before or reappearing in places where they had been eradicated years ago.

Some of these diseases are listed next.

Foot-and-Mouth Disease. Foot-and-mouth disease (FMD) is the most feared of all the animal diseases worldwide because of its ease and speed of spread and the susceptibility of all clovenhooved animals. It has long been established in livestock populations of the Mideast causing abortions, deaths, weight loss, mastitis, and lowered milk production. Europe has managed to keep these Asian strains at bay in Turkey, however.

Prior to World War II, FMD swept through Europe every 6 to 8 years, severely reducing livestock productivity. Finally in the 1950s, better quality vaccine was introduced and stemmed the spread of the epizootics. Outbreaks that occurred in the early 1960s, and again in 1973, were halted by teams from the United Nations working with the affected countries and regions. Then, in March 1981, FMD was confirmed in France, in Brittany, and near Cherbourg. Within a few days it broke out on the Isle of Jersey and the Isle of Wight in Great Britain. Just prior to that positive diagnosis came in from the south of France, from Portugal, and from Spain--outbreaks that most likely originated in Spain. Austria, free of FMD for six years, was infected during the winter of 1981 from imported Asian buffalo meat.

After 10 years of freedom from the disease and despite a vigilant quarantine program, Denmark discovered an outbreak of FMD on the Isle of Fyn just last March. The origin is unknown, but suspicions are that it came from one of the Eastern bloc countries where severe FMD outbreaks were reportedly occurring just prior to Denmark's infection. The loss Denmark has suffered from those outbreaks is estimated at almost a quarter of a billion dollars. Most of the loss has come from the cessation of fresh pork sales to the U.S., Canada, and Japan, although sixteen other countries were also compelled to embargo Danish meats and animals. The cost of eradication alone was also significant.

U.S. policy prohibits imports of fresh, frozen, or chilled meat products, or swine, or ruminants for a full year from any country that experiences an FMD outbreak. We had to refuse United Kingdom exports for a year following their 1981 outbreaks. We also refused farm products from across the Canadian border in 1952 following an outbreak there--which also curtailed big game hunting for U.S. hunters up north. (Certified "clean" animals, however, may be brought in from FMD-infected countries through the Harry S. Truman Animal Import Center on Fleming Key, Florida, after three months of high-security quarantine and testing at the importer's expense; or they may be shipped directly to an approved zoo.)

Some call this "politics," aimed at protecting the home meat industry. U.S. meat producers do gain a competitive edge from such an embargo, certainly, but that is not the aim. The productivity of more than 200 million fully susceptible cattle, swine, sheep, and goats is at stake. The collective worth of that stock is estimated at $24 billion --if they stay healthy. Healthy animals with their high-quality protein will keep America healthy, help alleviate worldwide hunger, and help keep the U.S. trade balance healthy.

If FMD were to invade the U.S. and enter a major market, it could spread to over a dozen states within 24 hours. By the end of the first year, direct losses would be an estimated $3.6 billion--indirect losses to allied industries and curtailed exports could climb to $10 billion.

Rinderpest and Peste de Petits Ruminants. Another serious, almost 100% fatal among cattle and resurging as a serious problem in Africa, is rinderpest. As with FMD, the United States (by law) cannot import cattle or other ruminants from rinderpest-infected countries (unless, as with FMD, they are brought in through the Truman Center on Fleming Key or are destined for an approved zoo.)

Rinderpest was essentially eliminated from Africa in 1975, following a decade of intensive and widespread vaccination. A team of international cooperators trudged from West to East Africa, vaccinating some 80 million head of cattle at a cost of $30 million. Unfortunately, though, the people became complacent once the disease had been suppressed. As the livestock populations were built up again, the young were not all vaccinated and were allowed contact with remaining carriers and wild animals.

So, the incidence is climbing again, especially in West Africa, and the virus is spreading east. Egypt is affected now, and the Arab Gulf States are reporting such frequent outbreaks that it appears the disease has become enzootic.

Peste de petits ruminants is a virus quite similar to rinderpest and readily infects goats and sheep in West Africa. It is actually classified as a strain of rinderpest that has lost its ability to infect cattle under natural conditions.

Rift Valley Fever. Another disease that has moved up from Kenya through the Sudan into the Sinai is Rift Valley fever. RVF primarily affects sheep but may also affect goats. It is also a serious human health problem. An epidemic in Egypt a few years ago killed nearly 600 people, along with a large number of livestock. The disease can spread rather easily. A human body incubating the virus could carry it anywhere in the world or a mosquito could carry it while hitchhiking on an airplane.

Israel has been vaccinating its animals against RVF for self-protection as well as to help curb further spread of the disease around the Eastern Mediterranean Basin.

Because of grave concern over Rift Valley fever, the U.S. Department of Agriculture has developed diagnostic capability of its own and is making preliminary vaccine studies. Work on possible domestic insect vectors is also planned.

African Swine Fever. Over the past 20 years, swine-producing nations have faced a rising risk of African swine fever infection. ASF has been established for many years in wart hog, bush hog, and giant forest pig populations in tropical Africa, south of the equator. Other animals may not have the same tolerance to the virus: in 1909 Europeans tried bringing domestic hogs into Africa, and the hogs were dead from the virus in a matter of a few days.

The first appearance of ASF in Europe was in Portugal in 1957 where the disease was at first mistaken for classic hog cholera and finally recognized as ASf. More than 16,000 pigs became infected before the virus was recognized as ASF and finally stamped out. In 1959, the disease was found in Spain, and in 1960 reappeared in Portugal. France was invaded by ASF three times in the 1960s and 1970s, though the French managed to eradicate it each time at an early stage--a tribute to their veterinary surveillance system.

Italy became infected in 1967 through the Rome airport. More than 100,000 swine were slaughtered before the disease was wiped out. Finally, about four years ago, Malta and Sardinia became infected. In Malta, the entire swine population was slaughtered to achieve eradication--at a cost of some $50 million (U.S.). Italy is still working on eradication of the disease on Sardinia.

The Western Hemisphere managed to avert an invasion of the dread ASF until 1971 when the virus hit Cuba. Approximately half a million pigs died or were destroyed before the disease was eradicated. Then, another Cuban outbreak occurred in 1980 and was successfully eradicated in about six weeks. ASF appeared in Brazil 5 years ago and is still there today. Positive diagnosis came from the Dominican Republic in 1978 and in Haiti early in 1979.

The U.S., Canada, and Mexico are particularly and acutely concerned about African swine fever in the Western Hemisphere. All three countries are free of the disease but are working together in the Dominican Republic toward eradication. The program used a hard strategy: they completely depopulated the country's swine farms after which they restocked with healthy, imported pigs. Not a trace of the virus has reappeared since the restocking was completed. The Haitian effort repeated the model program (and first of its kind in the West) completed in the adjacent Dominican Republic last year. Equal success is anticipated in Haiti.

Success is crucial to the three big cooperators. Mexico's Yucatan Peninsula is just a short hop across the

Caribbean from the Dominican Republic; Canada has trade interests all over Latin America and the Caribbean; and the threat of ASF lies right at the doorstep to the United States. The U.S. strictly prohibits swine imports from ASF-infected countries, but the virus could enter through contaminated food waste from ships or planes, through farm soil still clinging to a returning traveler's shoe, or as a result of illegal immigration. The investment in such an eradication program is sizable. Direct costs for the Dominican Republic--shared by the UN's Food and Agriculture Organization, the Inter-American Development Bank, and the United States--amounted to approximately $20 million. The bill for Haiti should run about the same.

The cost of chronically battling a disease like ASF can be far higher, however. Spain, for example, has spent roughly $315 million fighting it over the past 20 years. In addition, exports from Spain (along with Portugal and Brazil) have been restricted because of ASF-infected status.

Heartwater Disease. Latin America was jolted last year by the diagnosis of heartwater disease in a goat on Guadaloupe, an island in the eastern Caribbean. This was its first appearance in the Western Hemisphere and the implications are serious. We are cooperating with other countries in the region to make certain that it does not spread.

Heartwater is native to southern Africa and Madagascar and may be a major cause of animal deaths reported in West Africa. We have no simple and reliable means of diagnosing the disease; a blood test, especially, is badly needed.

Contagious Bovine Pleuropneumonia. Contagious bovine pleuropneumonia (CBPP) recently broke out in France for the first time in years. The disease may have been smouldering in the Basque region in Spain for some time. CBPP was the first major exotic disease to invade U.S. shores back in the 1800s. Cattle ranching had become so productive after the Civil War that a surplus of more than a quarter million head was available for export to England. When pleuropneumonia showed up in the animals, however, Britain closed off the market. That was the spark that set off the first animal disease eradication program in the United States.

Goat Diseases. In the United States evidence is accumulating of two possibly serious diseases of goats: contagious caprine pleuropneumonia and caprine arthritis encephalitis. CCPP is well established in much of the developing world--the regions that depend heavily on goats. The true impact of the disease is clouded as yet, largely because of common and widespread disagreement among diagnosticians. More research is needed to develop definitive diagnostic tests and effective vaccines.

Caprine arthritis encephalitis was only recently recognized in the United States and may exist yet undiscovered in other countries also. Research may give us some answers to this relatively new entity.

<u>Trypanosomiasis</u>. A full third of Africa--taking in most of the wide band of savannah and forest across the middle of the continent--is plagued by infestations of the tsetse fly. A blood parasite of animals as well as people, the tsetse fly transmits trypanosomiasis, or animal sleeping sickness, from animal to animal. If the tsetse-infested areas could be reclaimed for cattle production, it is estimated that annual meat production in Africa could double.

<u>Bacterial Disease, Intestinal Parasites, and Others</u>. The more dramatic virus diseases, such as FMD, ASF, rinderpest, bovine pleuropneumonia, and external parasites that transmit East Coast fever and trypanosomiasis, are the most highly visible of the livestock plagues worldwide.

Much more research remains to be done on bacterial diseases and intestinal parasites, particularly with small ruminants. Also more research is needed to understand the incidence and significance of the clostridial diseases--tetanus, anthrax, blackleg, and malignant edema--and paratuberculosis, and brucellosis. In the U.S., eleven states are free of cattle brucellosis now and another 25 have an extremely low incidence. Eradication of the disease requires only time. The damage done by different strains of the disease worldwide, however, is not entirely known. Since all strains can infect people as well as animals, the public health problem is always important. Anaplasmosis and babesiosis also are economically significant blood-parasite diseases in many tropical areas of the world, but the actual extent of the damage they do is not fully appreciated, much less known.

Another virus disease--one less dramatic in its toll, but gaining more and more visibility--is bluetongue. Bluetongue affects cattle, sheep, goats, and wild ruminants. It is particularly damaging to sheep, with a mortality rate running as high as 50% among affected animals; in cattle and goats, bluetongue lowers reproductive ability. Currently, U.S. cattle cannot be exported directly to Europe because of bluetongue in the South and southwestern states. Canada has a competitive edge over the United States in the European livestock trade--with the exception of the Holstein dairy breed--because the small gnat that carries the disease does not appear that far north.

Like caprine arthritis encephalitis in the United States, the possibility of diseases yet undiscovered still remains--diseases that may have been smouldering since the first goat herds on record were tended some 10,000 years B.C.

THE NEED FOR STRONGER CONTROLS

Proper vaccines, a solid border inspection and quarantine program, and accurate diagnosis could take care of many of the most severe disease problems in the world.

Vaccine

The alarming resurgence of rinderpest in Africa shows what can happen when a successful vaccination program breaks down. On the other hand, a laboratory opened in Mali just 5 years ago has been researching, tracing, and producing vaccine for several diseases rampant in West African livestock --rinderpest and bovine pleuropneumonia. The Central Veterinary Laboratory (CVL), as it is called, is also coordinating regional vaccination programs, and CVL veterinarians and technicians are being welcomed now by Malian herdsmen as they drive their cattle back and forth for the grazing season. The incidence of disease appears to be declining, at least locally.

Quarantine

Contributing to the rinderpest problem in Africa was the failure to keep sick animals from the healthy. Quite simply, quarantine can prevent the spread of many infectious diseases. For example, double fencing, as practiced in some parts of Africa, effectively protects domestic swine from African swine fever.

Border Inspection

Strict import controls over meat, feed, and animal products and a tight border guard are essential to keeping disease out. Years ago when most travel was by boat, a virus disease could probably not survive a transocean trip. Today a few hours on a jet plane are little threat to virus survival.

Customs and Agriculture inspectors at all 82 U.S. ports-of-entry inspect as much luggage and as many carry-on bags and travelers as possible. The goal of 100% inspection is a difficult one to achieve, however. International travel rose 8% in the United States last year, and 6% worldwide. Also more and more cargo is containerized every year.

Diagnosis

Research must proceed and develop simpler and more effective means of diagnosing heartwater disease, contagious caprine pleuropneumonia, and other diseases. Clinical signs of certain diseases are changing--and we must keep up with them. Genetic research must continue so that animal science can continue to breed for resistance to disease. Biological, chemical, ecological, and other means of controlling insect vectors are unfolding through field and laboratory trials.

CONCLUSION

Each country or region of the world will have its own formula for prioritizing the danger that different animal diseases present. Generally speaking, the most dangerous diseases are those that are spread not only by animal-to-animal contact (contagious diseases) but whose virus, mycoplasma, bacteria, or rickettsia may be transmitted by live vectors, inanimate fomites, or meat scraps. The most dangerous diseases do significant economic damage to producers and exporters and carry a human health hazard, such as brucellosis or Rift Valley fever.

Most important before the animal disease situation in the Third World can improve, veterinary services will have to be strengthened. Developing countries have nearly 50% of the world's total livestock population but less than 20% of the world's veterinary forces. A number of developing nations, in fact, have little or no organized veterinary structure at all. Certain nations in Africa, in fact, have only one or two veterinarians working the entire country; and in others they are often young, with little or no field experience. The expertise of technicians, if it can be found, is usually limited--gained through in-service training. Veterinary technicians have no reliable means of communication or transportation--yet national animal health policy decisions are based on their input. Thus, the global data on disease prevalence, productivity losses, and effectiveness of control measures on which many veterinary directorates rely is woefully deficient.

It becomes clear why diseases such as contagious bovine pleuropneumonia, foot-and-mouth, and rinderpest still exist and spread. With existing reliable diagnostic tests for both, as well as effective vaccines, eradication of CBPP and rinderpest would require little more than organization and finance, backed up by the government's strong commitment. With heavier investment in livestock health programs--and greater cooperation among nations--the great toll that diseases take can be curbed. The people <u>can</u> be fed.

40

IMPACT OF ANIMAL DISEASES IN WORLD TRADE

Harry C. Mussman

IMPACT OF ANIMAL DISEASES ON WORLD TRADE

Animal diseases are an important factor inhibiting world trade and hampering the free movement of both live animals and animal-derived products. The foot-and-mouth disease outbreak in Denmark in early 1982 provides a recent example of the impact of animal disease on trade. At one point, Denmark's export trade in meat and dairy exports was suffering badly--$7 million was lost each week because of the outbreak. The U.S.--as well as several other countries--placed Denmark on the list of countries from which animals and animal products cannot be imported until free status is regained. Overall costs of the outbreak will run in excess of one billion dollars.

Another example closer to home is our bluetongue situation. In 1980, the European Economic Community banned animal imports from the U.S. because this cattle and sheep disease was found in a portion of our country.

U.S. COMPETITIVE EXPORT POSITION

Despite the bluetongue situation and other domestic diseases such as brucellosis, tuberculosis, and leukosis, the U.S. export position remains competitive. In 1981, our animal and animal-product exports had a market value of $3.24 billion; and, in the same year, we had a $307 million trade surplus in these commodities, the first since 1977.

A U.S. animal export health certificate enjoys high credibility. One reason is that we have eradicated 12 major animal diseases that still plague many other countries of the world--diseases such as foot-and-mouth, rinderpest, hog cholera, and contagious bovine pleuropneumonia. Only a half-dozen or so other countries can claim freedom from these diseases. Largely because of our strict animal import health requirements and procedures, we are successful in keeping animal diseases out of our country.

IMPORT PROCEDURES

Livestock destined for this country must first be examined, tested, and certified by government veterinary officials as being healthy and meeting U.S. requirements in the country of origin. The foreign exporter must obtain health certification papers from his government and, in a majority of cases, an import permit in advance from us. At the U.S. port of entry the livestock must be examined again, this time by our veterinarians. This examination sometimes includes further testing and port-of-entry isolation, depending upon the kind of animal and the country it came from.

Although we are strict in our import controls, we try to meet the needs of American importers, particularly U.S. livestock breeders needing new bloodlines and exotic breeds of cattle. We recently opened a specialized import-quarantine facility--the Harry S. Truman Animal Import Center, at Fleming Key, Florida. Imported cattle are carefully tested and held there in quarantine for 3 months. They are mingled with a select "sentinel" group of susceptible U.S. cattle and swine to make sure they will present no disease threat to U.S. livestock.

Because some swine diseases, such as African swine fever, could devastate our swine herds, we do not accept swine imports from most of the world. Imports of sheep and goats are also severely limited because of the threat of scrapie, which has an incubation period of up to four years or more.

Restrictions on horse imports are generally less stringent than those required for swine, cattle, sheep, and goat. Still, incoming horses are tested for such diseases as dourine, glanders, equine infectious anemia (EIA), equine piroplasmosis, and contagious equine metritis (CEM).

Because of Venezuelan equine encephalomyelitis, horses from all Western Hemisphere countries--except Canada and Mexico--are quarantined for at least a week. Horses from countries known to have African horse sickness are quarantined for two months.

Contagious equine metritis is a recently identified disease of breeding horses. It has been found in parts of Europe, Japan, and Australia. As a result, horses cannot be freely imported from these countries.

Stallions and mares can be imported only after extensive treatment and negative culturing of the genitalia, both in the country of origin and again in the U.S. while under quarantine.

PROCEDURES FOR IMPORTS FROM CANADA AND MEXICO

The entry procedures for animals from Mexico and Canada are generally less strict than those for animals from overseas nations because Mexico and Canada have animal disease situations relatively similar to ours. The exception is hog cholera in Mexico, and for that reason we do not import swine from that country. In the case of other animals, however, entry quarantine and advance import permits are not required for either Mexico or Canada.

The U.S. has 16 crossing points on the Mexican border and 43 on the Canadian border where APHIS veterinarians examine and process animal imports. Entry procedures vary. For example, cattle from Mexico must be dipped in a pesticide solution as a precaution against cattle fever ticks and scabies.

ANIMAL SEMEN, EMBRYOS

Since animal semen is as much a potential disease threat as live animals, it must be subjected to strict standards for collection, handling, and shipping. These standards are spelled out in agreements between USDA and the foreign countries involved. While the technology for testing animal semen is well researched and established, the same cannot be said for embryos, the newest practical method for exporting animals. More research is needed to determine the diseases to which embryos are immune.

Our restrictions on animal imports may seem extreme to some. But we have a major responsibility for maintaining the health of our livestock--for example, a $9 billion swine industry, a $35 billion cattle industry, and a $10 billion poultry industry. Considering what is at stake, our restrictions are reasonable.

EXPORT PROCEDURES

Just as we insist that only healthy livestock be imported into this country, we have an obligation to see that only healthy animals are exported to other countries.

We ensure the health status of our exported animals in two ways. First, we establish our own health rules for exports; second, we cooperate fully in meeting the import rules of receiving nations.

All animals we export are subjected to special testing and certification requirements to indicate freedom from certain diseases found in the United States: bluetongue for cattle, sheep, and goats; brucellosis for cattle and goats; anaplasmosis for cattle; equine infectious anemia for horses; and pseudorabies for swine.

Our own export rules are necessary because we do have some health problems, i.e., EEC bluetongue of concern to foreign importers. We want to avoid damaging our position in the world market by making sure the animals and animal products we export are free of disease.

The animal health requirements of foreign countries vary, reflecting the particular animal disease problems and danger they face in their part of the world. Testing and certification are performed by private veterinarians accredited by us (usually at the shipper's expense) who normally conduct their tests on the farm or cattle ranch. Adult dairy and breeding cattle and goats must be tested and found free of brucellosis and tuberculosis within specified time limits before shipment. An animal health certificate is endorsed by the APHIS area veterinarian after the private accredited veterinarian completes his testing. The endorsement certifies that the private veterinarian is qualified to conduct the examination and tests.

Once the certificate is endorsed, livestock can move to a port of embarkation. There they must rest for 5 hours at a USDA-approved facility while the animals and paperwork are given a final check by an APHIS veterinarian. If the animals are healthy, if they are properly identified, and if the health certificate is in order, they are loaded on a ship or aircraft under APHIS supervision for export.

COMPLEXITY OF FOREIGN REQUIREMENTS

The health requirements imposed by foreign governments can be quite complex, so whenever possible we work out agreements with these countries. We do all we can to negotiate standard requirements, but we have over 150 agreements with some 70 foreign governments, and they can be changed on short notice. It is nearly impossible for an exporter or an examining veterinarian to keep track of the different rules for every overseas livestock shipment. To avoid delay and frustration, the exporter and the accredited veterinarian are urged to contact the APHIS Veterinary Services office in their state before attempting the process livestock or animal products.

Each APHIS Veterinary Services office keeps a current file of foreign animal import health requirements. Each area office of APHIS Veterinary Services also has a veterinarian assigned to work with exporters and their veterinarians. His job is to check the export health tests and certifications and place the final endorsements on the health papers before the livestock can leave this country.

CANADIAN, MEXICAN EXPORTS

As with our imports, our exports to Canada and Mexico are handled more simply and quickly than those to overseas nations. Exported livestock do not need an APHIS veterinary examination at the port of export. Once the health tests, certification, and APHIS endorsements are completed, the animals move directly to the border where they are examined by the Canadian or Mexican officials. Canada and Mexico are by far our biggest customers for exported livestock (and poultry as well), so it is important to devote some special attention to health matters on shipments across our borders.

Exports to Mexico move with few special problems. However, Canada has some requirements that exceed our own export rules. The Canadians are particularly concerned about chemical residues in cattle shipped to slaughter, so feed additives and antibiotics should be withdrawn from livestock within the recommended time limits. This is the responsibility of the exporter.

CLOSE COORDINATION REQUIRED

Successful U.S. exports, particularly those to overseas nations, require close coordination between the exporter, the private veterinarian, APHIS officials, the forwarder, the broker, the insurance underwriter, and the carrier. Among the most common causes of costly delay is the failure to conduct all required tests and failure to allow enough time for completion of the tests at the diagnostic laboratory.

Even if a plane or ship is waiting, animals cannot move to the port of embarkation unless APHIS endorses the health papers. Therefore, exporters should be aware of all the requirements and plan to allow sufficient time for testing when making plans to ship animals to another country.

INTERNATIONAL INVOLVEMENT

The U.S., along with its major partners, is actively involved in international organizations such as the Food and Agriculture Organization (FAO), the Office of International Epizootics (OIE), and the General Agreement on Tariff and Trade (GATT) in dealing with animal diseases worldwide and taking steps to assure the expeditious movement of healthy livestock and livestock products. We are actively involved in international health programs because the more we can do to reduce diseases worldwide, the more freely our own animals will move in international markets--an advantage of the U.S. exporter.

CONCLUSION

APHIS does all it can to help the stockman with his exports. We safeguard his markets by making sure no diseased animal gets out; and we make sure no diseased animal gets in to infect his livestock. We are against the unduly restrictive animal health import requirement that functions as a nontariff barrier and, more often than not, serves to protect a country's livestock industry--more from foreign competition than from foreign animal diseases.

The U.S. strongly supports "free trade" and endeavors, whenever possible, to make it a dominant principle in world trade, which includes trade in live animals and animal products. This is compatible with our free-enterprise tradition. If we adhere to this principle and keep our disease defenses strong, our livestock should remain the healthiest in the world. And export opportunities for U.S. stockmen should expand as world trade in animals expands.

41
HOW TO TELL WHICH IS THE LAME LEG

William C. McMullen

Early recognition of lameness is vital to the success of treatment in many cases. If you are sharp enough to notice the early pain signs of tendon fibers being torn faster than they are being replaced, you can reduce the stress and load on the tendon by reducing the amount and frequency of exercise. This would prevent a full-blown, incapacitating "bowed tendon," and a possible 6-months or longer layoff.

You must be alert to such signs when looking at a standing or moving horse; you want to know if he is normal or if he's "off." How do you tell if he is "off?"

Let's look at the horse in the stall or crosstie. A horse that is lame in the foot will try to place the foot so that direct pressure is less and it doesn't hurt as much. Consequently, he stands with the foot placed forward of the normal vertical position. This is called pointing. Horses with severe navicular disease or a coffin-bone fracture are good examples. Keep in mind that horses with a subtle lameness may stand normally and only show signs at a trot.

A horse that is very sore in the knee and in back of the fetlock or heel will stand with the knee forward and the heel raised off the ground. If he is sore in the shoulder, he may place the foot and limb back of the normal vertical position. It is normal for a horse to shift his weight from one limb to the other, just like we do if just standing around talking. But if, while idling, he rests the same rear limb, then suspect a sore rear limb.

If the horse has an abscess, fracture, joint infection, severe strains or sprains, the way he stands should lead you to the lame leg that should be examined more closely by visual inspection, palpation, and manipulation.

What about the horse with a mild lameness, who is "just a little off?" To make a subtle lameness more apparent, I like to see the horse move at a slow trot on a hard, smooth surface, preferably a nonslip rough-textured concrete or asphalt. He should be wearing a halter, not a bridle. If he does not work well on a longe line, have someone trot him on as loose a shank as his temperament will allow. This is especially important in front-leg problems because a horse will tell you by the nod of his head which leg hurts. With rear-leg lameness, however, head nod is not evident unless the lameness is severe. What is head nod? Normally, if a horse is trotting alongside a fence the same height as his head, the top of the fence and his head will stay even, but if he is lame, his head drops down several inches lower than the fence every other beat or each time the sound front foot hits the ground. The distance the head goes down is proportionate to the severity of the lameness and can vary from barely perceptible to 18 in. or more. When the lame front foot hits the ground, the horse's head comes up higher than the fence as he tries to reduce the painful impact by counterbalancing with the weight of his head. To observe for lameness, stand 25 ft away so that you can see the horse's head and feet at the same time. If the horse's nod is pretty consistent, synchronize your head motion with the nods of the horse and then watch to see which front foot hits the ground when your head goes down. That will be his sound front and the other front foot is the lame one. On hard surfaces, you may get a clue by listening to the hooves hit; the sound foot will hit harder and make a louder noise if he is shod or barefooted on all fours.

To determine lameness, have the horse trot in circles both to the right (clockwise) and to the left (counterclockwise). Ninety-five percent of the time lameness shows up better (more head nod) if the lame leg is on the inside of the circle. A horse that is only slightly lame will not show anything trotting straightaway down the road but will show a lameness when he is trotted in circles.

A horse that is equally lame in each front foot or leg shows a shortened length of stride and a choppy gait. He tries to hit the ground easily (egg walking, ice walking). Sometimes a horse will show head nod for left-front lameness if in a left circle and then head nod for a right-front lameness when reversed. In a few cases, I resort to making one foot numb with local anesthetic. If the horse quits limping, I know I've deadened the sore spot.

One other thing that helps to determine either a front or rear lame leg (especially if pain is coming from the fetlock), is to see how much the lower back part of the fetlock drops on impact. If it hurts, he will try to keep it from moving and the affected fetlock will not drop as far down when compared to the descent of the fetlock of the opposite leg.

Lameness in a rear limb is a different situation and is always more difficult for beginners to notice. Head nod is not helpful in rear-leg lameness. When a horse is lame enough in a rear leg to have a head nod, the affected limb already has an obviously swollen hock or a bad wound.

So what do you look for to detect rear-limb lameness before it is severe? Have the horse trotted away from you, view from behind; the sound rear leg pushes up and forward harder than the lame one. Still from behind, watch the croup muscles on either side as he trots. In a normal sound horse these muscles move up and down rhythmically like 2 pistons, bunching up equally first on one side and then the other as each leg does its equal share in propelling the body forward. If he is lame in the right rear, the croup (gluteal) muscles on the left sound side will bunch up higher than do those on the right. Next, watch him trot--viewed from one side. In most cases, he will show a shorter stride in the lame rear foot.

Because joint pain is increased when joints are flexed, a horse lame in the hock or stifle (knee or shoulder in front) will try to reduce the pain by decreasing joint flexion. When these joints don't bend enough, the foot is not lifted up as high and does not clear the ground as well and the horse may drag his toe(s) and/or stumble. Do not confuse lameness with the toe drag of a lazy, uncollected horse or one that badly needs a foot trim. Assessing a rear-leg lameness while the horse trots in a circle is more difficult, but a subtle problem may not show otherwise.

Flexion tests on a slightly lame horse may identify the problem. For the rear leg, lift the suspect leg (for instance, the right rear) from the ground as you stand close to his right flank facing the rear. Flex (bend) the hock as tightly as possible by clasping both hands around the cannon and pulling it up so that the cannon is horizontal to the ground. Hold this position for 90 seconds and have the person at the halter jog him off at the same instant you drop the leg. If he is lame in any of the joints in the leg, he should show an increase in lameness for the first 4 or 5 strides. This hock flexion or "spavin test" is void if the horse leaps off at a lope or puts his foot down and stands about 2 seconds before jogging off.

This same technique can be use defectively on the knee (front leg) or fetlock. The knee can be folded completely so that the back of the cannon is up against the back of the forearm. A fetlock flexion test will more specifically incriminate the fetlock if the hoof is lifted only slightly off the ground. With one hand on the front of the hoof wall, push the bulbs of the heel up toward the fetlock until it is tight. Hold it tight for 45 seconds before releasing. Immediately after releasing have the horse jogged off.

To test the shoulder, grasp the leg around the knee and lift up and forward as high as possible. All horses resent this, so do the same thing on the opposite leg for comparison. Then grasp the leg around the pastern and pull to the rear trying to touch the toe to the stifle. No normal horse will allow you to do it but he should let you get to within 12 inches of touching without showing pain. While the foot is close to the stifle, have someone press his knuckles firmly at the point of the shoulder from the front. This puts pressure on a bursa (bursitis) at the shoulder.

Tapping with a hammer on the walls and soles of the feet or squeezing with hoof testers or an oversized "Channel lock" pliers often will produce a diagnostic pain response if the horse is foundered or has a sole abscess or coffin-bone fracture. If you wonder if his reaction on a suspect foot is significant, use the same pressure on the other foot or sometimes a rear foot (if a front foot is suspect) for comparison.

Visual observation of swelling or atrophy (shrinkage) should be at a distance of 8 ft to 10 ft. In doing this, compare the suspect area with the same spot on an opposite leg. If you detect swelling, touch it with the backs of your fingers to see if it is warmer than normal. Apply firm digital pressure both when the limb is bearing weight and when off the ground to detect tender areas. Repeat pressure on the other leg to be sure the horse is not "just goosey." A knowledge of anatomy and knowing just where to put pressure on tendons, ligaments, and muscles, and not on nerves, is very helpful.

As with any medical problem, a history of what has happened up to the time of the appearance of lameness is equally as important as the examination. In general, I like to know for what events the horse is used. For instance, a jumper is much more likely to have back problems than is a western pleasure horse or most any kind of horse. In some cases, you have to put him through his routine (take some jumps, gallop a quarter of a mile at speed, or do whatever it is that the horse does) before signs of lameness will be recognizable. Quick stops and working downhill usually make hock and stifle problems more apparent.

An important part of both the history and the examination involves shoeing. If a lameness appears within 1 to 7 days after shoeing, each nail should be pressed with hoof testers to determine if it has been driven into the sensitive lamina ("quicked") or if a nail is right up against the sensitive lamina ("close nailed"). A good horseshoer is hardly ever guilty of this, but some of those thin-walled jumpy critters really make it difficult. Checking for a problem horseshoe nail is simple and quick. A horse with flattened soles (normal is concave) is easily "pressured" by a regular shoe if trimmed down a routine amount. The soles will get sore after the next workout. This can be prevented by using a hollow ground shoe that is beveled toward the center on the sole side.

If a horse gets lame several days after shoeing, I always ask if shoeing was changed in any way--the heel raised or lowered more than 2° to 4°, or raised on the inside to make him move better. In other words, was there some type of corrective shoeing? If a change is dramatic, new stresses result and the new pull on ligaments and joint capsules causes pain.

In getting the horse's history, I ask if the horse has had any kind of shots and where. An abscess in a muscle can cause a lameness just as easily as can a sole abscess. A little detective work often pays off in discovering lameness!

As a veterinarian, after I decide which leg (or legs) is involved, I examine it by palpation, visual observation, and manipulation. If I can determine the specific area that is the origin of the soreness, I will usually x-ray the part. If the specific area cannot be determined for sure, I usually start at the heel and numb it out with a local anesthetic (lidocaine or mepivacaine) injected over a nerve. If the horse quits limping after the injection, I know that the deadened area is the sore spot. If he doesn't quit limping, I inject the nerve higher up the leg and numb out his entire foot--and so on up the leg. In some cases, joints are injected with anesthetic to see if the pain will disappear or at least be significantly relived. When the pain is blocked, then I x-ray that part.

Several new but expensive techniques are being used to help in lameness work like the thermovision machine that makes a color photo showing hot or cold spots on a leg; or diagnostic ultrasound that can be used to detect hairline fractures; or radioactive uptake to spot bone lesions; or electromyograph to evaluate nerve and muscle function. These are reserved for the extremely difficult cases and, for the most part, are available only at universities.

The point of this paper has been to help you recognize pain as an early warning sign from your horse so that he can get rest and the proper treatment to prevent irreparable damage. I hope it will help in dealing with any such problems you may encounter in the future.

42

BIOLOGY AND CONTROL OF INSECT PESTS OF HORSES

R. O. Drummond

Although virtually no horses are used today in U.S. farming and there are considerably fewer in ranching, the total number of horses has nevertheless increased dramatically in recent years. This increase has occurred solely among "pleasure" horses, a category that includes racing horses, special breeds, and ordinary pets. Often these horses are housed close to human dwellings, and the flies and other pests found around horses can become a nuisance and a concern to humans.

Horses are parasitized by many of the same insect pests that attack cattle; in addition, they have several ectoparasites not found on other livestock. These pests include flies, bots, lice, ticks, and mites that bite, irritate, annoy, injure, transmit disease to, suck blood from, and detract from the appearance of horses. Most insect pests of horses can be controlled by the proper use of safe and effective insecticides (and other practices) that will not be hazardous or dangerous to horses, the applicators, or the environment, or create illegal residues of insecticides in horsemeat. This article contains information about the biology of a number of pests of horses, describes accepted techniques for their control, and lists precautions for the safe use of insecticides on horses.

EXTERNAL PARASITES

Flies

Horses are parasitized by a variety of flies, some of which are bloodsucking and commonly called "biting" flies (stable flies, horse flies, deer flies, mosquitoes, and gnats); some of which are nonbiting (house flies, face flies, blow flies, and screwworm flies); and some of which are nonfeeding (bot flies).

Biting flies. The stable fly, Stomoxys calcitrans (L.), the most common biting fly of horses, is prevalent around stables and other areas where horses are held. Adult flies, often called "dog flies," are vicious biters that are the size of a house fly, and they visit horses and other large animals (or often humans) one or more times a day to fill themselves with blood. The rest of the time they rest on walls, rafters, ceilings, and other structures that house horses. Female stable flies lay eggs on moist decayng matter, especially manure-contaminated hay and feed, and maggots develop.

The first and most important steps in control of the stable fly are the sanitary practices of removing spoiled hay and feed and spreading manure and other organic matter out on fields to dry. Dry material is not attractive to females for laying eggs, and it does not support growth of larvae. Also, breeding materials can be treated with sprays or granules of insecticides. Adult stable flies can be controlled by spraying resting surfaces with long-lasting residual insecticides to kill stable flies when they rest on the treated surfaces. Insecticidal baits used to attract and kill house flies are not effective against stable flies. Insecticides and repellents are applied as sprays, washes, wipes, aerosols, and mists to kill or repel stable flies on horses. Since stable flies prefer to feed on the legs and lower extremities of horses, these areas should be treated thoroughly. Usually the effectiveness of such treatments is short-lived, and horses must be retreated at short intervals if they are to be protected against attack by stable flies.

Horse flies, deer flies, mosquitoes, gnats, black flies, and a number of other biting flies that attack horses are especially difficult to control. Occasionally mosquitoes, gnats, and black flies attack in such massive numbers that they interfere with breathing, and a horse may suffocate. Horn flies are easily killed, and effective treatments may last for a week or more. Horse flies often inflict large and presumably painful feeding lesions on horses. Usually horse flies and other biting flies attack for only brief periods of the year, but one or more species may be active from spring to fall. Insecticides and repellents can be applied to horses to protect them from horse flies, but, as with treatments for control of stable flies, the period of protection is usually very short.

Nonbiting flies. The face fly, Musca autumnalis (De Geer), a recent introduction into the U.S., is of particular interest to those caring for horses because of its annoying habit of landing on the mouth and eyes of horses. The house fly-sized adults are attracted to the moisture around the eyes and nostrils and may gather in large numbers--25-100 flies/face. Horses continually fight these flies and thus do not graze normally. Female face flies lay eggs on and larvae develop in freshly dropped cow manure.

Control of face flies is difficult because insecticides do not readily adhere to the moist areas and short hair on the face of the horse. Frequent applications of insecticidal dusts, oils, smears, or wipes may give some control for short periods of time, but highly effective, long-lasting measures are not yet available.

The house fly, Musca domestica (L.), does not suck blood but may be found on horses. House flies may be a problem to horse owners because they can breed in large numbers in horse manure and decaying organic matter around stables. The flies can become a nuisance to people who live near stables.

As with stable flies, sanitary practices are the most effective method of control of house flies. Wastes around stables should be removed to eliminate breeding sites. In addition to treating breeding sites with insecticides, buildings can be sprayed to kill house flies when they rest on treated surfaces. Insecticide-treated baits also are used to control house flies. Insecticides applied directly to horses may provide short-term control.

The screwworm fly, Cochliomyia hominivorax (Coquerel), has been eradicated from the U.S. and is currently the subject of a highly successful eradication campaign in Mexico. Although screwworm are eradicated from the U.S., horse owners along the Mexico border should continually examine their horses for wounds. If a wound contains maggots, some should be collected and sent to the Screwworm Eradication Program, P.O. Box 969, Mission, Texas 78572, for identification. All wounds should be treated with an effective screwworm spray, smear, or dust of insecticide to kill maggots and protect the wound from reinfestation.

Horses may have infestations of other types of larvae in wounds, but these larvae are usually secondary invaders that live on dead and decaying flesh and do not tear and destroy flesh as does the primary screwworm. Such larvae should be controlled by treating wounds with insecticidal sprays, dusts, or smears.

Nonfeeding flies. Horses are infested by 3 species of bot flies: the horse bot fly, Gasterophilus intestinalis (De Geer); the throat bot fly, G. nasalis (L.); and the nose bot fly, G. haemorrhoidalis. Although each species has a slightly different life cycle, all have certain elements in common: The nonfeeding adult females usually dart in front of horses and attach eggs to hair on the forelegs, heart-breast area, throat, jaws, or lips. Larvae hatch from these eggs, enter the horse's mouth, and either live in pockets in the gums near the teeth or burrow into the tongue; later they emerge from these sites into the mouth area and are swallowed by the horse. Larvae then attach by means of strong, recurved mouth hooks to the wall of the horse's stomach or small intestine. When larvae are fully grown (about 3/4 to 1 inch long), they detach and are passed out with the manure. Some G. haemorrhoidalis larvae may re-attach around the anus and appear to be hemorrhoids. The full-grown larvae pupate in the manure or soil. After 2 to 4 weeks the nonfeeding adult flies emerge from the pupae.

The egg-laying activities of bot flies may frighten horses, which may injure themselves and their riders. Also, the small larvae in the gums, mouth, and tongue produce irritation and damage of these tender parts. The greatest damage is done by the bots attached to the wall of the stomach or intestine. The wall of the stomach may be perforated, and digestion may be impaired. Infested animals may show signs of colic or other gastric disturbances.

Control of bot flies may include removing eggs from hair or encouraging to hatch prematurely by sponging the horse's skin and the eggs with warm (at least 120° F) water. However, horses are more commonly treated for bots with drugs and insecticides administered orally. Some treatments must be carefully administered under the supervision of a veterinarian; others can be added to the horse's feed for one day so the horse treats itself. Effective treatments cause bots to detach from the stomach and intestine and pass out with the manure. Treatments may have to be administered several times during the warm months when flies are active, but one treatment after frost in the fall should keep animals free of bots during the winter.

Lice

Horses are infested with 2 species of lice: the horse biting louse, Bovicola equi (Denny), and the horse sucking louse, Haematopinus asini (L). Lice live constantly on the horse where they either suck blood or feed on skin scales, debris, or hair. Female lice attach eggs (called nits) to hair. Infestations of lice cause irritation, scratching, rubbing, and biting of horses--often patches of hair are rubbed off. Lice are more prevalent in the winter than in the summer; in fact, many lice are lost from a horse when it sheds its winter hair coat.

Control of lice on horses is usually accomplished by thoroughly dipping, spraying, or dusting the horse with an approved insecticide. A second application may be necessary to kill lice that hatched from eggs surviving the first treatment.

Ticks

Horses that graze in pastures are attacked by a variety of ticks. These 8-legged relatives of spiders may do much harm to horses because ticks suck blood, transmit diseases, cause paralysis, and create unthrifty and unsightly horses as a result of massive infestations and constant irritation ("tick worry").

Soft ticks. Horses are infested by only one species of "soft ticks," a large group of ticks characterized by the wrinkled, leathery texture of their "skin." This species is the spinose ear tick, Otobius megnini (Duges), which lives deep in the ears of horses and other animals. Adults of this species are free living (they do not feed) and are found in protected places such as cracks and crevices of stables and fences, under salt troughs, etc. Females are bred and lay eggs in these protected places. The tiny 6-legged larvae hatch from eggs, seek horses, attach deep in the animals' ears, feed for a short period, and molt to the spiny-appearing nymph that may feed for several months. This species may be found on horses in most states but is common in the southwestern states.

Spinose ear ticks are usually controlled by thoroughly treating the ears of horses with insecticide dusts, low-pressure sprays, aerosols, and smears. Effective treatments will provide adequate control for a month or longer. Attempts to control the adult ticks in the environment are generally unsuccessful.

Hard ticks. Most ticks in the U.S. have a hard covering on all or part of the back and thus are called "hard" ticks. These ticks are 1-host or 3-host ticks. The 1-host tick attaches to a host as a larva (or seed tick), engorges on blood, molts to the next stage (the nymph), engorges as a nymph, and then molts to become the adult male or female. The adults mate on the host; the females completely fill with blood, detach, drop to the ground, find a secluded spot, and lay eggs from which larvae will hatch. All the molting and engorging therefore takes place on a single host. The 3-host tick feeds on a host as a larva, drops to the ground when fully fed, and molts on the ground to the nymph. The nymph finds another host, engorges fully, drops off the host, and molts on the ground to become the adult male or female. The adults find a third host and mate; then the female engorges, drops off, and lays eggs.

Each region of the U.S. has a separate group of hard ticks that attack horses. The Pacific Coast tick, Dermacentor occidentalis (Marx), is found on the Pacific Coast west of the coastal mountains. The Rocky Mountain wood tick, D. andersoni (Stiles), is generally distributed in the northern Rocky Mountain States. The American dog tick, D. variabilis (Say), is found distributed over the eastern half of the U.S. The blacklegged tick, Ixodes scapularis (Say), is found in the southcentral U.S. The lone star tick, Amblyomma americanum (L.), is found throughout the southeastern one-third of the U.S. The Gulf Coast tick, Amblyomma maculatum (Koch), is limited to the South Atlantic and Gulf Coast States though large populations are found in eastern Oklahoma and surrounding states.

Two species of ticks are of special interest to horse owners. The winter tick, D. albipictus (Packard), is found on horses in the northern tier of states and as far south as Texas. Large infestations of this 1-host species that appear on horses in the winter when feed is scarce can cause an unthrifty condition in horses called "water belly." Engorged females drop off horses in the winter and lay egg masses (as many as 5000 eggs/mass). Larvae hatch from eggs and then enter a quiescent state throughout the spring and summer. In the fall, they become active and attach to horses, cattle, and other large animals.

The tropical horse tick, Anocentor nitens (Neumann), is another 1-host species whose primary host is the horse. This tick is limited in its distribution in the U.S. to Alabama, Georgia, Florida, and Texas. It is of special interest in the southeast because it transmits equine piroplasmosis, a protozoan blood disease of horses. All stages of this tick may be found deep in the ears of horses, and some ticks may be found in the false nostrils of horses, but with heavily infested horses, some ticks may be found in the mane, tail, and other parts of the body. Because all stages of this tick attach in the ears and false nostrils of horses, it is necessary to treat those areas with insecticides in oil in order to achieve good control.

In general, control of hard ticks on horses is accomplished by the application of insecticides to horses by dipping, spraying, or dusting the animals thoroughly. Also, insecticides may be sponged or wiped on horses, though thorough treatment is necessary to kill the small ticks.

Because some ticks spend considerable periods off the host and thus are found in the grass, soil, or debris in pastures and around stables, these areas can be treated with approved insecticides for control of ticks.

Mites

Horses are infrequently infested with several species of mange, itch, or scab mites that live on or burrow into the skin of horses and cause considerable irritation, itching, thickening of the skin, and loss of hair. Infested horses should be treated thoroughly with dips or sprays of insecticides. Several treatments may be necessary to kill the infestation so the lesions can heal.

USE OF INSECTICIDES ON HORSES

Treatments

The first line of defense against most insect pests of horses is the application of insecticides to the horses. Since insecticides should be used when they have maximum effectiveness, the horse owner must know which pest is attacking his horse. It is advisable to collect specimens of the pests and show them to a local county agent, agricultural advisor, or similar official who has the specialized knowledge to make the identification. These officials also have the latest information concerning the seasonal appearance of pests, species found in a given location, officially approved methods and materials available for their control, recommended times of treatment, and a variety of facts about local pests, their biology, and their control.

Special care should be exercised when treating horses. Most animals react violently to the noise and action of spray dispensed by high-pressure sprayers. Therefore, low-pressure, handpumped sprayers are best suited for application of sprays to horses. Horses are especially sensitive to the application of sprays and aerosols about the head. It is a good practice to confine the horse in a chute or other area to make sure that it will not injure itself or people if it reacts strongly to the application. Horses have a very sensitive skin and may react dermally to the solvents and other ingredients in an insecticide formulation. Therefore, it is <u>absolutely essential</u> that horses be treated only with those formulations approved for application to horses. Other formulations may cause burning, blistering, or cracking of the skin, and may be detrimental to the appearance and well-being of the animals.

Precautions

The insecticides that can be used to kill insect pests of horses can also be toxic to the horses and to the humans who apply them. In addition, these insecticides can create illegal residues in tissues of treated horses and can be destructive to the environment if they are not used and handled in a safe and correct manner. The following are a few precautions to follow when choosing and applying insecticides for the control of insect pests of horses:

- Use only those insecticides recommended and approved for use on horses by a recognized authority, usually a government official such as an agricultural agent or advisor.
- Use a formulation of the insecticide that is approved and designed for use on horses. In dipping vats, use only those formulations designed specifically for dipping vats.
- Follow the label directions <u>exactly</u>. The label contains all the information about dilution, time of retreatment of animals, antidotes for poisoning, methods of disposing of unused insecticide, and other important facts.
- Avoid treating horses in cold, stormy weather, and avoid treating stressed, overheated, or sick animals.
- Be sure that spraying equipment is clean, working properly, and provides sufficient agitation to allow for thorough mixing of insecticides.
- Be aware of safe practices when mixing and applying insecticides. Wear protective clothing; do not smoke, drink, or eat while applying insecticides. Do not contaminate feed or water troughs.
- Learn to recognize signs of insecticide poisoning in livestock (and humans) to avoid delaying antidotal measures.

- Store all insecticides in original containers. Do not store insecticides with food or where they can be reached by children, animals, or unauthorized persons.
- Avoid contamination of the environment: dispose correctly of all containers, unused concentrate, and used diluted insecticides.

43
EQUINE HEALTH MAINTENANCE PROGRAMS

R. Gene White

Preventive medicine is not a new concept in veterinary medicine. Large-animal veterinarians have applied preventive health programs in the food-animal industry for years. Such programs have proved to be very profitable for the cow-calf feedlot and swine industries and poultry producers have used this concept very effectively.

All segments of the horse population have grown rapidly in the recent years. Along with this has come increased numbers of brood-mare farms, pleasure-horse owners, and training stables. Horse ownership brings with it a responsibility, both in terms of the economics of the investment as well as getting the most out of the horses for the dollars invested. Owners should protect this investment with a preventive medicine program by working with a veterinarian who understands the goals and objectives of each particular operation. The veterinarian may be called to see the pleasure horse where there is only one horse involved or to the breeding farms and to the racing stables where there are many. The owner should select a veterinarian with whom he feels comfortable and who understands his type of operation. The veterinarian should take time to discuss owner problems with him on a consultant basis, or during a routine farm call.

Client education should be continual and lead to a higher treatment success rate because of earlier problem recognition by the client. Horse owners may range from the 4-H member who is just starting in a horse club to the sophisticated trainer. The veterinarian has to constantly be on his toes to see that the proper level of client education is made available to each type of owner.

A preventive medicine program for pleasure- or show-horse owners often begins with the examination of the horse when purchased. Many problems can be avoided if the purchaser will allow his veterinarian to examine the horse prior to purchase. This is especially true for the pleasure-horse owner. There are so many levels of sophistication in the horse industry that the veterinarian may do well to visit with his horse-owning clients and learn as much as possible from them.

Preventive medicine programs will help to keep the brood-mare farm in operation, to keep a horse on the track much longer, and to enable the 4-H or the pleasure-horse owner to enjoy his/her animal much more.

The arrival of a new horse on a farm provides a good point to begin the discussion of a preventive medicine program. On large farms, or breeding farms, it is wise to have an isolation barn for shipping and receiving horses. This barn should be located far enough away from the main horses to reduce transmission of any disease that might have been brought onto the farm. Personnel visiting the newly arrived animal should be particularly careful about going to the other animals on the farm or using tack from one animal on another.

Each new arrival should receive a physical examination to assess its general health. The horse should be in isolation for approximately 30 days with a final health check before mixing with the other animals on the farm. During this isolation, a Coggins test for equine infectious anemia should be run. It would be wise to check for leptospirosis while you have blood drawn.

Horses that are purchased and brought to a new facility have frequently been exposed to respiratory diseases that are contagious and that may be capable of causing abortions.

IMMUNIZATION PROGRAMS

Immunization programs must be designed according to location of the farm and special problems that may be geographic in nature.

Tetanus. Tetanus is a highly fatal infectious disease caused by the toxin of Colostridium tetani. The disease causes the horse to have a very hard, rigid muscle, with increased sensitivity to pain, noise, and convulsions in horses of all ages. The most common route of infection is by wound contamination with the Colostridium tetani spores. These spores are present in most areas where horses are kept. The organisms grow in deep puncture wounds where there is little or no oxygen. After the organism starts growing in the deep wound, a potent exotoxin is produced and acts principally on the central nervous system. This toxin is spread through the system by the blood and by movement along nerves that supply the infected area. This produces a tetanic contraction of the muscle fibers after normal sensory stimulation. The severity and rate of progression of the clinical signs depend upon the amount of toxin and the size, age, and immune status of the infected animal.

In many cases, a slightly stiffened animal is noted. The animal is reluctant to feed off the ground and may overreact to external noises. The jaw muscles may be affected early. Retracted lips, flared nostrils, protruding third eyelid and erect ears are observed. Horses with tetanus are unable to eat or drink in the normal manner. They have been described as having a "saw horse" stance. After an affected horse falls, it is generally unable to regain its feet. Death usually occurs after 5 to 7 days and is often caused by paralysis of the respiratory muscles and pneumonia. If the animal does not die, recovery is gradual and takes about 6 weeks. If recovery occurs, it is usually complete and without side effects.

Some horses are being saved using modern treatment regimes for tetanus. First of all the wound or the route of infection should be found and debrided as soon as possible. High levels of penicillin and tranquilizers should be administered. Keep the animal in a dark quiet place. The ears can be plugged with cotton to cut down on noises. The animal may have to be fed by stomach tube, and fluids should be given to prevent dehydration.

Vaccinations for tetanus are necessary on all horse farms, regardless of size, and should be done on an annual basis. Mares that are to foal should receive their annual tetanus booster approximately 30 days prior to foaling. This results in a high colostrial antibody titer that may be passed on to the foal, thereby eliminating the need for antitoxin in the foals at birth. Foals may be vaccinated with tetanus toxoid as early as the day after birth. However, if the mare has been adequately immunized, passive immunity received through the colostrum may persist for 2 to 6 months. Active immunization is generally started at 2 to 3 months of age. If a foal is born to a mare that has not been immunized against tetanus, the foal should receive antitoxin or start immediately upon tetanus toxoid. If they receive tetanus antitoxin, tetanus toxoid should follow prior to 2 months of age. Two doses of toxoid should be administered 4 weeks apart. When injury occurs, a booster dose of tetanus toxoid should be given. When an unvaccinated horse is injured, tetanus antitoxin may be given to provide passive immunity for approximately 2 weeks. However, the use of tetanus antitoxin has been known to transmit serum hepatitis.

<u>Strangles (Distemper)</u>. This disease is caused by a bacteria called Strepcoccus equi. More than one strain of this bacteria exists, which may account for some variable clinical signs and variable responses to vaccination. Clinical signs produced by S. equi infection can vary; it may be confined to the upper respiratory tract or it can cause a systemic disease. The horses may run a temperature of 102° to 106° F. Varying degrees of depression and a nasal discharge that may range from a liquid to a purulent exudate may be seen. Affected animals may cough and have difficulty in breathing. The organism quite frequently localizes and causes abscesses in the lymph nodes under the jaws and just behind the mandible, or they can form anywhere in the respiratory tract or abdominal area.

This disease is most commonly transmitted from one horse to another through infected secretions such as nasal discharges. Once the disease establishes itself on a farm or a ranch, it is often persistent and morbidity in susceptible horses is generally quite high. Death in most uncomplicated cases is quite low. Pneumonia can occur as a result of this infection.

The course of the disease can vary from mild to extremely virulent. If abscesses occur in the abdominal area, death can occur from rupture of the abscesses into the abdominal cavity. A positive diagnosis of this infection can only be confirmed by isolation of the bacteria from the lesions. Horse owners and veterinarians become suspicious of strangles because of the clinical signs. However, other organisms are capable of producing the same type lesions.

Horses with strangles should be encouraged to eat. They should be fed a soft diet until they can eat normally. Antibiotic therapy is quite frequently disappointing due to the inability to get antibiotics to the organisms in the abscess. The organism is usually susceptible to penicillin.

If recognized early, an outbreak may be contained by isolation of infected horses and strict hygiene by the handlers. Vaccines have been developed that may be of some benefit, but their efficacy has been questioned. Difficulty in vaccine developments stems from the relatively low ability of the organism to stimulate antibody production. Local reaction and abscesses at the injection sight can occur from use of vaccines presently available. If using the vaccine, completely sterile techniques should be utilized. The area of injection should be cleansed and swabbed with organic iodine and the vaccine injected deep into the muscle of the neck using a new sterile syringe and needle. There can also be secondary abscesses due to a Streptococcus zooepidemicus. This organism can be involved with strangles or the S. equi abscesses. S. zooepidemicus is also sensitive to penicillin but it is very difficult to get the antibiotic therapy deep into the abscessed material. The owner and the veterinarian must be aware of these potential failures and adverse reactions if use of the vaccine is elected. Before embarking upon a vaccination program for S. equi, consult your veterinarian for the incidence of the disease in your area.

Viral Infections

Nine viruses have been associated with disease of the equine respiratory tract.

Influenza. The terms influenza or "flu" have been used indiscriminately for many years in the literature. Equine influenza is produced by a myxovirus and was first documented in 1956. Influenza is caused in horses by two subtypes of the myxoviruses (A-equi-1 and A-equi-2).

The influenza virus attacks the cells lining the respiratory tract. These cells have a cilia or whip-like projections on the borders that normally prevent bacteria, viruses, and dust from entering the respiratory tract. When these cells are damaged, this allows secondary bacterial invasion. Uncomplicated cases of influenza require 3 weeks for regeneration of the normal ciliated cells. Influenza virus infects the entire lining of the respiratory tract and is not limited to the upper portion of the tract.

Clinical signs of equine influenza vary greatly. Young horses with no prior exposure are most susceptible. The disease may spread explosively through a susceptible herd. Infection usually produces fever that may be as high as 106° F. Most affected animals have some degree of depression and loss of appetite. A dry, nonproductive cough may last for several weeks. Secondary bacterial infections are common. Exercise should be minimized for 3 weeks. Antibiotics should be used to keep down secondary bacterial infection.

Influenza is best prevented by proper vaccination prior to exposure. Killed-virus vaccine contains both strains of the influenza virus. A primary series of two intramuscular injections followed by a single annual booster is recommended. There has been some indication that horses subjected to high degrees of infection should be vaccinated more often.

The vaccination program for any disease should be individualized for specific populations depending upon the risk. Foals should receive a primary immunization series beginning at 3 months of age. All horses in a particular population should be vaccinated, if possible, to maintain a high level of immunity. Horses are most likely to catch the "flu" when they are in groups, such as horse shows, race tracks, horse sales, and boarding stables.

Rhinopneumonitis. Rhinopneumonitis is a disease of the respiratory system that occurs most often in young horses. This virus, an equine herpes virus #1 (EHV--1), may also cause viral abortions and nervous disorders. All of these signs may be present in a disease outbreak. This virus cannot survive long outside the host.

The respiratory farm of rhinopneumonitis is usually mild and limited to weanlings and yearlings. It is one of the most common upper respiratory tract infections in young horses. The incubation is 2 to 10 days, and the spread of this disease is often rapid with high morbidity and low death loss. The initial signs include a watery nasal discharge and a fever up to 106° F. Other symptoms that may be observed include depression, being off feed, coughing, and a slight swelling of the lymph nodes. In uncomplicated cases in young horses, recovery usually requires 5 to 7 days. The disease may go completely unnoticed unless a secondary bacterial infection or other forms of the disease show up.

In pregnant mares, the rhinopneumonitis virus can cause abortion. If a vaccinated mare is subjected to a massive exposure of herpes virus it can be carried to the fetus and cause death in the fetus with abortion even though the mare is carrying antibodies.

Treatment is usually unnecessary unless the disease is complicated by a secondary bacterial infection. Modified live and killed vaccines are available to prevent the respiratory disease. Foals should be vaccinated initially at 3 months of age and then every 3 months until 2 years of age. Older horses should be vaccinated less frequently. The neurological forms or nerve forms may be seen in foals but occur more commonly in older horses.

Encephalomyelitis (Sleeping sickness). Equine encephalomyelitis is caused by a virus that has recently been named togavirus. This is the agent of Western equine encephalomyelitis (WEE), Eastern equine encephalomyelitis (EEE) and Venezuelan equine encephalomyelitis (VEE). However, as recently as 1971, VEE threatened to spread into the U.S. It was of sufficient importance to warrant the declaration of a national state of emergency and the mobilization of considerable resources of the Department of Defense.

These viruses are not only equine pathogens, they may cause disease in humans.

In nature, WEE and EEE viruses are maintained between outbreaks in reservoir hosts that probably include certain birds, rodents, and reptiles. The numbers can be multiplied many fold by the rapid transfer from bird to mosquito to bird. Many species of birds can become infected but do not usually become ill. The number of virus particles in infected horses is so low that further infection of feeding mosquitoes usually does not occur. Outbreaks of EEE and WEE tend to occur in mid to late summer when weather conditions favor breeding, longevity, and mobility of mosquito-vector populations.

In South and Central America and Florida, the intermediate host for VEE appears to be wild rodent and mosquito cycles. The natural history of an outbreak of VEE may resemble WEE and EEE. However, the horse with VEE is capable of increasing virus numbers, and the disease can be transmitted from one horse to another by the mosquito.

Often the first things noticed in a horse with an encephalomyelitis is a change in behavior. Docile animals may become irritable and bite their handler; other horses may seem sleepy or fail to respond to their owner's call. The horse may develop signs of compulsive walking, often in a circle, blindness, and leaning against a wall or a fence. Death rates range from 70% to 95% for EEE, 19% to 15% for WEE, and 49% to 90% for VEE. A four-fold rise in antibody titer between acute and convalescent serum samples is considered positive, although a very high acute titer in an unvaccinated horse is probably sufficient to establish a diagnosis.

Treatment for equine encephalomyelitis is largely supportive; fluids may be required during the severe stages of the disease to keep the animal from dehydration.

The EEE and WEE viruses will probably never be eradicated from U.S. because reservoirs exist in many areas and do not depend upon horse infection for their maintenance. Continual vigilance and conscientious immunization programs will be necessary to minimize and contain these diseases in horses.

Vaccines can be obtained that contain both EEE and WEE, or all three are available. Choice of a particular product depends upon the prevalence or likely occurrence of the three diseases.

Horses should be vaccinated at least a month before the anticipated risk period. This time will change in different geographic areas of the country. The vaccines will not produce a long lasting immunity. In areas where mosquitoes are prevalent year round, the horses should be vaccinated every 6 months. An outbreak of VEE requires quarantine of the horses with reporting to federal authority.

We have discussed in general some of the diseases of horses and how vaccination programs can be utilized. A preventive health program combined with attention to nutrition and enteric diseases can go a long way toward maintaining the health and well-being of your horses.

Part 10

MARKETING, ECONOMICS, AND COMPUTER TECHNOLOGY

44

OFFICIAL AQHA RECORDS FOR MARKETING

Ronald Blackwell

Horses are a little different from the other species of livestock associated with the International Stockmen's School. They are not a source of food, in this country anyway, and few, if any of them, are raised commercially.

However, the purebred horse business is big business. And there are tens of thousands of people in the U.S. that earn a living from horses. There are major breeding operations all across the country. It is estimated by the American Horse Council, an organization in Washington, D.C. that represents all horse breeds, that there are 8.5 million horses in the U.S. and that land, equipment, horses, and other related items easily represent a $15 billion dollar investment by America's 3.2 million horse owners.

Although some equine breed registries have been around for more than 100 years, there is none larger than the American Quarter Horse Association. We were organized in 1940 and since then have registered more than 1.8 million horses. In 1981, 148,000 horses were registered by AQHA--that number is more than all the other breed registries combined, as illustrated in the statistics computed by the American Horse Council (figure 1). We have more than 130,000 dues-paying members from all 50 states and 67 foreign countries, plus 7,900 members of the American Junior Quarter Horse Association.

For those of you not familiar with quarter horses, we bill them as the world's most versatile horse. Quarter horses can do about anything that needs to be done on horseback. They are the world's fastest horse up to distances of a quarter-of-a-mile; they are unsurpassed as a using horse on ranches; and they perform in 17 different events in the show arena, in Western and English tack. Because of their gentle disposition, they are sought after as a pleasure horse by thousands of weekend horse enthusiasts.

Along with this multitude of uses comes the necessity to keep accurate, official records on the accomplishments of our horses. I am proud to say that AQHA's system of official record keeping on its horses is second to none in the industry. Representatives of every major horse breed, plus many livestock associations, have studied and often copied

REGISTRATION FIGURES
Major American Light Horse Breeds, 1960–1981

	1960	1968	1973	1978	1980	1981	No. of Living Horses in Registry*
Appaloosa	4,052	12,389	20,357	17,802	25,384	18,277	356,500
Anglo & Half Arab	2,200	9,800	13,222	15,396	14,257	10,035	175,000
Arabian	1,610	6,980	12,266	16,669	19,725	20,300	180,000
Hackney	459	656	466	619	595	801	11,500
Morgan	1,069	2,134	3,052	3,519	4,537	4,785	59,000
Paint	-	2,390	4,331	7,622	9,654	9,411	60,000
Palomino	657	1,262	1,580	2,002	1,548	1,334	43,000
Pinto	250	2,258	2,270	1,870	1,502	1,520	NA
POA	525	1,759	1,550	1,080	1,272	1,124	23,000
Quarter Horse	37,000	57,000	87,568	119,287	137,090	147,787	1,646,069
Saddlebred	1,600	3,500	4,011	3,801	3,879	3,855	60,000
Standardbred	7,100	10,200	11,393	14,000	14,691	17,458	200,000
Tennessee Walker	2,623	8,492	7,116	6,200	6,673	5,037	NA
Thoroughbred	12,901	22,700	26,760	31,000	33,170	37,499	200,000
TOTAL	72,046	141,520	195,942	240,867	273,977	279,223	3,014,069

*Registry estimates *(Compiled by the American Horse Council)*

Figure 1.

our record-keeping system. Without a doubt, our records are the basis of marketing quarter horses to the general public. Let me explain how our system works.

Basically, there are four categories of records within our system: 1) show and race records on the individual horses themselves, 2) production records (which we call get of sire) and produce records on stallions and mares, 3) ownership records, and 4) breeder records. The last two kinds of records both include the people, farms and ranches, and corporations that are in our business.

Our records are kept by means of an IBM 370-148 system that features tape and disc units. It has a capacity of more than 3.8 billion characters. Many of our records are updated "on line" while others are updated each cycle, which is every other working day. At our present rate of growth, which is about 8%, this system should handle our needs through the next 5 years.

Let's look at each type of record beginning with the individual show and performance records, also known as a master registration record. I have two examples to show you. The first is Kaweah Bar, a race horse who was twice a world champion and five times voted champion gelding.

Should someone request a master registration record on Kaweah Bar, it would look something like this (figure 2).

First is a listing of basic information on the horse including registration number, color, sex, year foaled, sire, dam, breeder, and current or last recorded owner. A summary of the horse's show record, if any, is followed by a race record summary including starts, wins, places and shows, and total earnings on the track. This is followed by an accounting of the horse's performance in stakes races, in which it placed at least third. And finally, any special awards the horse has won are listed. Every horse who has started in any AQHA-approved race would have a similar type of master registration record available from AQHA.

Now let's look at a master registration record on one of our show horses--Van Decka (figure 3). This horse has quite an incredible list of accomplishments. After the basic information is listed, the adult show record summary is given and you can see this horse has points in six events. He also has a youth show record; therefore, a breakdown of points earned by each youth that exhibited the horse is listed. And again, any awards he had received, including placings at either of our world championship shows, or any association-approved awards (such as AQHA champion, superior, register of merit) are listed. This horse has quite an impressive record in our youth division under two different riders. The accomplishments under each youth exhibitor is linked separately.

Other types of records are the get of sire and produce of dam records. We have these available in two formats for get of sire: one that lists all the foals and one that lists performing foals only.

In our examples here (figure 4) we are using the sire of Kaweah Bar, which was Alamitos Bar, and Decka Center, the sire of Van Decka. The information listed is in basically the same form as for individual master registration records for any get that have raced and includes a summary of the racing career, awards, and stakes races. For sires of show horses, such as Decka Center, sire of Van Decka, a breakdown of points earned by get in either halter or performance events is listed followed by a summary of awards earned.

Our records can go a step further and give a detailed race or show record. For Kaweah Bar, his detailed race record is similar to a master performance record. However, his entire racing history in a race-by-race accounting is detailed. Information includes date of race, track, distance, type of race, how the horse finished, and the margin, earnings, grade, time, and odds the horse paid, if there was pari-mutuel wagering. All of this is followed by a summary of his racing history.

Similar information is kept on show horses, with a show-by-show and class-by-class accounting of the horse's showing activity. The information kept includes date, place of show, the grade of the show (A, B, C, D), the name of the

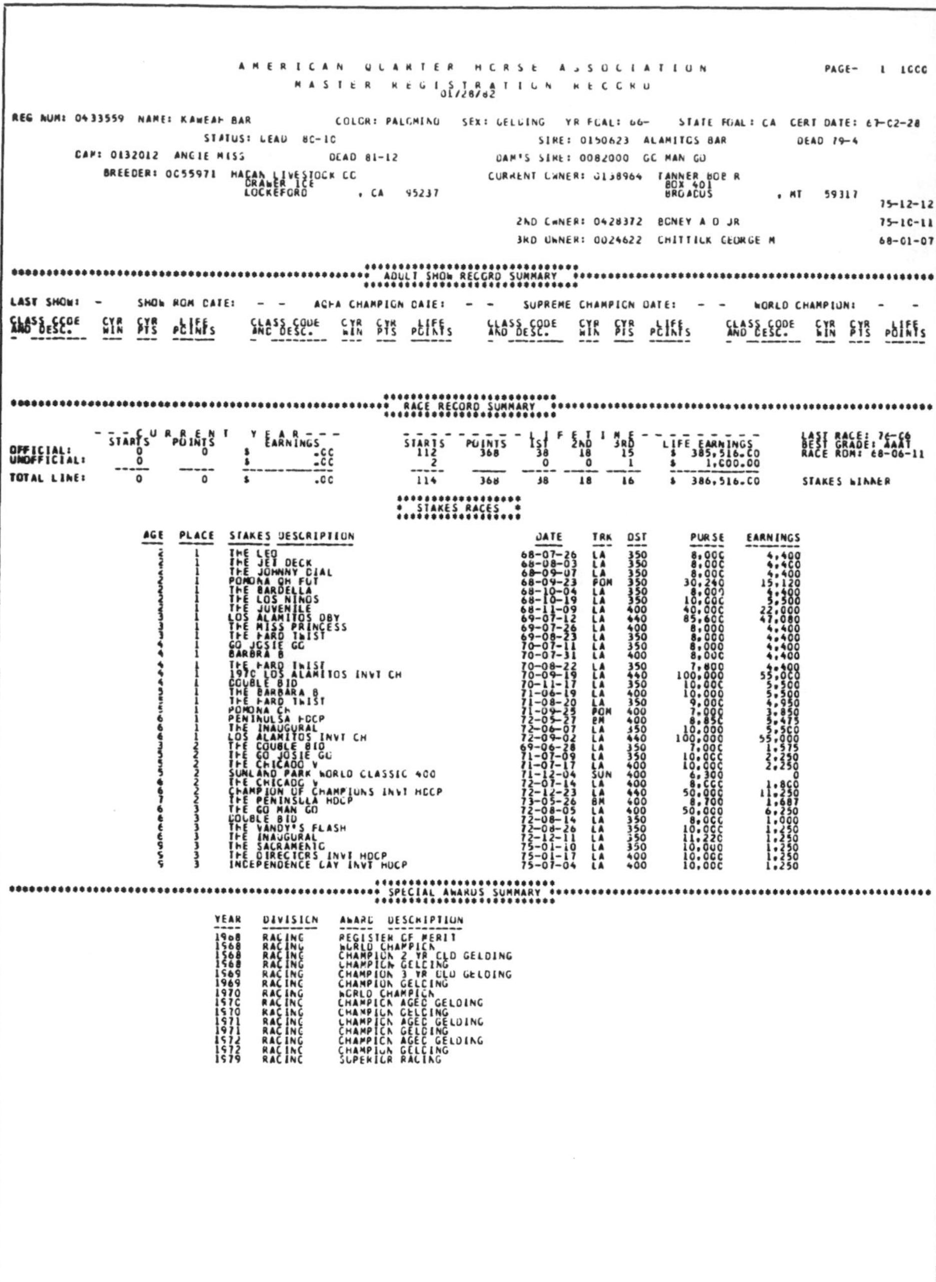

AMERICAN QUARTER HORSE ASSOCIATION PAGE- 1 1000
MASTER REGISTRATION RECORD
01/28/82

REG NUM: 0433559 NAME: KAWEAH BAR COLOR: PALOMINO SEX: GELDING YR FOAL: 66- STATE FOAL: CA CERT DATE: 67-02-28
STATUS: DEAD 80-10 SIRE: 0150623 ALAMITOS BAR DEAD 79-4
DAM: 0132012 ANGIE MISS DEAD 81-12 DAM'S SIRE: 0082000 GO MAN GO
BREEDER: 0055971 HAGAN LIVESTOCK CO CURRENT OWNER: 0138964 TANNER BOB R
DRAWER ICE BOX 401
LOCKEFORD , CA 95237 BROADUS , MT 59317
75-12-12
2ND OWNER: 0428372 BONEY A D JR 75-10-11
3RD OWNER: 0024622 CHITTICK GEORGE M 68-01-07

***** ADULT SHOW RECORD SUMMARY *****

LAST SHOW: - SHOW ROM DATE: - - AQHA CHAMPION DATE: - - SUPREME CHAMPION DATE: - - WORLD CHAMPION: - -

CLASS CODE AND DESC.	CYR WIN	CYR PTS	LIFE POINTS	CLASS CODE AND DESC.	CYR WIN	CYR PTS	LIFE POINTS	CLASS CODE AND DESC.	CYR WIN	CYR PTS	LIFE POINTS	CLASS CODE AND DESC.	CYR WIN	CYR PTS	LIFE POINTS

***** RACE RECORD SUMMARY *****

	CURRENT YEAR STARTS	POINTS	EARNINGS	LIFETIME STARTS	POINTS	1ST	2ND	3RD	LIFE EARNINGS
OFFICIAL:	0	0	$ .00	112	368	38	18	15	$ 385,516.00
UNOFFICIAL:	0		$ .00	2		0	0	1	$ 1,000.00
TOTAL LINE:	0	0	$ .00	114	368	38	18	16	$ 386,516.00

LAST RACE: 76-06
BEST GRADE: AAAT
RACE ROM: 68-06-11
STAKES WINNER

***** STAKES RACES *****

AGE	PLACE	STAKES DESCRIPTION	DATE	TRK	DST	PURSE	EARNINGS
2	1	THE LEO	68-07-26	LA	350	8,000	4,400
2	1	THE JET DECK	68-08-03	LA	350	8,000	4,400
2	1	THE JOHNNY DIAL	68-09-07	LA	350	8,000	4,400
2	1	POMONA QH FUT	68-09-23	POM	350	30,240	15,120
2	1	THE BARDELLA	68-10-04	LA	350	8,000	4,400
2	1	THE LOS NINOS	68-10-19	LA	350	10,000	5,500
2	1	THE JUVENILE	68-11-09	LA	400	40,000	22,000
3	1	LOS ALAMITOS DBY	69-07-12	LA	440	85,600	47,080
3	1	THE MISS PRINCESS	69-07-26	LA	400	8,000	4,400
3	1	THE HARD TWIST	69-08-23	LA	350	8,000	4,400
4	1	GO JOSIE GO	70-07-11	LA	350	8,000	4,400
4	1	BARBRA B	70-07-31	LA	400	8,000	4,400
4	1	THE HARD TWIST	70-08-22	LA	350	7,800	4,400
4	1	1970 LOS ALAMITOS INVT CH	70-09-19	LA	440	100,000	55,000
4	1	DOUBLE BID	70-11-17	LA	350	10,000	5,500
5	1	THE BARBARA B	71-06-19	LA	400	10,000	5,500
5	1	THE HARD TWIST	71-08-20	LA	350	9,000	4,950
5	1	POMONA CH	71-09-25	POM	400	7,000	3,850
6	1	PENINULSA HDCP	72-05-27	BM	400	8,850	5,475
6	1	THE INAUGURAL	72-06-07	LA	350	10,000	5,500
6	1	LOS ALAMITOS INVT CH	72-09-02	LA	440	100,000	55,000
3	2	THE DOUBLE BID	69-06-28	LA	350	7,000	1,575
5	2	THE GO JOSIE GO	71-07-09	LA	350	10,000	2,250
5	2	THE CHICADO V	71-07-17	LA	400	10,000	2,250
5	2	SUNLAND PARK WORLD CLASSIC 400	71-12-04	SUN	400	6,300	0
6	2	THE CHICADO V	72-07-14	LA	400	8,000	1,800
6	2	CHAMPION OF CHAMPIONS INVT HDCP	72-12-23	LA	440	50,000	11,250
7	2	THE PENINSULA HDCP	73-05-26	BM	400	8,700	1,687
6	3	THE GO MAN GO	72-08-05	LA	400	50,000	6,250
6	3	DOUBLE BID	72-08-14	LA	350	8,000	1,000
6	3	THE VANDY'S FLASH	72-08-26	LA	350	10,000	1,250
6	3	THE INAUGURAL	72-12-11	LA	350	11,220	1,250
9	3	THE SACRAMENTO	75-01-10	LA	350	10,000	1,250
5	3	THE DIRECTORS INVT HDCP	75-01-17	LA	400	10,000	1,250
9	3	INDEPENDENCE DAY INVT HDCP	75-07-04	LA	400	10,000	1,250

***** SPECIAL AWARDS SUMMARY *****

YEAR	DIVISION	AWARD DESCRIPTION
1968	RACING	REGISTER OF MERIT
1968	RACING	WORLD CHAMPION
1968	RACING	CHAMPION 2 YR OLD GELDING
1968	RACING	CHAMPION GELDING
1969	RACING	CHAMPION 3 YR OLD GELDING
1969	RACING	CHAMPION GELDING
1970	RACING	WORLD CHAMPION
1970	RACING	CHAMPION AGED GELDING
1970	RACING	CHAMPION GELDING
1971	RACING	CHAMPION AGED GELDING
1971	RACING	CHAMPION GELDING
1972	RACING	CHAMPION AGED GELDING
1972	RACING	CHAMPION GELDING
1979	RACING	SUPERIOR RACING

The master registration record of Kaweah Bar shows all current information such as name, number, breeder, etc., and then goes to the show record summary (if he had one), followed by the race record summary and then the list of stake races in which he either won or placed. The final item on this record is the special awards summary.

Figure 2.

AMERICAN QUARTER HORSE ASSOCIATION — PAGE- 1 1000

MASTER REGISTRATION RECORD

01/28/82

REG NUM: 0523063 NAME: VAN DECKA COLOR: BAY SEX: GELDING YR FOAL: 67- STATE FOAL: OK CERT DATE: 68-07-31

SIRE: 0361940 DECKA CENTER

DAM: 0066558 VANESSA DEE DEAD 80-6 DAM'S SIRE: 0043628 VANDY DEAD 00-

BREEDER: 0072413 JOHNSON JOHNNY, 111 N PORT PLAZA SUITE 203, ST LOUIS, MO 63141

CURRENT OWNER: 0228872 JOHNSON KIM, 13704 CLAYTON RD, ST LOUIS, MO 63011 — 76-04-26

2ND OWNER: 0229049 JOHNSON CHERYL — 72-03-13

3RD OWNER: 0197901 JOHNSON CHERYL &/OR BETTY — 71-11-22

ADULT SHOW RECORD SUMMARY

LAST SHOW: 76-01 SHOW ROM DATE: 71-02-27 AQHA CHAMPION DATE: 72-04-16 SUPREME CHAMPION DATE: - - WORLD CHAMPION: 00- 00-

CLASS CODE AND DESC.	CYR WIN	CYR PTS	LIFE POINTS
A HALTER	0	0	33.0
O WS PLES	0	0	14.0
K REINING	0	0	1.0
P HUNTSEAT	0	0	28.0
L W RIDING	0	0	29.0
M TRAIL	0	0	5.0

RACE RECORD SUMMARY

	CURRENT YEAR STARTS	CURRENT YEAR POINTS	CURRENT YEAR EARNINGS	LIFETIME STARTS	LIFETIME POINTS	1ST	2ND	3RD	LIFE EARNINGS
OFFICIAL:	0	0	$ .00	3	0	0	0	0	$ 25.00
UNOFFICIAL:	0		$ .00	0		0	0	0	$ .00
TOTAL LINE:	0	0	$ .00	3	0	0	0	0	$ 25.00

LAST RACE: 65-10 BEST GRADE: B RACE ROM: - -

YOUTH SHOW RECORD SUMMARY

YOUTH CODE: A YOUTH NAME: JOHNSON CHERYL YOUTH CITY: ST LOUIS YOUTH STATE: MO

LAST SHOW: 76-02 SHOW ROM DATE: 72-04-29 AQHA CHAMPION DATE: 72-07-01 SUPREME CHAMPION DATE: - - WORLD CHAMPION: - -

CLASS CODE AND DESC.	CYR POINTS	LIFE POINTS
A HALTER	.0	142.0
M TRAIL	.0	62.0
B SHMNSHIP	.0	655.0
N WS HORSE	.0	688.0
K REINING	.0	29.0
O WS PLES	.0	723.0
L W RIDING	.0	123.0
P HUNTSEAT	.0	452.0

YOUTH CODE: B YOUTH NAME: JOHNSON KIM YOUTH CITY: ST LOUIS YOUTH STATE: MO

LAST SHOW: 78-07 SHOW ROM DATE: 76-06-06 AQHA CHAMPION DATE: 77-05-29 SUPREME CHAMPION DATE: - - WORLD CHAMPION: - -

CLASS CODE AND DESC.	CYR POINTS	LIFE POINTS
A HALTER	.0	37.0
N WS HORSE	.0	344.0
B SHMNSHIP	.0	218.0
O WS PLES	.0	403.0
L W RIDING	.0	42.0
P HUNTSEAT	.0	128.0
M TRAIL	.0	120.0
S HT ST EQ	.0	45.0

SPECIAL AWARDS SUMMARY

YEAR	DIVISION	AWARD DESCRIPTION	
1971	OPEN	REGISTER OF MERIT	
1972	OPEN	AQHA CHAMPION	
1974	OPEN	WORLD CHAMPION SR WESTERN RIDING	9TH PLACE
1972	YOUTH -A	REGISTER OF MERIT	
1972	YOUTH -A	AQHA CHAMPION	
1973	YOUTH -A	SUPERIOR HALTER	
1973	YOUTH -A	SUPERIOR SHOWMANSHIP	
1973	YOUTH -A	SUPERIOR WESTERN HORSEMANSHIP	
1973	YOUTH -A	SUPERIOR WESTERN RIDING	
1973	YOUTH -A	SUPERIOR WESTERN PLEASURE	
1973	YOUTH -A	SUPERIOR HUNT SEAT	
1973	YOUTH -A	AQHA PERFORMANCE CHAMPION	
1973	YOUTH -A	HIGH POINT HALTER GELDING	5TH PLACE
1973	YOUTH -A	HIGH POINT SHOWMANSHIP	4TH PLACE
1973	YOUTH -A	HIGH POINT WESTERN HORSEMANSHIP	3RD PLACE
1973	YOUTH -A	HIGH POINT REINING	9TH PLACE
1973	YOUTH -A	HIGH POINT WESTERN RIDING	
1973	YOUTH -A	HIGH POINT WESTERN PLEASURE	8TH PLACE
1973	YOUTH -A	HIGH POINT HUNT SEAT	3RD PLACE
1973	YOUTH -A	ALL AROUND	3RD PLACE
1973	YOUTH -A	WORLD CHAMPION WESTERN HORSEMANSHIP	7TH PLACE
1973	YOUTH -A	WORLD CHAMPION WESTERN RIDING	8TH PLACE
1973	YOUTH -A	WORLD CHAMPION HUNT SEAT	10TH PLACE
1974	YOUTH -A	SUPERIOR TRAIL HORSE	
1974	YOUTH -A	HIGH POINT HALTER GELDING	7TH PLACE
1974	YOUTH -A	HIGH POINT SHOWMANSHIP	
1974	YOUTH -A	HIGH POINT WESTERN HORSEMANSHIP	
1974	YOUTH -A	HIGH POINT WESTERN RIDING	2ND PLACE
1974	YOUTH -A	HIGH POINT WESTERN PLEASURE	3RD PLACE
1974	YOUTH -A	HIGH POINT TRAIL HORSE	
1974	YOUTH -A	HIGH POINT HUNT SEAT	
1974	YOUTH -A	ALL AROUND	
1974	YOUTH -A	WORLD CHAMPION AGED GELDING	5TH PLACE
1974	YOUTH -A	WORLD CHAMPION SHOWMANSHIP	
1974	YOUTH -A	WORLD CHAMPION WESTERN HORSEMANSHIP	6TH PLACE
1974	YOUTH -A	WORLD CHAMPION TRAIL HORSE	6TH PLACE
1975	YOUTH -A	HIGH POINT SHOWMANSHIP	
1975	YOUTH -A	HIGH POINT WESTERN HORSEMANSHIP	2ND PLACE
1975	YOUTH -A	HIGH POINT WESTERN PLEASURE	
1975	YOUTH -A	HIGH POINT HUNT SEAT	4TH PLACE
1975	YOUTH -A	ALL AROUND	
1975	YOUTH -A	WORLD CHAMPION SHOWMANSHIP	3RD PLACE
1975	YOUTH -A	WORLD CHAMPION WESTERN RIDING	6TH PLACE
1975	YOUTH -A	WORLD CHAMPION WESTERN PLEASURE	4TH PLACE
1975	YOUTH -A	WORLD CHAMPION HUNT SEAT	5TH PLACE
1976	YOUTH -B	REGISTER OF MERIT	
1976	YOUTH -B	SUPERIOR SHOWMANSHIP	
1976	YOUTH -B	SUPERIOR WESTERN HORSEMANSHIP	
1976	YOUTH -B	SUPERIOR WESTERN PLEASURE	
1976	YOUTH -B	AQHA PERFORMANCE CHAMPION	
1976	YOUTH -B	HIGH POINT WESTERN HORSEMANSHIP	8TH PLACE
1977	YOUTH -B	AQHA CHAMPION	
1977	YOUTH -B	SUPERIOR TRAIL HORSE	
1977	YOUTH -B	SUPERIOR HUNT SEAT	
1977	YOUTH -B	HIGH POINT SHOWMANSHIP	2ND PLACE
1977	YOUTH -B	HIGH POINT HUNT SEAT EQUITATION	2ND PLACE
1977	YOUTH -B	HIGH POINT WESTERN HORSEMANSHIP	
1977	YOUTH -B	HIGH POINT WESTERN RIDING	5TH PLACE
1977	YOUTH -B	HIGH POINT WESTERN PLEASURE	
1977	YOUTH -B	HIGH POINT TRAIL HORSE	2ND PLACE
1977	YOUTH -B	HIGH POINT HUNT SEAT	2ND PLACE
1977	YOUTH -B	ALL AROUND	2ND PLACE
1977	YOUTH -B	WORLD CHAMPION SHOWMANSHIP	9TH PLACE
1977	YOUTH -B	WORLD CHAMPION WESTERN PLEASURE	7TH PLACE

The master registration record of Van Decka gives the same information as that of Kaweah Bar, but in this case the emphasis is on the show arena.

Figure 3.

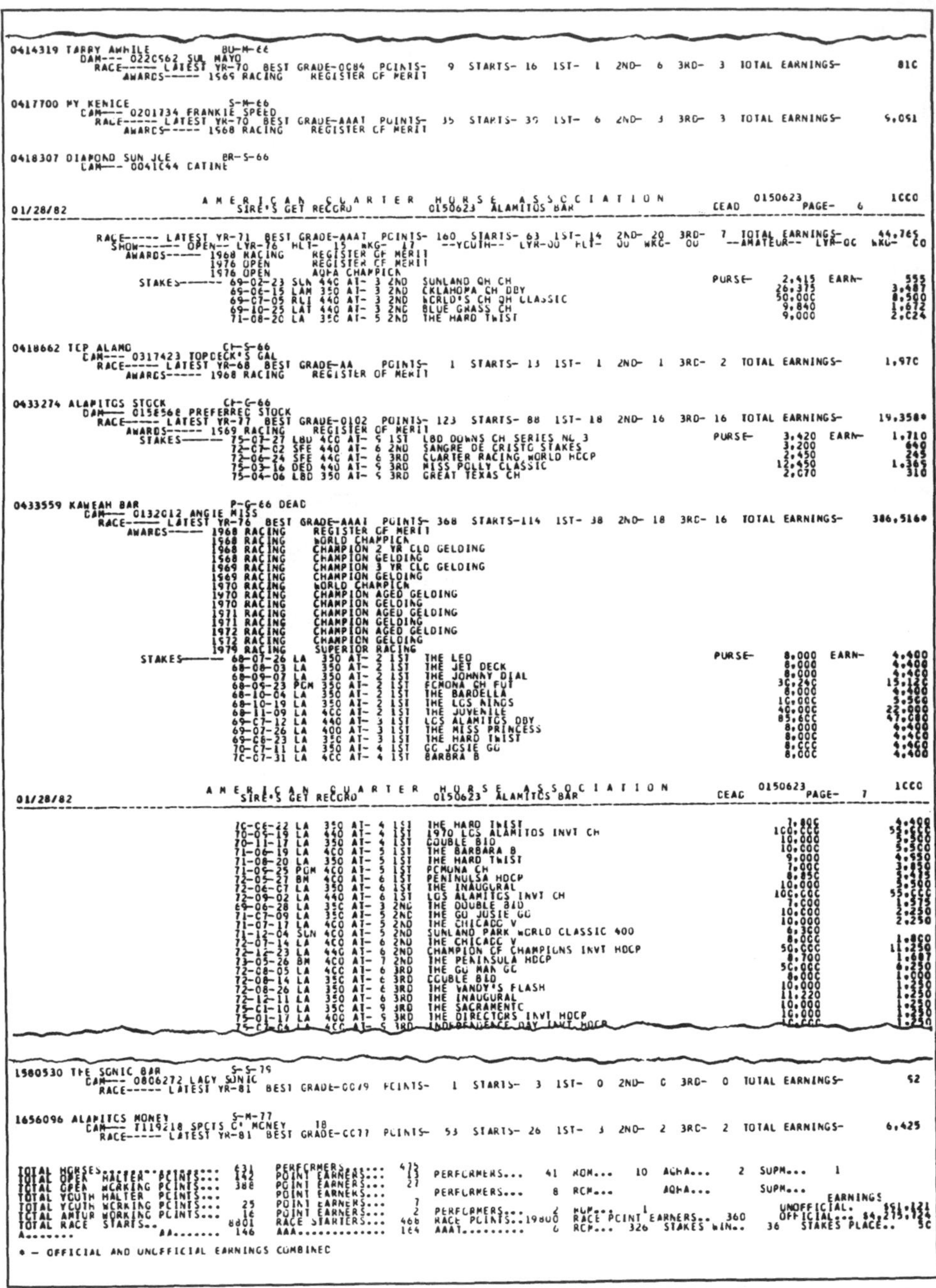

```
0414319 TARRY AWHILE             BU-M-66
        DAM--- 0220562 SUL MAYO
          RACE----- LATEST YR-70  BEST GRADE-0084  POINTS-    9  STARTS- 16  1ST-  1  2ND-  6  3RD-  3  TOTAL EARNINGS-        810
              AWARDS----- 1969 RACING     REGISTER OF MERIT

0417700 MY KENICE                S-M-66
        DAM--- 0201734 FRANKIE SPEED
          RACE----- LATEST YR-70  BEST GRADE-AAAT  POINTS-   35  STARTS- 39  1ST-  6  2ND-  3  3RD-  3  TOTAL EARNINGS-      9,051
              AWARDS----- 1968 RACING     REGISTER OF MERIT

0418307 DIAMOND SUN JOE          BR-S-66
        DAM--- 0041044 CATINE

                       A M E R I C A N   Q U A R T E R   H O R S E   A S S O C I A T I O N        DEAD 0150623          1000
01/28/82                     SIRE'S GET RECORD              0150623  ALAMITOS BAR                         PAGE-   6
-------------------------------------------------------------------------------------------------------------------------------------

          RACE----- LATEST YR-71  BEST GRADE-AAAT  POINTS- 160  STARTS- 63  1ST- 14  2ND- 20  3RD-  7  TOTAL EARNINGS-     44,765
           SHOW------ OPEN-- LYR-76  HLT-  15  WKG-  17    --YOUTH--  LYR-00  HLT-  00  WKG-  00   --AMATEUR--  LYR-00  WKG-  00
              AWARDS----- 1968 RACING     REGISTER OF MERIT
                          1976 OPEN       REGISTER OF MERIT
                          1976 OPEN       AQHA CHAMPION
               STAKES------ 69-02-23 SLN 440 AT- 3 2ND  SUNLAND QH CH                         PURSE-    2,415  EARN-      555
                            69-06-15 LAM 350 AT- 3 2ND  OKLAHOMA CH DBY                                26,375              3,487
                            69-07-05 RLI 440 AT- 3 2ND  WORLD'S CH QH CLASSIC                          50,000              8,500
                            69-10-25 LAT 440 AT- 3 2ND  BLUE GRASS CH                                   9,840              1,672
                            71-08-20 LA  350 AT- 5 2ND  THE HARD TWIST                                  9,000              2,024

0418662 TOP ALAMO                CH-S-66
        DAM--- 0317423 TOPDECK'S GAL
          RACE----- LATEST YR-68  BEST GRADE-AA    POINTS-    1  STARTS- 13  1ST-  1  2ND-  1  3RD-  2  TOTAL EARNINGS-      1,970
              AWARDS----- 1968 RACING     REGISTER OF MERIT

0433274 ALAMITOS STOCK           CH-G-66
        DAM--- 0155568 PREFERRED STOCK
          RACE----- LATEST YR-77  BEST GRADE-0102  POINTS- 123  STARTS- 88  1ST- 18  2ND- 16  3RD- 16  TOTAL EARNINGS-     19,358*
              AWARDS----- 1969 RACING     REGISTER OF MERIT
               STAKES------ 75-07-27 LBD 400 AT- 9 1ST  LBD DOWNS CH SERIES NO 3             PURSE-    3,420  EARN-    1,710
                            72-07-02 SFE 440 AT- 6 2ND  SANGRE DE CRISTO STAKES                         3,200                640
                            72-06-24 SFE 440 AT- 6 3RD  QUARTER RACING WORLD HDCP                       2,450                245
                            75-03-16 DED 440 AT- 9 3RD  MISS POLLY CLASSIC                             12,450              1,365
                            75-04-06 LBD 350 AT- 9 3RD  GREAT TEXAS CH                                  2,070                310

0433559 KAWEAH BAR               P-G-66 DEAD
        DAM--- 0132012 ANGIE MISS
          RACE----- LATEST YR-76  BEST GRADE-AAAT  POINTS- 368  STARTS-114  1ST- 38  2ND- 18  3RD- 16  TOTAL EARNINGS-    386,516*
              AWARDS----- 1968 RACING     REGISTER OF MERIT
                          1968 RACING     WORLD CHAMPION
                          1968 RACING     CHAMPION 2 YR OLD GELDING
                          1968 RACING     CHAMPION GELDING
                          1969 RACING     CHAMPION 3 YR OLD GELDING
                          1969 RACING     CHAMPION GELDING
                          1970 RACING     WORLD CHAMPION
                          1970 RACING     CHAMPION AGED GELDING
                          1970 RACING     CHAMPION GELDING
                          1971 RACING     CHAMPION AGED GELDING
                          1971 RACING     CHAMPION GELDING
                          1972 RACING     CHAMPION AGED GELDING
                          1972 RACING     CHAMPION GELDING
                          1979 RACING     SUPERIOR RACING
               STAKES------ 68-07-26 LA  350 AT- 2 1ST  THE LEO                               PURSE-    8,000  EARN-    4,400
                            68-08-03 LA  350 AT- 2 1ST  THE JET DECK                                    8,000              4,400
                            68-09-07 LA  350 AT- 2 1ST  THE JOHNNY DIAL                                 8,000              4,400
                            68-09-23 PCM 350 AT- 2 1ST  POMONA CH FUT                                  30,240             15,120
                            68-10-04 LA  350 AT- 2 1ST  THE BARDELLA                                    8,000              4,400
                            68-10-19 LA  350 AT- 2 1ST  THE LOS NINOS                                  10,000              5,500
                            68-11-09 LA  400 AT- 2 1ST  THE JUVENILE                                   40,000             22,000
                            69-07-12 LA  440 AT- 3 1ST  LOS ALAMITOS DBY                               85,600             47,080
                            69-07-26 LA  400 AT- 3 1ST  THE MISS PRINCESS                               8,000              4,400
                            69-08-23 LA  350 AT- 3 1ST  THE HARD TWIST                                  8,000              4,400
                            70-07-11 LA  350 AT- 4 1ST  GO JOSIE GO                                     8,000              4,400
                            70-07-31 LA  400 AT- 4 1ST  BARBRA B                                        8,000              4,400

                       A M E R I C A N   Q U A R T E R   H O R S E   A S S O C I A T I O N        DEAD 0150623          1000
01/28/82                     SIRE'S GET RECORD              0150623  ALAMITOS BAR                         PAGE-   7
-------------------------------------------------------------------------------------------------------------------------------------

                            70-08-22 LA  350 AT- 4 1ST  THE HARD TWIST                                  7,800              4,400
                            70-09-19 LA  440 AT- 4 1ST  1970 LOS ALAMITOS INVT CH                     100,000             55,000
                            70-11-17 LA  350 AT- 4 1ST  DOUBLE BID                                     10,000              5,500
                            71-06-19 LA  400 AT- 5 1ST  THE BARBARA B                                  10,000              5,500
                            71-08-20 LA  350 AT- 5 1ST  THE HARD TWIST                                  9,000              4,950
                            71-09-25 PCM 400 AT- 5 1ST  POMONA CH                                       7,000              3,850
                            72-05-27 BM  400 AT- 6 1ST  PENINULSA HDCP                                  8,850              5,475
                            72-06-07 LA  350 AT- 6 1ST  THE INAUGURAL                                  10,000              5,500
                            72-09-02 LA  440 AT- 6 1ST  LOS ALAMITOS INVT CH                          100,000             55,000
                            69-06-28 LA  350 AT- 3 2ND  THE DOUBLE BID                                  7,000              1,575
                            71-07-09 LA  350 AT- 5 2ND  THE GO JOSIE GO                                10,000              2,250
                            71-07-17 LA  400 AT- 5 2ND  THE CHICADO V                                  10,000              2,250
                            71-12-04 SLN 400 AT- 5 2ND  SUNLAND PARK WORLD CLASSIC 400                  6,300
                            72-07-14 LA  400 AT- 6 2ND  THE CHICADO V                                   8,000              1,800
                            72-12-23 LA  440 AT- 6 2ND  CHAMPION OF CHAMPIONS INVT HDCP                50,000             11,250
                            73-05-26 BM  400 AT- 7 2ND  THE PENINSULA HDCP                              8,700              1,687
                            72-08-05 LA  400 AT- 6 3RD  THE GO MAN GO                                  50,000              6,250
                            72-08-14 LA  350 AT- 6 3RD  DOUBLE BID                                      8,000              1,000
                            72-08-26 LA  350 AT- 6 3RD  THE VANDY'S FLASH                              10,000              1,250
                            72-12-11 LA  350 AT- 6 3RD  THE INAUGURAL                                  11,220              1,250
                            75-01-10 LA  350 AT- 9 3RD  THE SACRAMENTO                                 10,000              1,250
                            75-01-17 LA  400 AT- 9 3RD  THE DIRECTORS INVT HDCP                        10,000              1,250
                            75-07-04 LA  400 AT- 9 3RD  INDEPENDENCE DAY INVT HDCP                     10,000              1,250

1580530 THE SONIC BAR            S-S-79
        DAM--- 0806272 LADY SONIC
          RACE----- LATEST YR-81  BEST GRADE-0079  POINTS-    1  STARTS-  3  1ST-  0  2ND-  0  3RD-  0  TOTAL EARNINGS-         52

1656096 ALAMITOS MONEY           S-M-77
        DAM--- T119218 SPOTS O' MONEY     18
          RACE----- LATEST YR-81  BEST GRADE-0077  POINTS-   53  STARTS- 26  1ST-  3  2ND-  2  3RD-  2  TOTAL EARNINGS-      6,425

TOTAL HORSES.................   631   PERFORMERS.......   475
TOTAL OPEN  HALTER  POINTS...   142   POINT EARNERS....    13   PERFORMERS...   41  ROM...   10  ACHA...   2  SUPM...   1
TOTAL OPEN  WORKING POINTS...   388   POINT EARNERS....    27
TOTAL YOUTH HALTER  POINTS...         POINT EARNERS....         PERFORMERS...    8  ROM...        AQHA...      SUPM...
TOTAL YOUTH WORKING POINTS...    25   POINT EARNERS....     7                                                     EARNINGS
TOTAL AMTUR WORKING POINTS...    16   POINT EARNERS....     2   PERFORMERS...    2  ROM...    1            UNOFFICIAL.     $51,121
TOTAL RACE  STARTS..           8801   RACE STARTERS....   466   RACE POINTS..19800  RACE POINT EARNERS..  360  OFFICIAL... $4,275,124
A.......               AA.......146   AAA..............   164   AAAT.........    6  ROM...  326  STAKES WIN..  36  STAKES PLACE..  50

* - OFFICIAL AND UNOFFICIAL EARNINGS COMBINED
```

This portion of the sire's get record of Alamitos Bar shows how Kaweah Bar appears on that record. A short summary follows the name of the horse, and then all his awards and stake races are listed.

Figure 4.

judge, the horse's placing, the number of horses that competed, the name of the class, and points earned.

Two other types of records are available, those being breeder and owner records. One problem we encounter in this area is identity. Oftentimes, people will assemble a broodmare band and fail to transfer all of the mares into the same name. All of our owner and breeder records are linked to our system by identification numbers. Should someone purchase a horse and transfer it into his name, then purchase another horse, transfer it into his and his wife's name, he creates two separate ownerships linked to our system by two separate identification numbers. Two breeders' records will then be on file, should foals be registered under each ownership.

Owner and breeder records are similar to our other records. Each horse owned or bred by the individual is listed, along with performance records, if any.

As I stated earlier, our records are very well received by our owners, breeders, and prospective owners. There are several reasons for this. The most important is accessibility. Within our office we have a records research section that consists of four ladies who answer all inquiries regarding records of horses. Since all of our horses are "on line" on the computer, a simple phone call to our office can determine whether any horse has a race or show record, has produced any foals with any records, etc.

Written requests for copies of AQHA records are processed upon receipt and normally mailed the next working day. We are proud of our records and also proud that most of our records are free to anyone requesting them. There are cases where charges are assessed for reproducing and mailing large numbers of records. For instance, the first 19 pages of individual performance (show or race) records are free. If 20 or more pages are requested, a charge of 25 cents per page is assessed.

Likewise, the first 19 pages of get of sire and produce of dam records are free. Twenty or more pages are 50 cents per page.

As an example, a get of sire record on Alamitos Bar, who has 631 registered foals, amounts to 42 pages. The cost for his sire record is currently $21. A produce of dam record on Angie Miss, the dam of Kaweah Bar is free, as she has produced nine foals, which amounts to two pages. Actually, all AQHA records are maintained free of charge. We simply charge in some cases to reproduce the records.

If you don't think our records are popular with our people, drop by our office sometime and observe our records research section in action. When they are not communicating on the phone, which is not very often, they are handling written requests.

Our records are extremely important in the marketing of American quarter horses. They are also a permanent part of any horse's history. As the leading breeder of AQHA champions, Howard Pitzer of Ericson, Nebraska, recently stated

"Our aim always has been to build records, to accumulate AQHA points first, and everything else second. Futurities are getting more and more popular...as time goes on. People forget about a futurity win, but points go down in the books, and they are in the books of our association so you can study them when you go looking for a stallion or mare."

If someone is considering purchasing a horse, a quick phone call to AQHA can verify any claims made by the seller as to the horse's official records.

Also, all claims made in advertisements in The Quarter Horse Journal are checked by our staff before they are allowed in the ad. This works two ways. First, it assures accuracy and uniformity among all ads. And secondly, we oftentimes find that the horses in question actually have a better record than the advertiser was aware.

Also, many types of monthly and year-end-leaders lists are maintained by AQHA, such as show and contest leaders in each of our divisions: youth, amateur, and open; leading money-earning race horses; leading breeders in many different categories of both show and race horses; and several others.

Simply stated, AQHA records are the key to marketing and promotion of American quarter horses and the people who breed them. It has become one of AQHA's primary day-to-day functions and will always be an integral part of our association.

```
01/28/82            A M E R I C A N   Q U A R T E R   H O R S E   A S S O C I A T I O N        0361940          1000
                            SIRE'S GET RECORD          0361940  DECKA CENTER                           PAGE-   2
---------------------------------------------------------------------------------------------------------------------

0517799 SAILING DECK            S-G-67
      DAM---- 0310400 TONEY BEGGAR
        RACE----- LATEST YR-69  BEST GRADE-D     POINTS-   0  STARTS-  3  1ST-  0  2ND-  0  3RD-  0  TOTAL EARNINGS-

0523062 DECKA PEARL             B-M-67
      DAM--- 0228698 MISS BAYOU CREEK
        RACE----- LATEST YR-70  BEST GRADE-A     POINTS-   0  STARTS-  7  1ST-  1  2ND-  2  3RD-  1  TOTAL EARNINGS-      722
             STAKES------ 70-05-08 END 350 AT- 3 3RD  GARFIELD DOWNS SPRING DBY                   PURSE-   1,512  EARN-   302

0523063 VAN DECKA               B-G-67
      DAM---- 0066558 VANESSA DEE
        RACE----- LATEST YR-69  BEST GRADE-B     POINTS-   0  STARTS-  3  1ST-  0  2ND-  0  3RD-  0  TOTAL EARNINGS-       25
          SHOW------ OPEN-- LYR-76  HLT-  33  WKG-  77    --YOUTH--  LYR-78  HLT- 179  WKG-4036    --AMATEUR--  LYR-00  WKG-  00
          AWARDS------ 1971 OPEN      REGISTER OF MERIT
                       1972 OPEN      AQHA CHAMPION
                       1974 OPEN      WORLD CHAMPION SR WESTERN RIDING          9TH PLACE
                       1972 YOUTH   A REGISTER OF MERIT
                       1972 YOUTH   A AQHA CHAMPION
                       1973 YOUTH   A SUPERIOR HALTER
                       1973 YOUTH   A SUPERIOR SHOWMANSHIP
                       1973 YOUTH   A SUPERIOR WESTERN HORSEMANSHIP
                       1973 YOUTH   A SUPERIOR WESTERN RIDING
                       1973 YOUTH   A SUPERIOR WESTERN PLEASURE
                       1973 YOUTH   A SUPERIOR HUNT SEAT
                       1973 YOUTH   A AQHA PERFORMANCE CHAMPION
                       1973 YOUTH   A HIGH POINT HALTER GELDING                  5TH PLACE
                       1973 YOUTH   A HIGH POINT SHOWMANSHIP                     4TH PLACE
                       1973 YOUTH   A HIGH POINT WESTERN HORSEMANSHIP            3RD PLACE
                       1973 YOUTH   A HIGH POINT REINING                         9TH PLACE
                       1973 YOUTH   A HIGH POINT WESTERN RIDING
                       1973 YOUTH   A HIGH POINT WESTERN PLEASURE                8TH PLACE
                       1973 YOUTH   A HIGH POINT HUNT SEAT                       3RD PLACE
                       1973 YOUTH   A ALL AROUND                                 3RD PLACE
                       1973 YOUTH   A WORLD CHAMPION WESTERN HORSEMANSHIP        7TH PLACE
                       1973 YOUTH   A WORLD CHAMPION WESTERN RIDING              8TH PLACE
                       1973 YOUTH   A WORLD CHAMPION HUNT SEAT                  10TH PLACE
                       1974 YOUTH   A SUPERIOR TRAIL HORSE
                       1974 YOUTH   A HIGH POINT HALTER GELDING                  7TH PLACE
                       1974 YOUTH   A HIGH POINT SHOWMANSHIP
                       1974 YOUTH   A HIGH POINT WESTERN HORSEMANSHIP
                       1974 YOUTH   A HIGH POINT WESTERN RIDING                  2ND PLACE
                       1974 YOUTH   A HIGH POINT WESTERN PLEASURE                3RD PLACE
                       1974 YOUTH   A HIGH POINT TRAIL HORSE
                       1974 YOUTH   A HIGH POINT HUNT SEAT
                       1974 YOUTH   A ALL AROUND
                       1974 YOUTH   A WORLD CHAMPION AGED GELDING                5TH PLACE
                       1974 YOUTH   A WORLD CHAMPION SHOWMANSHIP
                       1974 YOUTH   A WORLD CHAMPION WESTERN HORSEMANSHIP        6TH PLACE
                       1974 YOUTH   A WORLD CHAMPION TRAIL HORSE                 6TH PLACE
                       1975 YOUTH   A HIGH POINT SHOWMANSHIP
                       1975 YOUTH   A HIGH POINT WESTERN HORSEMANSHIP            2ND PLACE
                       1975 YOUTH   A HIGH POINT WESTERN PLEASURE
                       1975 YOUTH   A HIGH POINT HUNT SEAT                       4TH PLACE
                       1975 YOUTH   A ALL AROUND
                       1975 YOUTH   A WORLD CHAMPION SHOWMANSHIP                 3RD PLACE
                       1975 YOUTH   A WORLD CHAMPION WESTERN RIDING              6TH PLACE


01/28/82            A M E R I C A N   Q U A R T E R   H O R S E   A S S O C I A T I O N        0361940          1000
                            SIRE'S GET RECORD          0361940  DECKA CENTER                           PAGE-   3
---------------------------------------------------------------------------------------------------------------------

                       1975 YOUTH   A WORLD CHAMPION WESTERN PLEASURE            4TH PLACE
                       1975 YOUTH   A WORLD CHAMPION HUNT SEAT                   5TH PLACE
                       1976 YOUTH   B REGISTER OF MERIT
                       1976 YOUTH   B SUPERIOR SHOWMANSHIP
                       1976 YOUTH   B SUPERIOR WESTERN HORSEMANSHIP
                       1976 YOUTH   B SUPERIOR WESTERN PLEASURE
                       1976 YOUTH   B AQHA PERFORMANCE CHAMPION
                       1976 YOUTH   B HIGH POINT WESTERN HORSEMANSHIP            8TH PLACE
                       1977 YOUTH   B AQHA CHAMPION
                       1977 YOUTH   B SUPERIOR TRAIL HORSE
                       1977 YOUTH   B SUPERIOR HUNT SEAT
                       1977 YOUTH   B HIGH POINT SHOWMANSHIP                     2ND PLACE
                       1977 YOUTH   B HIGH POINT HUNT SEAT EQUITATION            2ND PLACE
                       1977 YOUTH   B HIGH POINT WESTERN HORSEMANSHIP
                       1977 YOUTH   B HIGH POINT WESTERN RIDING                  5TH PLACE
                       1977 YOUTH   B HIGH POINT WESTERN PLEASURE
                       1977 YOUTH   B HIGH POINT TRAIL HORSE                     2ND PLACE
                       1977 YOUTH   B HIGH POINT HUNT SEAT                       2ND PLACE
                       1977 YOUTH   B ALL AROUND                                 2ND PLACE
                       1977 YOUTH   B WORLD CHAMPION SHOWMANSHIP                 9TH PLACE
                       1977 YOUTH   B WORLD CHAMPION WESTERN PLEASURE            7TH PLACE

        RACE------ LATEST YR-81  BEST GRADE-0067  POINTS-   0  STARTS-  2  1ST-  0  2ND-  0  3RD-  0  TOTAL EARNINGS-

1585775 DECKA JINA              S-M-77
      DAM---- 0512096 VENN'S JINA
        RACE----- LATEST YR-80  BEST GRADE-0053  POINTS-   0  STARTS-  1  1ST-  0  2ND-  0  3RD-  0  TOTAL EARNINGS-


TOTAL HORSES..................  257   PERFORMERS.......   136
TOTAL OPEN  HALTER  POINTS....  255   POINT EARNERS....     6   PERFORMERS...   12  ROM...    3  AQHA...   1  SUPM...
TOTAL OPEN  WORKING POINTS....  127   POINT EARNERS....     7
TOTAL YOUTH HALTER  POINTS....  263   POINT EARNERS....     5   PERFORMERS...    3  ROM...    3  AQHA...   3  SUPM...
TOTAL YOUTH WORKING POINTS.... 4064   POINT EARNERS....     3                                                  EARNINGS
TOTAL AMTUR WORKING POINTS....        POINT EARNERS....         PERFORMERS...       ROM...                UNOFFICIAL.   $15,767
TOTAL RACE  STARTS..           1526   RACE STARTERS....   134   RACE POINTS.. 1915  RACE POINT EARNERS..  77   OFFICIAL...  $345,646
A.......               AA.......   36   AAA.............    28   AAAT.........       ROM...   65  STAKES WIN..   5  STAKES PLACE..     5

* - OFFICIAL AND UNOFFICIAL EARNINGS COMBINED
```

This portion of Decka Center's sire's get record shows how Van Decka appears on the record. The example is simply a portion of the record, and it is spliced to also show how the ending appears.

Figure 5.

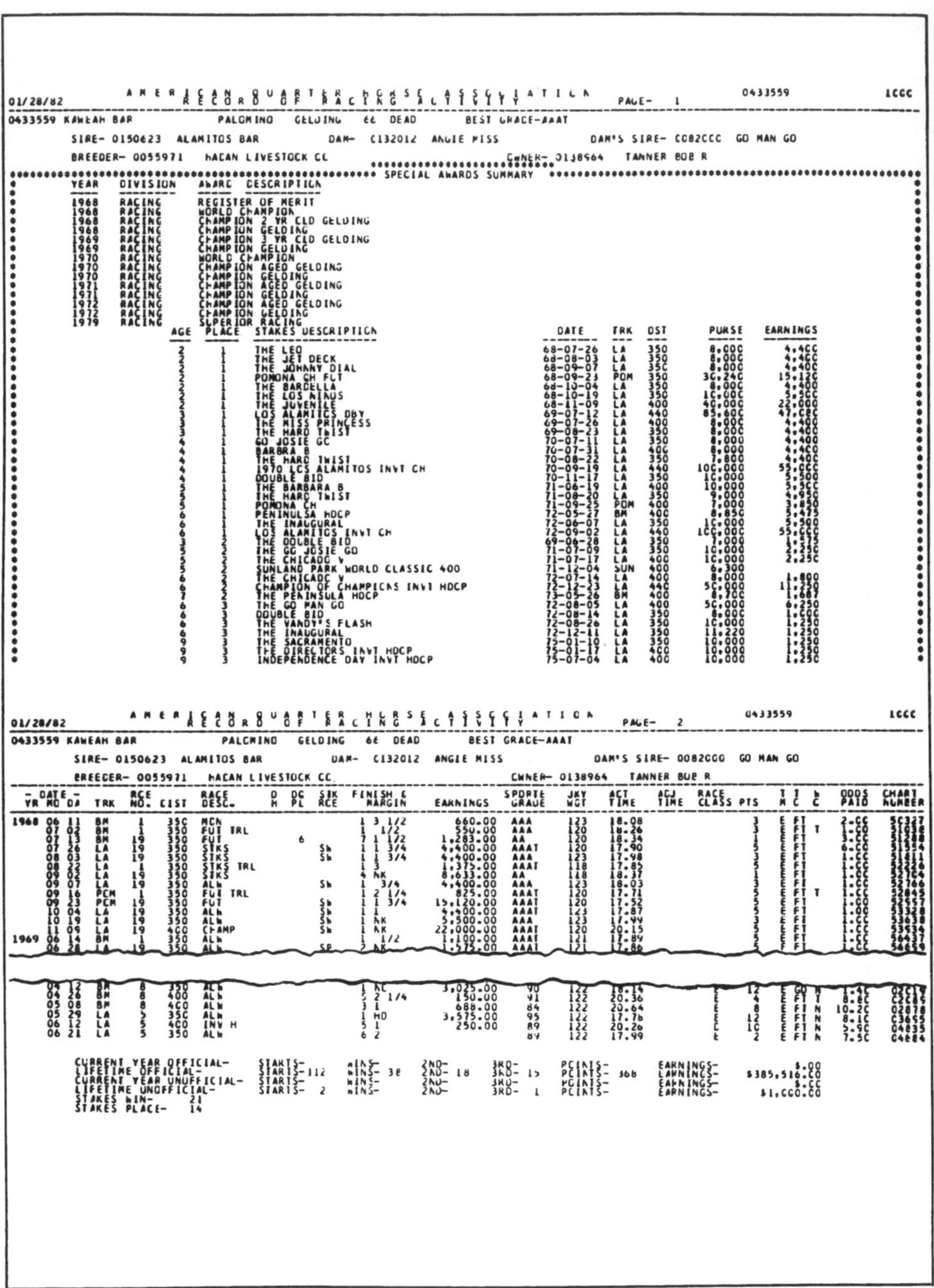

01/28/82 AMERICAN QUARTER HORSE ASSOCIATION RECORD OF RACING ACTIVITY PAGE- 1 0433559 1CCC

0433559 KAWEAH BAR PALOMINO GELDING 66 DEAD BEST GRADE-AAAT

SIRE- 0150623 ALAMITOS BAR DAM- C132012 ANGIE MISS DAM'S SIRE- C082CCC GO MAN GO

BREEDER- 0055971 HACAN LIVESTOCK CC OWNER- 0138964 TANNER BOB R

SPECIAL AWARDS SUMMARY

YEAR	DIVISION	AWARD DESCRIPTION
1968	RACING	REGISTER OF MERIT
1968	RACING	WORLD CHAMPION
1968	RACING	CHAMPION 2 YR CLD GELDING
1968	RACING	CHAMPION GELDING
1969	RACING	CHAMPION 3 YR CLD GELDING
1969	RACING	CHAMPION GELDING
1970	RACING	WORLD CHAMPION
1970	RACING	CHAMPION AGED GELDING
1970	RACING	CHAMPION GELDING
1971	RACING	CHAMPION AGED GELDING
1971	RACING	CHAMPION GELDING
1972	RACING	CHAMPION AGED GELDING
1972	RACING	CHAMPION GELDING
1979	RACING	SUPERIOR RACING

AGE	PLACE	STAKES DESCRIPTION	DATE	TRK	DST	PURSE	EARNINGS
2	1	THE LEO	68-07-26	LA	350	8,COC	4,4CC
2	1	THE JET DECK	68-08-03	LA	350	8,00C	4,4CC
2	1	THE JOHNNY DIAL	68-09-07	LA	35C	8,000	4,40C
2	1	POMONA CH FUT	68-09-23	POM	350	3C,24C	15,12C
2	1	THE BARCELLA	68-10-04	LA	350	8,00C	4,400
2	1	THE LOS NINOS	68-10-19	LA	350	1C,00C	5,5CC
2	1	THE JUVENILE	68-11-09	LA	400	40,00C	22,000
3	1	LOS ALAMITCS DBY	69-07-12	LA	440	85,60C	47,C8C
3	1	THE MISS PRINCESS	69-07-26	LA	400	8,00C	4,400
3	1	THE HARD TWIST	69-08-23	LA	350	8,00C	4,400
4	1	GO JOSIE GC	70-07-11	LA	350	8,000	4,400
4	1	BARBRA B	70-07-31	LA	40C	8,000	4,4C0
4	1	THE HARD TWIST	70-08-22	LA	350	7,800	4,40C
4	1	1970 LCS ALAMITOS INVT CH	70-09-19	LA	440	10C,000	55,CCC
4	1	DOUBLE BID	70-11-17	LA	350	1C,000	5,500
5	1	THE BARBARA B	71-06-19	LA	400	10,000	5,5CC
5	1	THE HARD TWIST	71-08-20	LA	350	9,000	4,95C
5	1	POMONA CH	71-09-25	POM	400	7,000	3,850
6	1	PENINULSA HDCP	72-05-27	BM	400	8,85C	5,475
6	1	THE INAUGURAL	72-06-07	LA	350	1C,000	5,500
6	1	LOS ALAMITCS INVT CH	72-09-02	LA	440	1CC,00C	55,CCC
3	2	THE DOUBLE BID	69-06-28	LA	350	7,000	1,575
5	2	THE GC JOSIE GO	71-07-09	LA	350	10,000	2,25C
5	2	THE CHICADC V	71-07-17	LA	400	1C,000	2,25C
5	2	SUNLAND PARK WORLD CLASSIC 400	71-12-04	SUN	400	6,300	
6	2	THE CHICADC V	72-07-14	LA	400	8,000	1,800
6	2	CHAMPION OF CHAMPICNS INVT HDCP	72-12-23	LA	44C	5C,000	11,250
7	2	THE PENINSULA HDCP	73-05-26	BM	400	8,70C	1,687
6	3	THE GO MAN GO	72-08-05	LA	400	5C,000	6,250
6	3	DOUBLE BID	72-08-14	LA	350	8,00C	1,C0C
6	3	THE VANDY'S FLASH	72-08-26	LA	350	1C,000	1,250
6	3	THE INAUGURAL	72-12-11	LA	350	11,220	1,250
9	3	THE SACRAMENTO	75-01-10	LA	350	10,000	1,250
9	3	THE DIRECTORS INVT HDCP	75-01-17	LA	40C	10,000	1,250
9	3	INDEPENDENCE DAY INVT HDCP	75-07-04	LA	40C	10,000	1,25C

01/28/82 AMERICAN QUARTER HORSE ASSOCIATION RECORD OF RACING ACTIVITY PAGE- 2 0433559 1CCC

0433559 KAWEAH BAR PALOMINO GELDING 66 DEAD BEST GRADE-AAAT

SIRE- 0150623 ALAMITOS BAR DAM- C132012 ANGIE MISS DAM'S SIRE- 0082C00 GO MAN GO

BREEDER- 0055971 HACAN LIVESTOCK CC. OWNER- 0138964 TANNER BOB R

YR	MO	DA	TRK	RCE NO.	DIST	RACE DESC.	D H	DC PL	STK RCE	FINISH & MARGIN	EARNINGS	SPORTE GRADE	JKY WGT	ACT TIME	ADJ TIME	RACE CLASS	PTS	T M	T C	W C	ODDS PAID	CHART NUMBER
1968	06	11	BM	1	35C	MCN				1 3 1/2	660.00	AAA	123	18.08			3	E	FT		2.CC	5C327
	07	02	BM	1	350	FUT TRL				1 1/2	550.00	AAA	120	18.26			3	E	FT	T	1.C0	51038
	07	13	BM	19	350	FUT		6		7 1 1/2	1,283.00	AA	120	18.34			1	E	FT		3.CC	51288
	07	26	LA	19	350	STKS			SW	1 1 3/4	4,400.00	AAAT	120	17.90			5	E	FT		6.C0	51554
	08	03	LA	19	350	STKS			SW	1 1 3/4	4,400.00	AAA	123	17.98			3	E	FT		1.CC	51811
	08	22	LA	1	350	STKS TRL				1 3	1,375.00	AAAT	118	17.85			5	E	FT		1.CC	52226
	09	02	LA	19	350	STKS				4 NK	8,633.00	AA	118	18.37			1	E	FT		1.CC	52764
	09	07	LA	19	350	ALW			SW	1 3/4	4,400.00	AAA	123	18.03			3	E	FT		1.CC	52766
	09	16	PCM	1	350	FUT TRL				1 2 1/4	825.00	AAAT	120	17.71			5	E	FT	T	1.CC	52845
	09	23	PCM	19	350	FUT			SW	1 1 3/4	15,120.00	AAAT	120	17.52			5	E	FT		1.C0	52557
	10	04	LA	19	350	ALW			SW	1 1	4,400.00	AAAT	123	17.87			5	E	FT		1.00	53328
	10	19	LA	19	350	ALW			SW	1 NK	5,500.00	AAA	123	17.99			3	E	FT		1.CC	53638
	11	09	LA	19	4C0	CHAMP			SW	1 NK	22,000.00	AAAT	120	20.15			5	E	FT		1.CC	53534
1969	06	14	BM	1	350	ALW				1 1/2	1,100.00	AAAT	121	17.89			5	E	FT		1.CC	56437
	06	28	LA	19	350	ALW			SP	2 NK	1,575.00	AAAT	121	17.86			5	F	FT		1.CC	56659
	04	12	BM	8	350	ALW				1 NC	3,025.00	90	122	18.14			12	E	GD	N	1.4C	02C19
	04	26	BM	8	400	ALW				5 2 1/4	150.00	91	122	20.36		E	4	E	FT	T	8.8C	C2C85
	05	08	BM	8	4C0	ALW				3 1	688.00	84	122	20.64		E	8	E	FT	N	10.2C	02878
	05	29	LA	5	35C	ALW				1 HD	3,575.00	95	122	17.76		E	12	E	FT	N	8.1C	C3655
	06	12	LA	5	4C0	INV H				5 1	250.00	89	122	20.26		C	1C	E	FT	N	5.9C	04835
	06	21	LA	5	350	ALW				6 2		89	122	17.99		E	2	E	FT	N	7.5C	C4884

CURRENT YEAR OFFICIAL-	STARTS-	WINS-	2ND-	3RD-	POINTS-	EARNINGS- $.00
LIFETIME OFFICIAL-	STARTS- 112	WINS- 38	2ND- 18	3RD- 15	POINTS- 368	EARNINGS- $385,516.C0
CURRENT YEAR UNOFFICIAL-	STARTS-	WINS-	2ND-	3RD-	POINTS-	EARNINGS- $.CC
LIFETIME UNOFFICIAL-	STARTS- 2	WINS-	2ND-	3RD- 1	POINTS-	EARNINGS- $1,CC0.C0
STAKES WIN- 21						
STAKES PLACE- 14						

This is a portion of Kaweah Bar's racing record. Awards the horse earned lead off the record, followed by a list of all stake races in which he placed. The remainder of the record is the list of races in which he ran, and is the same as what has been available in the past. At the end of the record is a summary of the horse's racing activity.

Figure 6.

01/28/82	AMERICAN QUARTER HORSE ASSOCIATION RECORD OF SHOW ACTIVITY	PAGE- 1	0523063	1000

0523063 VAN DECKA BAY GELDING 67 BEST GRADE-E

SIRE- 0361940 DECKA CENTER DAM- 0066558 VANESSA DEE DAM'S SIRE- 0043626 VANDY

BREEDER- 0072413 JOHNSON JOHNNY OWNER- 0228872 JOHNSON KIM

SPECIAL AWARDS SUMMARY

YEAR	DIVISION	AWARD DESCRIPTION	
1971	OPEN	REGISTER OF MERIT	
1972	OPEN	AQHA CHAMPION	
1974	OPEN	WORLD CHAMPION SR WESTERN RIDING	9TH PLACE
1972	YOUTH -A	REGISTER OF MERIT	
1972	YOUTH -A	AQHA CHAMPION	
1973	YOUTH -A	SUPERIOR HALTER	
1973	YOUTH -A	SUPERIOR SHOWMANSHIP	
1973	YOUTH -A	SUPERIOR WESTERN HORSEMANSHIP	
1973	YOUTH -A	SUPERIOR WESTERN RIDING	
1973	YOUTH -A	SUPERIOR WESTERN PLEASURE	
1973	YOUTH -A	SUPERIOR HUNT SEAT	
1973	YOUTH -A	AQHA PERFORMANCE CHAMPION	
1973	YOUTH -A	HIGH POINT HALTER GELDING	5TH PLACE
1973	YOUTH -A	HIGH POINT SHOWMANSHIP	4TH PLACE
1973	YOUTH -A	HIGH POINT WESTERN HORSEMANSHIP	3RD PLACE
1973	YOUTH -A	HIGH POINT REINING	9TH PLACE
1973	YOUTH -A	HIGH POINT WESTERN RIDING	
1973	YOUTH -A	HIGH POINT WESTERN PLEASURE	8TH PLACE
1973	YOUTH -A	HIGH POINT HUNT SEAT	3RD PLACE
1973	YOUTH -A	ALL AROUND	3RD PLACE
1973	YOUTH -A	WORLD CHAMPION WESTERN HORSEMANSHIP	7TH PLACE
1973	YOUTH -A	WORLD CHAMPION WESTERN RIDING	8TH PLACE
1973	YOUTH -A	WORLD CHAMPION HUNT SEAT	10TH PLACE
1974	YOUTH -A	SUPERIOR TRAIL HORSE	
1974	YOUTH -A	HIGH POINT HALTER GELDING	7TH PLACE
1974	YOUTH -A	HIGH POINT SHOWMANSHIP	
1974	YOUTH -A	HIGH POINT WESTERN HORSEMANSHIP	
1974	YOUTH -A	HIGH POINT WESTERN RIDING	2ND PLACE
1974	YOUTH -A	HIGH POINT WESTERN PLEASURE	3RD PLACE
1974	YOUTH -A	HIGH POINT TRAIL HORSE	
1974	YOUTH -A	HIGH POINT HUNT SEAT	
1974	YOUTH -A	ALL AROUND	
1974	YOUTH -A	WORLD CHAMPION AGED GELDING	5TH PLACE
1974	YOUTH -A	WORLD CHAMPION SHOWMANSHIP	
1974	YOUTH -A	WORLD CHAMPION WESTERN HORSEMANSHIP	6TH PLACE
1974	YOUTH -A	WORLD CHAMPION TRAIL HORSE	6TH PLACE
1975	YOUTH -A	HIGH POINT SHOWMANSHIP	
1975	YOUTH -A	HIGH POINT WESTERN HORSEMANSHIP	2ND PLACE
1975	YOUTH -A	HIGH POINT WESTERN PLEASURE	
1975	YOUTH -A	HIGH POINT HUNT SEAT	4TH PLACE
1975	YOUTH -A	ALL AROUND	
1975	YOUTH -A	WORLD CHAMPION SHOWMANSHIP	3RD PLACE
1975	YOUTH -A	WORLD CHAMPION WESTERN RIDING	6TH PLACE
1975	YOUTH -A	WORLD CHAMPION WESTERN PLEASURE	4TH PLACE
1975	YOUTH -A	WORLD CHAMPION HUNT SEAT	5TH PLACE
1976	YOUTH -B	REGISTER OF MERIT	
1976	YOUTH -B	SUPERIOR SHOWMANSHIP	
1976	YOUTH -B	SUPERIOR WESTERN HORSEMANSHIP	
1976	YOUTH -B	SUPERIOR WESTERN PLEASURE	
1976	YOUTH -B	AQHA PERFORMANCE CHAMPION	

SPECIAL AWARDS SUMMARY

YEAR	DIVISION	AWARD DESCRIPTION	
1976	YOUTH -B	HIGH POINT WESTERN HORSEMANSHIP	8TH PLACE
1977	YOUTH -B	AQHA CHAMPION	
1977	YOUTH -B	SUPERIOR TRAIL HORSE	
1977	YOUTH -B	SUPERIOR HUNT SEAT	
1977	YOUTH -B	HIGH POINT SHOWMANSHIP	2ND PLACE
1977	YOUTH -B	HIGH POINT HUNT SEAT EQUITATION	2ND PLACE
1977	YOUTH -B	HIGH POINT WESTERN HORSEMANSHIP	
1977	YOUTH -B	HIGH POINT WESTERN RIDING	5TH PLACE
1977	YOUTH -B	HIGH POINT WESTERN PLEASURE	
1977	YOUTH -B	HIGH POINT TRAIL HORSE	2ND PLACE
1977	YOUTH -B	HIGH POINT HUNT SEAT	2ND PLACE
1977	YOUTH -B	ALL AROUND	2ND PLACE
1977	YOUTH -B	WORLD CHAMPION SHOWMANSHIP	9TH PLACE
1977	YOUTH -B	WORLD CHAMPION WESTERN PLEASURE	7TH PLACE

INDEX	YR	MO	FRST	LAST	CITY AND STATE		GRADE	JUDGE	PLACE	NO IN CLASS	CLASS DESC AND CODE		HALTER	WORKING
1031031	1968	08	18	18	IOLA	KS	B	BLISS F E	2	13	HALTER	A		
1067031	1968	09	07	07	STANLEY	KS	B	COOK JACK	1	8	HALTER	A		
1070031	1968	09	08	08	LEE'S SUMMITT	MO	C	BUSH BILLY C	1	5	HALTER	A		
1089031	1968	08	24	24	LEE'S SUMMIT	MO	B	MOON KEITH	2	10	HALTER	A		
1117031	1968	08	25	25	SEDALIA	MO	B	DANIELS JIM	2	10	HALTER	A		
1123031	1968	08	31	31	CARTHAGE	MO	B	REMKES J R	1	5	HALTER	A		
1136031	1968	09	15	15	FRUITLAND	MO	C	SCOTT CLYDE	1	5	HALTER	A		
					GRAND CHAMPION			RESV. CHAMPION	CURRENT YEAR ***		TOTALS ***		0.0	0.0
0772011	1969	07	18	18	CARTHAGE	MO	B	HUNTER JAMES 'ROY'			RESERVE	A		
0772032	1969	07	18	18	CARTHAGE	MO	B	HUNTER JAMES 'ROY'	1	3	HALTER	A		
0801011	1969	07	19	19	JOPLIN	MO	B	MILLER H S			RESERVE	A		
0801032	1969	07	19	19	JOPLIN	MO	B	MILLER H S	1	5	HALTER	A	1.0	
0815011	1969	07	20	20	NEOSHO	MO	C	FOSTER J C			RESERVE	A		
0815032	1969	07	20	20	NEOSHO	MO	C	FOSTER J C	1	5	HALTER	A	1.0	
					GRAND CHAMPION			RESV. CHAMPION 3	CURRENT YEAR ***		TOTALS ***		2.0	0.0
0043010	1970	02	28	28	ELLISVILLE	MS	A	WOOD OLIVER			GRAND	A	1.0	
0043033	1970	02	28	28	ELLISVILLE	MS	A	WOOD OLIVER	1	10	HALTER	A	2.0	
0043421	1970	02	28	28	ELLISVILLE	MS	A	WOOD OLIVER	6	34	WS PLES	O		1.0
0048010	1970	03	07	07	ELLISVILLE	MS	B	HANKS RUBE			GRAND	A	2.0	
0048033	1970	03	07	07	ELLISVILLE	MS	B	HANKS RUBE	1	8	HALTER	A	1.0	
0062340	1975	02	09	09	GOODLETTSVILLE	TN	B	REHL ROBERT P DR	1	5	REINING	K		1.0
0082360	1975	02	09	09	GOODLETTSVILLE	TN	B	REHL ROBERT P DR	1	6	W RIDING	L		1.0
1297440	1975	08	02	02	SALEM	IL	A	JOHNSON ROBERT L	2	26	HUNTSEAT	P		4.0
1943360	1975	11	28	28	WEST PALM BEACH	FL	A	MCMURTRIE EDWARD	2	10	W RIDING	L		1.0
					GRAND CHAMPION			RESV. CHAMPION	CURRENT YEAR ***		TOTALS ***		0.0	5.0
0046342	1976	01	27	27	PHOENIX	AZ	A	RYDBERG L OMER	1	24	W RIDING	L		4.0
0046422	1976	01	27	27	PHOENIX	AZ	A	RYDBERG L OMER	5	55	WS PLES	O		2.0
					GRAND CHAMPION			RESV. CHAMPION	CURRENT YEAR ***		TOTALS ***		0.0	6.0
					GRAND CHAMPION 7			RESV. CHAMPION 16	LIFE ***		TOTALS ***		33.0	77.0
					GRAND CHAMPION 7			RESV. CHAMPION 16	LIFE ***		TOTALS ***		0.0	0.0

CLASS	CYR. WINS	CYR. PTS.	LIFE PTS.	CLASS	CYR. WINS	CYR. PTS.	LIFE PTS.	CLASS	CYR. WINS	CYR. PTS.	LIFE PTS.	CLASS	CYR. WINS	CYR. PTS.	LIFE PTS.
HALTER	00	000	33.0	REINING	00	000	1.0	W RIDING	00	000	29.0	TRAIL	00	000	5.0
WS PLES	00	000	14.0	HUNTSEAT	00	000	28.0								

A show record on Van Decka begins with a list of all awards earned, and then goes into the regular show record which has been available before. This spliced record shows the beginning and ending, and shows the summary that is still available at the end.

Figure 7.

45

AGNET: A NATIONAL COMPUTER SYSTEM FOR CATTLEMEN

Harlan G. Hughes

In the early 1970s, two professors at the University of Nebraska conceived the idea of an agricultural computer system designed specifically for farmers and ranchers. They developed the computer system now known across the country as AGNET--The Agricultural Computer Network. In 1977, the governors of five states (Nebraska, South Dakota, North Dakota, Montana and Wyoming) jointly funded a pilot project to test if farmers and ranchers in their respective states would use a computer system to make better management decisions. AGNET has now developed so that over 400,000 calls a year are being made to the AGNET computer. AGNET is, indeed, a management tool for agriculture.

Wyoming now has computer terminals in all 23 county extension offices, and county extension agents are now receiving training on how to use these terminals with their farmer and rancher clientele.

AGNET is one of three computers in the Wyoming computer center. This operator controls the AGNET computer from the central station. If we have done the job right, the operator should not have to do much. Due to the speed of the computer, we prefer that the machine do as much of its own operation as possible. This operator, however, can and does take over control of the machine whenever necessary.

AGNET is a mass storage system. Behind the dark windows in AGNET storage units are stacks of phonograph-like records used for storage of data and programs. All of AGNET's programs are stored on disks so that when you type in the name of a program, the computer can immediately go to the appropriate disk and find the requested program. We do not have to wait for an operator to mount a tape or to do any manual intervention. AGNET is one of the largest mass storage systems in the world.

Farmer advisors on the AGNET payroll are very special persons to AGNET. George, one of the advisors, is a real character whom I wish everyone could met. George's role is to help make sure that what we have on AGNET will work for farmers and ranchers. I have heard George say, "Harlan, that is the dumbest #%&"* thing I have ever heard!" Or I have heard George say, "That may be well and fine in your ivory tower, but out on the farm we do not have that kind of data." We have two half-time farmers on the AGNET payroll and they play a very important and unique role in the design and operation of the total AGNET system.

AGNET is equipped so that we can have over 200 telephone calls coming in to the computer at one time. We are now averaging a phone call into AGNET every four minutes, seven days a week, 24 hours a day. That is over 400,000 phone calls a year.

The AGNET computer in Nebraska is located in the basement of the State Capitol. By design AGNET is not on a university campus computer (and probably never will be on a campus computer) because of our computer needs and demands. A university computer is set up for research and administrative data processing. We need a service-oriented computer center that can consult us before the system is changed or shut down. Our users are not computer science PhD's and become frustrated with computer down time or off time. AGNET is often our user's first contact with a computer and since they are paying for the computer time, we place some stringent demands on the computer center.

THE FULL PARTNER STATES AND THE STAFF

In 1977, five states previously mentioned became full partners in the AGNET system. The best way to describe a full partner state is to say that each state has a member on the AGNET Board of Directors.

In July of 1980, the state of Washington joined as a full partner state and in October 1980 the state of Wisconsin joined AGNET. In July of 1981, Wisconsin withdrew, which leaves six full-partner states in the AGNET system, but other states currently are considering partner status.

The concept behind AGNET is to share the development and operating costs among the full-partner states. There are approximately 17 people on the AGNET payroll. Of the 17 people, Wyoming is paying for two. Each state pools its resources with the other states so that each state can take advantage of the total efforts of the total 17 people.

I have a goal in life and it is to dissolve state boundaries when it comes to information dissemination and use. We are proving that states have information and computer programs to share across state lines, and as we share our extension resources, the winners are our clientele.

AGNET PROGRAM LIBRARY

The six partner states in the AGNET system have developed the world's largest agricultural and home economics computer-program library in the world. Today there are over 200 programs available to AGNET users. With a library of this size, no one is expected to use or even know how to use all the programs.

Our goal is not to have users able to use all the programs in the library, but rather to have a large enough library so that every user can find at least one program of interest. This large smorgasbord of programs means that users should be able to find several programs of special interest. Appendix A provides a partial list of the programs available on AGNET. I have grouped the programs by subject matter to facilitate user interests.

The AGNET library has been put together with approximately 35 man-years of programming effort. In addition, each program development is supervised by a subject matter extension specialist who is responsible for the content of the program. Each subject-matter program is owned by the subject matter specialist and not by AGNET.

AGNET is exceptionally well equipped for the livestock producer. There are livestock ration-formulation programs available for range cattle, feedlot cattle, hogs, sheep, and poultry. There are programs available that will let you simulate on paper what your cattle will do in the feedlot given a description of your cattle and the ration that you are going to feed them. There are livestock budgets and planning prices stored in selected programs.

AGNET also has programs for the crop farmer. Machinery-cost calculators and crop budgets are available. In addition, there are whole farm or ranch budgeting programs designed to help you make long-run business investment decisions. There are many, many more programs designed to help you make better management decisions.

HARDWARE USED TO ACCESS THE AGNET LIBRARY

Touch-tone telephone. The first computer terminal that I installed in a county extension office in 1972 was a touch-tone telephone. It cost us $14 per month. We used the number pad to send information to the computer and the computer sent back the information over the special loud speaker attached to the phone. We would send in the input numbers by typing them into the telephone. The computer would talk back and say, "Answer number 1 is 420." We printed the answer onto a preprinted form that explained the interpretation of the number. This touch-tone terminal served us very well as a low-cost computer terminal. Industry still uses this type of small, low-cost terminal.

Execuport terminal. It soon became evident that we would like to have terminals in our county extension offices that would print out the computer information. We now have five of the Execuports in the Wyoming AGNET inventory. These cost $1,400 for reconditioned terminals.

Texas Instrument 745 Terminal. We have installed small portable TI-745s in most of our extension offices. The TI-745 weighs 13 pounds, has a clamp-on lid and a handle. It is the size of a small briefcase and weighs about half as much. Wyoming agents transport their terminals all over their counties. The TI-745 costs approximately $1,400 new.

North Dakota's CRT Terminal. Terminals come in all sizes and shapes. The Animal Science Department at North Dakota has a CRT terminal with a TV screen where one can read the data. It also has a printer that can be used to generate a printed copy of the output when desired. These dual-purpose units cost more money, but the flexibility is convenient and does reduce paper costs.

A Decwriter terminal. A Decwriter terminal is used by the Department of Agriculture in Alberta, Canada. I used their terminal to check my electronic mail. Alberta Agriculture subscribes to the AGNET system. This terminal costs around $2,000.

Teletype 43 terminal. My secretary and I use Teletype 43 terminals. Obviously this is the terminal that I like best. The TT-43 gets used more hours than any of our terminals and is virtually a maintenance-free terminal. The only problem is that it is not portable. It weighs 45 pounds and has the terminal plus the telephone coupler and the paper to move. This terminal also costs approximately $1,400.

Terminal with TV screens. We have one special terminal that drives 23-inch TV screens for demonstration and teaching purposes. These are the same TV screens that you see in airports with flight schedules. We use these screens so that clientele and students can see exactly what we type on the terminal and exactly what the computer sends back to the screens. These screens have helped to promote AGNET in Wyoming. The screens work so well that I will not give a demonstration without these screens. The special terminal and the two screens cost approximately $4,000; therefore, we have only one in Wyoming.

MICROCOMPUTERS FOR FARM AND RANCH

Let's now boil this all down. What does it mean for you on the farm or ranch?

Agriculture is going to have some serious challenges in the 80s. During the 60s and 70s your challenge was production, but the challenge in the 80s is going to be financial management.

Yes, the computer has the potential to improve your financial management. Let me make a prediction. Those of you that will be farming in 1990 will be using computers. Those of you that do not want to use a computer will not be farming in 1990. I often hear, "No damn computer is going to tell me how to run my ranch!" I predict that that person won't be farming in 1990. Many will have retired and others will have gone out of business. Computers are going to become commonplace on U.S. farms and ranches during the 1980s. Producer owned microcomputers can be useful in relation to AGNET. As I travel around the country talking to farmers and ranchers, I hear them expressing interest in three applications of microcomputers. The three applications are:

- Business accounting.
- Herd performance reporting.
- Financial management.

In an accompanying paper, I have discussed microcomputers, their use, and purchase.

Information Networking on AGNET

If you have a telephone coupler for your micro, you can access the following from AGNET:

- Current commodity market prices.
- Current USDA, Foreign Ag and Wyoming news releases.
- Agricultural outlook and situation reports.
- Western Livestock Market Information Project livestock analyses.
- Hay for sale.
- Sheep for sale.
- Certified pesticide applicators in Wyoming.
- People interested in judging county and state fairs.
- Horticultural tips during the summer.
- Home-canning tips during the canning season.
- Emergency information such as drought tips, Mount St. Helen's emergencies, etc.

You can even use your micro to access the UPI and AP news services for news stories dealing with, for example, the Farm Bill or "beef." The AP and UPI news services are available from two commercial time-sharing companies. You can do all this today with your micro if it has a telephone coupler on it.

Marketing Information On AGNET

We are putting about 17 different daily market-price files on AGNET, including the futures opening and closing prices. We have Chicago, Kansas City, and Minneapolis futures going onto AGNET. In addition, we have both national and selected local cash markets. We are reporting local feeder-cattle sales in Wyoming, Northeastern Colorado, and Western Nebraska. Local grain and cattle markets are being put on weekly for Nebraska. Feedlot reports for the major cattle feeding states are going periodically. Export data is also going on weekly. AGNET is becoming a major source of market information for agricultural producers.

This appears to be the major reason for most of our producer subscriptions to AGNET. They want current market information.

SUMMARY

In Wyoming we are using the AGNET system to provide Wyoming farmers and ranchers with their first contact with computers for:

- Record keeping such as beef herd performance.
- Problem solving for computer-aided decision making.
- Information networking such as daily market information.
- Electronic mail to speed up the delivery of research and extension information to clientele.

Computerized Management Aids (CMAs) are not new to agriculture. They are just new to the west. Leading midwest farmers have been using CMAs for over 10 years.

APPENDIX A

PARTIAL LIST OF AGNET PROGRAMS AVAILABLE

Livestock Production Models on AGNET:

BEEF Simulation and economic analysis of feeder's performance.
BHAP/BHPP Beef herd performance program and beef herd analysis program.
COWCULL Package to help determine which dairy cow to cull and when.
COWGAME Beef genetic selection simulation game.
CROSSBREED Evaluates beef crossbreeding systems & breed combinations.
FEEDMIX Least cost feed rations for beef, dairy, sheep, swine, & poultry.
FEEDSHEETS Prints batch weights of rations including scale readings.
RANGERATION Ration balancer for beef cows, wintering calves, horses & sheep.
SWINE Simulation and economic analysis of feeder's performance.
TURKEY Simulation and economic analysis of turkey's performance.
VITAMINCHECK Checks the level of vitamins & trace minerals in swine diet.
WEAN Performance testing of weaning-weight calves.
YEARLING Performance testing of yearling-weight calves.

AG Engineering Models on AGNET:

BINDRY Predicts results of natural air & low temp. corn drying.
CONFINEMENT Ventilation requirements & heater size for swine confinement.
DRY Simulation of grain drying systems.
DUCTLOCATION Determines ducts to aerate grain in flat storage bldg.
FAN Determination of fan size and power needed for grain drying.
FUELALCOHOL Estimates production costs of ethanol in small-scale plants.
GRAINDRILL Calculates the lowest cost width for a grain drill for your farm.

PIPESIZE Computes most cost-effective size irrigation pipe to install.
PUMP Determination of irrigation costs.
SPRINKLER Examines feasibility of installing sprinkler irrigation.
STOREGRAIN
Cost analysis of on-farm and commercial grain storage.
TRACTORSELECT
Assists in determining suitability of tractors to enterprise.

Crop Production Models on AGNET:

BASIS Develops "historical basis" patterns for certain crops.
BESTCROP Provides equal return yield & price analysis between crops.
CROPINSURNACE
Analyzes whether to participate in crop insurance program.
FLEXCROP Forecasts yields based on amount of water available for crop growth.
IRRIGATE Irrigation scheduling.
RANGECOND Calculates the range condition and carrying capacity.
SEEDLIST Lists seed stocks for sale.
SOIL LOSS Estimates the computed soil-loss (tons/acre/year).
SOILSALT Diagnoses salinity & sodicity hazard for crop production.
SOYBEANPROD
Demonstration soybean production management model.

Home Economic Models on AGNET:

BEEFBUY Comparison of alternative methods of purchasing beef.
BUSPAK Package of financial analysis programs.
CARCOST Calculates costs of owning & operating a car or light truck.
DIETCHECK Food intake analysis.
DIETSUMMARY
Summary of analysis saved from DIETCHECK.
FIREWOOD Economic analysis of alternatives available with wood heat.
FOODPRESERVE
Calculates costs of preserving foods at home.
MONEYCHECK
Financial budgeting comparison for families.
PATTERN Helps select a commercial pattern size & type for figure.
STAINS Tells how to remove certain stains from fabrics.

4-H and Youth Models on AGNET:

CARCASS Scoring & tabulation of beef or lamb carcass judging contest.
FAIR Scoring and tabulation of judging contests.
JUDGELIST List of judges available for fairs and contests.
PREMIUM Compiles and summarizes fair premiums.

Farm and Ranch Planning Models on AGNET:

BUSPAK Package of financial analysis programs.
CALFWINTER Analyzes costs and returns associated with wintering calves.
COWCOST Examines the costs and returns for beef cow-calf enterprise.
CROPBUD Prints out select Wyoming crop budgets.
CROPBUDGET Analyzes the costs of producing a crop.
DAIRYCOST Analyzes the monthly costs and returns with milk production.
EWECOST Analyzes the costs & returns of sheep production enterprise.
FARMPROGRAM Analyzes USDA Acreage Reduction Program.
GRASSFAT Analyze costs and returns associated with pasturing calves.
LANDPAK Package of programs to assist in land management decisions.
LSBUDGETS Designed to print out stored livestock budgets.
MACHINEPAK Machinery analysis package.
PLANPAK Package of programs designed to help analyze and plan aspects of the business.
PLANTAX Income tax planning/management program.

Information Networking on AGNET:

CONFERENCE A continuing dialogue among users on a specific topic.
EWESALE Lists sheep for sale.
FAS Prints trade leads & commodity reports provided by USDA-FAS.
GUIDES Prints available reports of reference material information.
HAYLIST Lists hay for sale.
MAILBOX Used to send and receive mail.
NEWS Latest notifications about programs and user-related information.
NEWSRELEASE A program for rapid dissemination of news stories.

WHO IS — Retrieves name and company affiliation of individual users.

WYOPROGS — List of specialized Wyoming programs available only to Wyoming users.

Market Price Retrievals, Plotting, and Forecasting Models on AGNET:

CASHPLOT — Prints a plot of selected cash prices.

CORNPROJECT — Projects avg U.S. corn price for various marketing years.

MARKETCHART — Prints various charts on selected future and cash prices.

MARKETS — Various market reports and specialists' comments.

PRICEDATA — Prints selected historic cash and/or futures prices.

PRICEPLOT — Designed to plot market prices in graphic form.

Miscellaneous Programs on AGNET:

EDPAK — Demo programs illustrating computer-assisted instruction.

FILLIN — A "fill in the blank" quiz routine.

GAMES — Package of game programs.

INPUTFORMS — Prints available input forms.

JOBSEARCH — Matches abilities and interests to occupations.

MC — A multiple choice quiz routine.

MICROPROGRAM — Lists programs for microcomputers.

TESTPLOT — Standard analysis of variance.

TREE — Summarization of community forestry inventory.

46

RANCHER-OWNED MICROCOMPUTER SYSTEMS: WHAT'S AVAILABLE

Harlan G. Hughes

In the fall of 1977, Radio Shack started advertising the TSR-80 microcomputer for Christmas. This was the beginning of general-public awareness of the personal microcomputer. Another highly advertised microcomputer is the Atari which can be hooked up to a regular TV set, but the Atari is a game computer and, to my knowledge, has no agricultural programs available yet.

CURRENT MICROCOMPUTERS FOR RANCH AND FARM USE

There are two levels of microcomputers being considered by farmers. For the lack of any other terminology, I will use Level I and Level II as the classifications. Level I micros are the lowest cost and most popular systems. The three most common Level I micros in agriculture are the Radio Shack, Apple, and Pet Commodore.

Level I Hardware

Radio Shack Models I, II, & III. The Animal Science Division at the University of Wyoming has a Model I Radio Shack microcomputer. As is typical of most microcomputers, it has a keyboard, a TV screen, a disk unit, and a printer. Dr. Schoonover from Wyoming has developed a herd performance program for the Radio Shack Model I and III microcomputers. This program keeps track of the cow/calf information that ranches have been keeping on 3- x 5-inch cards. Once the data is inside the computer, management reports can be quickly printed out to help the rancher determine the cows to keep and the cows to cull. The same herd performance program that Dr. Schoonover has on the Radio Shack microcomputer is also on the AGNET system. We have several Wyoming ranchers currently using these herd performance programs.

The Radio Shack Model II has the disk drive built into the unit. Radio Shack refers to this as their small business machine.

The Radio Shack Model III has two disk drives built into the unit and presents pictures and graphs of your data.

The Model III can present a bar graph to show how ranch profits have changed the last 5 years. It has been suggested by some ranchers that a graph is purely academic if it represents ranch profits since profits have disappeared rather than changed.

Apple Computers. AGNET has an Apple computer with which we have one of our Teletype 43 AGNET terminals as a slave terminal. With proper connections, you can use your existing terminal as a slave printer on your microcomputer. Also, if you have a black and white TV you can back it up as the CRT on the Apple (and other brands as well). The resolution is not quite as clear as a regular monitor, but it is a cheaper way to get set up with a microcomputer.

Dr. Menkhaus, at the University of Wyoming, uses his Apple microcomputer in his Price Analysis class to teach undergraduate students how to use microcomputers.

The newest Apple is the Apple III. It has been out for about a year, but has had some technical troubles that has set its acceptance back. The Animal Science Division at Wyoming cancelled its order for the Apple III and ordered the Apple II Plus. This fast-growing company moved into a new product and forgot something called "quality control."

Pet Commodore Microcomputer. The Pet Commodore is being used by Alberta Agriculture in Canada and the Ag Economics Department at Wyoming. The Canadians have written a fair amount of agricultural software for the Pet and have been willing to share it with Wyoming so that we do have several decision aids for our Pet Commodore.

Word Processor on Screen. Micros also can be used for word processing. You can buy word processing programs for almost all micros that will let you use your micro to generate printed materials like letters and reports.

Word processing allows you to electronically add words, delete words, add paragraphs, move paragraphs, etc. When you have your paper like you want it, you can print out the letter or paper on the computer's printer. I now write all my papers on the word processor.

While word processing will not be a big thing for many farmers or ranchers, it might be of value to those of you that are officers of farm organizations. Dave Flintner, President of Wyoming Farm Bureau, could surely use word processing in his Farm Bureau business.

Level II Computers

The more common level II microcomputers that farmers are considering are: Northstar, Vector Graphics, Superbrain, Hewlett Packard, and Cromenco. There are also other brands but they tend to be less popular.

Northstar Microcomputer. One Level II microcomputer that is fairly popular is the Northstar. Country Side Data out of Utah is selling agricultural software for the Northstar computer.

Vector Graphics Microcomputer. Another Level II microcomputer is called the Vector Graphics. Homestead Computers out of Canada has several software packages for the Vector. In addition, Loren Bennett in California has a dairy-ration package for the Vector.

This microcomputer and others can be equipped with a "professional" printer that is used for word processing. If we had a letter typed with this type of printer, I could convince you that the letter was typed by my secretary on her IBM electric typewriter. Professional printers sell for around $3,000; however, if you are going to do word processing, a professional printer is preferred.

One purebred cattleman has a professional printer on his micro. He uses the word processor to write individual letters to his purebred cattle customers. He keeps a list of potential customers inside his computer. When he has a bull for sale, he then uses the word processor to generate and address personal letters to each customer. Each customer thinks the cattleman personally typed the letter to them. In reality, his microcomputer merged the names into the standard letter stored in the micro. This cattleman argues that this is a very cost-effective way to advertise his purebred cattle. The key is the professional printer and the word processing software.

Superbrain microcomputer. A Superbrain is used by South Dakota AGNET with disk drives that are built into the cabinet. This is extremely nice when you move the microcomputer around.

Hewlett Packard. Hewlett Packard recently announced the HP-125 as their small business machine. HP long has a reputation of producing high quality products, and we believe this is also true for their microcomputers. To date, I am not aware of any agricultural software available for the HP machines.

Cromenco Computer. Cromenco microcomputer is configured to be a fairly powerful microcomputer, yet there are several empty slots for future additions to meet your expanding needs. The Level II machines are considerably more flexible than the level I machines.

Comparing Level I and II Microcomputers

There are several differences in the Level II micros as compared to the Level I micros. The key differences are: 1) basic language compilers that are faster than Level I interpretors, 2) 80 character screens that make VISICALC and communications easier to use, 3) more standard operating systems such as CP/M (this means it is easier to exchange programs from one machine to another), 4) more error diagnostics for software and hardware, 5) and the S-100 buss (for more hardware exchangeability).

Hardware Accessories

Data cassette. In the past, we used cassettes for data and program storage. In fact, you can use your kids' cassettes and their tape recorder on your micro to record data and programs. While this is a very cheap storage device, by today's standards it is too slow and inflexible.

Floppy disk. The technology that has made microcomputers of value to agriculture is the floppy disk--a phonograph record with a paper covering around it. Instead of recording music on the disk, the micro records data and computer instructions on the disk. The floppy disk now provides the microcomputer with mass storage capability. Dr. Schoonover can store data for 500 cows in the beef program on one of these disks. If you have 1,000 cows, you simply use two disks. In fact, you can have as many of these disks as you want on the shelf. You just pull off the shelf the disk that you want and put it into your microcomputer.

Hard disk. The newest storage technology is the hard disk. Inside this little box is the ability to store 5 million characters of data. You could store all the management information that you would ever need or generate on your farm or ranch on one hard disk. Most farmers or ranchers do not have this kind of data storage need. The purebred cattleman I know with a Vector Graphics machine keeps all his pedigree information for his cow herd on the hard disk. He can go back to 1932 with his pedigree searches. He feels that the microcomputer has helped his purebred business out considerably.

Instructional Aids

A Radio Shack Teaching Center on our campus has 15 microcomputers hooked up to a sixteenth computer. The sixteenth computer can monitor the other 15 computers. Wyoming's Agricultural Extension Service needs one of these to bring 15 ranchers or farmers in for computer training. You learn more about microcomputers by hands-on experience than from lecture or books. Many high schools and vocational technical schools have such instructional centers but the university extension services are behind.

MICROCOMPUTERS FOR FARM AND RANCH

Let's now boil all this down--what do microcomputers mean for you on the farm or ranch? Agriculture is going to have some serious challenges in the 80s. During the 60s and 70s your challenge was production, but the challenge in the 80s is going to be financial management. And the computer has the potential to improve your financial management. Let me make a prediction. Those of you that will be farming in 1990 will be using computers, and those of you that do not want to use a computer will not be farming in 1990. I often hear, "No damn computer is going to tell me how to run my farm!" I predict that that person won't be farming in 1990. Many will have retired and others will have gone out of business. Computers are going to become commonplace on U.S. farms and ranches during the 80s.

As I travel around the country talking to farmers and ranchers, I hear them expressing interest in three applications of the microcomputer. The three applications are:

- Business accounting.
- Herd performance reporting.
- Financial management.

Top producers are recognizing that they need to keep better books. They are looking to the microcomputer as a means to make bookkeeping easier and more flexible. They want current cashflow situations several times during the year. Today's profit margins do not allow the management errors that you could get by with in the 70s.

Top ranchers know the benefits of good cow-calf records and they have been keeping them on the 3- x 5-inch cards; however, it takes a lot of time to sort them into useful management reports. A herd performance system fits well onto a microcomputer and once the data is in the computer, management reports can easily be printed out. We even know of one rancher that takes his micro right out to the scales and enters the calf weights as they are weighed. When the last calf is weighed, he pushes the button and identifies the cows to be immediately culled. By not having to wait for culling data, this rancher argues that the dollar amount saved from not rounding up cattle the second time will pay for his microcomputer.

Bankers are requiring more and more financial information before they will make loans to producers. Top producers are starting to see the potential of being able to use the microcomputer to help generate these needed reports: financial statements, profit and loss statements, cash flow projections, five-year plans, etc.

VISICALC - a financial management tool. One of the most powerful financial management tools available is VISICALC. It is designed so that you don't have to be a programmer to program your own financial management programs. There is nothing equivalent on AGNET! Since I don't know how to describe in words what VISICALC can do, I suggest that you stop into a computer store and ask for a VISICALC demonstration.

Disk oriented system. In order to have sufficient capacity to handle your agricultural applications, producers should buy a disk-oriented system. It should contain:

- Dual-disk drives.
- A good 80-column printer.
- 32K to 48K memory (the horsepower of the computer).
- 80-column screen (preferred over a 40-column screen).
- Telephone coupler.

The system will cost between $4,000 to $5,000 for the hardware and about $2,000 to $3,000 for programs (software) for your farm or ranch.

Telephone coupler. One of the extremely useful attachments that you can purchase for your microcomputer is a telephone coupler. This will allow you to use your micro as a terminal to large mainframe computers such as AGNET, TELEPLAN, and CMN. You can call the mainframe on the telephone and type in your information on your micro's keyboard and have the output printed out on your micro's printer. The cost of a phone coupler is around $300 and you can access:

- Current commmodity market prices.
- Current USDA, Foreign Ag, and Wyoming news releases.
- Agricultural outlook and situation reports.
- Western Livestock Market Information Project livestock analyses.
- Hay for sale.
- Sheep for sale.
- Certified pesticide applicators in Wyoming.
- People interested in judging county and state fairs.
- Horticultural tips during the summer.
- Home-canning tips during the canning season.
- Emergency information such as drought tips, Mount St. Helen's emergencies, etc.

You can even use your micro to access the UPI and AP news services such as news dealing with the Farm Bill and "beef." The AP and UPI news services are available from two commercial time-share companies. You can do all this today with your micro if it has a telephone coupler on it.

HOW TO BUY A SMALL COMPUTER

What should a farmer and rancher do if he is thinking about buying a small computer?

There are two newsletters that I recommend that you subscribe to on computers in agriculture. Successful Farming publishes one newsletter for $40.00 per year. They make useful evaluations of hardware and agricultural software.

The second newsletter is published by Doane-Western Agricultural Service out of St. Louis, Missouri. Their subscription rate is $48.00 per year. If you are seriously considering a microcomputer, I strongly recommend that you subscribe to one or both of these newsletters.

The second thing I recommend that you do if you are considering purchasing a computer is attend one of the computer seminars that are being held around the country. Almost every state extension service is holding these seminars specifically for farmers and ranchers interested in learning more about microcomputers and the potential agricultural applications. Contact your local county extension agent or extension advisor for information on these seminars.

Books and magazines on microcomputers and how to use and program them are also helpful when selecting a microcomputer. I strongly encourage farmers and ranchers who are thinking seriously about purchasing a computer to get one or two magazines or books on microcomputers. Farmers and ranchers read several agricultural-related magazines, so why not read at least one computer-related magazine.

I personally subscribe to BYTE. It is a good magazine to read to find out what kind of hardware is available and to learn the jargon of computers.

I also subscribe to the Personal Computing magazine. It has stories written by people who are familiar with microcomputers for people like you and me who are not familiar with microcomputers.

SUMMARY

Microcomputers are the new farm- and ranch-management tools and innovative producers are buying them. More and more farmers and ranchers are going to own one or more microcomputers.

If you buy a microcomputer, be sure and buy the telephone coupler so that you can access the agricultural information networks being set up across the country. You will need to spend around $4,000 to $5,000 for a microcomputer with enough horsepower and flexibility to do your farm or ranch applications. I assure you that we are going to see considerably more farm and ranch purchases in the next five years.

47

SIX STEPS FOR A CATTLEMAN TO TAKE IN BUYING A COMPUTER

Harlan G. Hughes

INTRODUCTION

Today's low profit margins and high interest rates place a premium on a cattleman's management-information system. Automation of that system lends itself to the microcomputer. Microcomputers represent a relatively new farm and ranch-management tool that farmers and ranchers are investigating. Purchasing one may prove to be one of the few profitable equipment purchases of the 1980s. One study indicates that as high as 64% of the producers interviewed were planning to buy a microcomputer as a management tool in the next five years. Twenty-seven percent indicated they would purchase a microcomputer in one to two years. These producers ranked business record keeping as the number one management function they wanted to perform on the microcomputer. The preparation of financial balance sheets and income and cash-flow statements ranked second. Breakeven analysis of individual enterprises and crop-production records ranked as the third and fourth management functions, respectively.

An Alberta, Canada, study of producers owning microcomputers indicated they were using the microcomputers for 1) farm planning, 2) financial record keeping, 3) physical record keeping, and 4) analysis of records (cash flow, breakeven analysis, and costs of production).

What kind of microcomputers do producers own? Sixty percent of the Canadian producers owned Radio Shack and the rest owned Apple, Pet Commodore, Vector Graphics, and others.

A recent Successful Farming magazine survey indicated that 46% of the respondents owned Apples, 34% owned Radio Shacks, 4% owned IBMs, 4% owned Commodore or Pet and the remaining percentage covered all other brands.

SIX STEPS FOR A COST-EFFECTIVE INVESTMENT

A producer-owned microcomputer should pass the same cost/benefit analysis as any other machinery investment.

Costs can be easily identified and documented; however, the benefit of improved management is considerably more difficult to document. What is clear, however, is that benefits received depend heavily on the preparation that the cattleman makes before purchasing the microcomputer.

Step 1

Before purchasing a microcomputer, study your management-information needs. Collection and analysis of management information requires time and money. You cannot afford to collect management information that you do not use or need. Some questions that you should ask are: What are the most important and significant decisions that I need to make? What information is needed to make these decisions? Can the generation of the needed information be scheduled? Can a microcomputer make this information collection easier? Studying your information requires some time and effort. It may well be worth your time to hire a consultant or visit with your university extension service and get a second opinion.

Step 2

Identify computer programs (software) that are available that might meet your management-information needs. As a cattleman you have four potential ways that you can obtain needed software. You can (1) buy it from a commercial vendor, (2) obtain it from the extension service, (3) hire it custom programmed, or (4) program it yourself.

If the software needed is available from a commercial vendor, this may well prove to be the most satisfactory method of acquiring software. Sometimes, however, you'll need software that is not available from a commercial vendor. The local extension service may have what is needed. Occasionally the only viable alternative is to hire a program custom-programmed or to program it yourself. Unless you have special training or a lot of spare time, I cannot recommend that you program the software on your own. Obtaining software tailored to your specific needs will be the most difficult and time-consuming task.

Step 3

Determine the hardware specifications required to execute the needed software. The size of the business affects the volume of management information needed and this, in turn, determines the size of hardware needed. Microcomputers come in different sizes (memory units), have different storage capabilities on the diskette (floppy disk), and have different add-on capabilities (80 column screens, upper and lower case characters, computer languages, CP/M operating systems, telephone modems, word processing software, etc.) Again, it is recommended that you contact a consultant or the extension service. Computer dealers are not necessarily the best information sources for determining specific hardware needs. Generally, they promote what they have to sell.

Step 4

Contact local hardware dealers and determine the viable hardware alternatives. Cattlemen should use the same criteria that they would use for any other equipment purchase: dealer knowledge of his own hardware, quality of the service department, apparent financial stability of the dealer's business and, in general, compatability with the dealer. Since cattlemen have purchased equipment before, they should feel reasonably comfortable with this step.

Step 5

Estimate the cost/benefit of the proposed computerized management-information system. A dealer can tell the purchaser exactly what the hardware will cost; and the cattleman already should have an estimate of what the software will cost. Remember that the cattleman-buyer can take investment credit and depreciation on computer hardware just like any other piece of machinery.

The clerical cost of collecting and processing the management information should also be included. This frequently is your time or that of your spouse. Collecting and typing data into the computer is time-consuming and boring. You might even consider hiring a person to be specifically responsible for the data processing of the management information.

While determining the cost/benefit, cost of the total management information system should be projected. A Michigan State University study indicates that it may cost $500 to $600 a year to process a producer's business records through his own microcomputer. Again, an outside consultant can be useful.

Estimating the dollar benefit of having a computerized management-information system is difficult for most cattlemen to do. Today's high costs of production and high interest rates do not leave much margin for management errors. Just preventing one management error a year may well pay for the microcomputer system. As could the ability to experiment with a decision on paper before implementation.

Step 6

The final step is to make the decision whether to set up a computerized management-information system. You should consider talking to other cattlemen that already own microcomputers. Many states are offering educational seminars for ranchers and farmers to learn more about how microcomputers can enhance a producer's decision-making process. The final decision rests with you the individual. There is no blank recommendation that will fit all situations. Microcomputers can, however, be an effective management tool.

Microcomputers are becoming a more common management tool for cattlemen. Innovative producers are purchasing microcomputers to enhance their personal management-information systems. This article summarizes six recommended steps that you should go through in making the decision to purchase a microcomputer. If these six steps are followed, you will have a higher probability for a successful experience with your first microcomputer.

REFERENCES

Engler, Verlyn, E. A. Unger and Bryan Schurle. 1981. The potential for microcomputer use in agriculture. Contribution 81-412-A. Department of Agricultural Economics, Kansas State Univ.

Nott, Sherrill. 1979. Feasibility of farm accounting on microcomputers. Agricultural Economics Report No. 336. Michigan State Univ.

Successful Farming. 1982. Successful Farming farm computer news. A special survey summary. Successful Farming.

1981. A survey of on-farm computer use in Alberta. Alberta Farm Management Branch, Olds, Alberta.

48

DIRECT DELIVERY OF MARKET INFORMATION THROUGH RANCHER-OWNED MICROCOMPUTERS: A RESEARCH REPORT

Harlan G. Hughes, Robert Price,
Doug Jose

Ranchers needs for marketing information have changed dramatically since the early 1970s. Increasing price variability, rapid inflation, higher interest rates, and closer ties to world supply-and-demand conditions for agricultural commodities have resulted in increased needs for short, intermediate, and long-run marketing information. Also, ranchers continually have fewer market outlets available so that they must do a better job of marketing their product. The net result is that many ranchers are unable to adequately evaluate marketing alternatives and, thus, are often unable to make good marketing decisions.

In late June 1981, Cooperative Agreement Number 12-05-300-522 was signed between the USDA Extension Service and the Colorado State University Cooperative Extension Service on behalf of the Western Livestock Marketing Information Project to give ranchers decision assistance. The agreement was to conduct a pilot study concerning the feasibility of direct electronic delivery of marketing and management information to farm and ranch families.

Ranchers base marketing decisions on information from both internal and external sources. Accounting records, herd performance records, and budgets are examples of internal information used. Market news, outlook reports, price forecasts, weather forecasts, and research reports are examples of external information. Internal and external information are required for almost all short, intermediate, and long-run marketing decisions.

Needs for short-run market information commonly relate to selling decisions. There are sometimes substantial risks associated with selling agricultural commodities today rather than waiting a few days, or vice versa. Short-run decisions are relatively simple to analyze in a budgeting sense as the costs are readily predictable. The difficult element is the probability of price increases and decreases.

Typically, university and government outlook specialists have not provided short-run market information. It has generally been left to the commodity brokerage firms and other private organizations to provide short-run market information. These sources tend to discount the risk and uncertainty aspects.

Intermediate-run needs for market information relate to such decisions as purchasing of stocker and feeder cattle, crop selections, fertilizer application, feed choice, and other decisions that do not result in immediate revenue. These decisions are generally more complex as the information needed to evaluate possible outcomes is more complicated and has more chance of error. University and government outlook specialists generally have been most active in providing intermediate-run information.

Examples of long-run market-information needs include land purchases, irrigation development, machinery selection, cattle herd expansion, and the construction of livestock production units. These decisions, although not made as frequently as the previous types, require significant information to allow for success of a farm business. Although farm management economists have devoted much time and effort to investment analysis, outlook specialists in the university and government realm generally have concentrated very little on this long-run arena.

Because of variability in agricultural prices and production, as well as high financing requirements, producers may risk bankruptcy before profits from an investment can be realized. Long-run market information can also be useful in assessing the amount of risk that a specific producer can afford when making investment decisions.

A comprehensive marketing-information system, used properly, could play a major role in stabilizing or increasing net ranch income during the 1980s. In the coming years, ranchers are going to need more marketing information, delivered faster, and available in an easy-to-use form. Computers can and should play a major role in such a marketing-information system and the associated educational needs. The rapid development of electronic technology also presents an exceptional opportunity for the Cooperative Extension Service to assume an even greater role in the delivery of timely market information.

DELIVERY OF MARKET INFORMATION IN THE WEST

The problem of delivering timely market information in the western U.S. is compounded by the vast geographical dispersion of producers. The extension specialists and county agents must travel extensively to accommodate the needs of farmers and ranchers. Most newspapers carry very little, if any, current market data. Farm magazines are major sources of intermediate-run market information, but timeliness of that information does not meet the standards necessary for decision making in today's economic environment.

The Western Livestock Marketing Information Project (WLMIP) was created over 25 years ago in recognition of the void that existed in the delivery of timely market information to livestock producers in the West. The proven record of WLMIP as a major source of useful intermediate-run market information for the region has been well documented (WLMIP, 1977; Bolen, 1949). However, the changing complexities of the livestock market, combined with the rapid growth in computer technology, present the need and opportunity for WLMIP to expand services. These opportunities include direct delivery of market information at the producer level, as well as increased service to professional economists and others in the West. It also presents the opportunity for WLMIP to go from almost exclusive emphasis on intermediate-run market information to expanding short-run information.

AGNET--AGRICULTURAL COMPUTER SYSTEM

AGNET is a time-sharing computer network headquartered in Lincoln, Nebraska. There are over 2500 subscribers to the network with a total yearly connecttime of over 75,000 hours. This averages out to 8.5 users per hour concurrently on a 24-hour 7-day-a-week basis. AGNET is being utilized for problem-solving and information networking. The system is very "user-friendly" and is designed for use by people with no computer background. Ranchers are allowed to subscribe to the AGNET System by paying variable costs associated with operating the system.

PREVIOUS STUDIES

In 1979 a survey was sent to state extension service administrators inquiring about the priority of marketing extension programs. (Watkins and Hoobler, 1980). Thirty-seven of the 44 state administrators returning the survey placed extension marketing programs in the range of "important" to "of highest importance." The following summary statement was taken from the report:

- "It is recommended that each state Cooperative Extension Service administration, in cooperation with their marketing specialists and representative clientele groups, examine the results of this national study, analyze their state's specific needs, determine where a cooperative effort is needed with other states, and develop plans for renewing and/or initiating programs to effectively manage the problems."

Brown and Collins (1978), University of Missouri, conducted a national study in 1977 on the information needs of large commercial farms. Their study revealed that:

- Commercial family farmers and ranchers perceive marketing information as their number one need.
- Extension and universities were rated the most important source of production technology, but only of minor importance as a source of marketing information.

- Farmers, agribusiness, extension, and the agricultural media all expressed the belief that marketing information is critical now and will continue to be critical in the future. They also agreed that present sources of market information are inadequate.

A joint USDA/NASULGC study (1968) committee recommended in 1968 "that extension increase its emphasis on marketing and farm-business management while reducing the percentage of effort in husbandry and production." The study goes on to say, "Extension should gradually shift towards giving more in-depth training to producers and to wholesaling information through supply firms."

Most extension marketing-program-appraisal studies generally include recommendations for experimentation with the latest electronic and computer innovations. For example, New York dairymen in a 1977 telephone survey felt that extension could improve its effectiveness by placing more emphasis on the use of the computer as an educational tool. (Ainsle et al., 1977).

In spite of the emphasis placed more than 10 years ago on changing extension priorities, little progress has been made to implement these program shifts. This is mainly due to extension administrators' reluctance to changing priorities of their extension programs. We are hearing the same priority requests coming from producers today as we did a decade ago. This paper reports on one pilot project that attempted to respond to some of these priority requests.

PILOT PROJECT

The state of Wyoming piloted a basic electronic market information system on the AGNET computer network during 1978-79 (Skelton, 1980). Four objectives of the pilot system were:

1. To collect price information of interest to Wyoming producers.
2. To provide county extension offices with the ability to retrieve market information that allowed them to put together today's, yesterday's, last week's, last month's, or last year's markets of interest for use by their producers.
3. To provide simple, down-to-earth interpretations of what market prices and associated outlook mean to Wyoming producers.
4. To provide price forecasts for extension personnel to use with producers in planning.

The Western Livestock Marketing Information Project piloted some initial work in computerized market-information delivery in 1980. Major livestock reports (Cattle on Feed, Hogs and Pigs, etc.) were placed on AGNET, complete with analysis and interpretation. WLMIP was instrumental in substantially increasing listings of producers with hay for sale and making these listings available to areas hardest hit by the drought of 1980. In addition, WLMIP served as a clearing house for drought conditions in many areas of the western plains region. This information was collected and transmitted throughout the region and forwarded to the office of the Secretary of Agriculture in Washington, D.C.

OBJECTIVES OF THIS STUDY

In an effort to provide an evaluation of electronic delivery of market information, a cooperative agreement was signed between the Colorado State University Extension Service on behalf of WLMIP and the USDA Extension Service. Subsequent cooperation was obtained from the University of Wyoming and the University of Nebraska. Four of the six objectives of the pilot study reported in this report are:

1. Research and develop mechanisms for direct producer access to AGNET via farmer-owned microcomputers and computer terminals.
2. Add current livestock market news information on AGNET for retrieval by producers and others.
3. Improve the documentation of marketing information and other pertinent information on AGNET and make it available to producers.
4. Evaluate the effectiveness and efficiency of this new delivery system in providing useful information to farmers.

RESULTS

Objective 1: Research and develop mechanisms for direct producer access to AGNET by farmer-owned microcomputers and computer terminals.

The technology of communication between computers of different brands and types is an involved science of its own. Different hardware requires different communication protocols and procedures. Much additional work is needed in this area that is receiving a lot of interest at the current time.

The importance of networking between microcomputers and mainframe computers housing networks such as AGNET is becoming increasingly obvious. As an example, of the 12 producers that participated in the pilot study, 9 accessed AGNET through the use of microcomputers. The other 3 used "dumb terminals" for communication. It is the authors' opinion that microcomputers will become more the norm in producer hardware than dumb terminals. The reason is that the microcomputer can also be used to solve on-the-farm types of production and marketing problems, handle production and accounting records, and handle other applications.

Current technology is readily available to enable a microcomputer to operate as a dumb terminal in communicating with AGNET. Generally, all that is required is a modem (telephone coupler) and software for the microcomputer, which is generally included with the hardware coupler. Such packages for microcomputers generally run in the price range of $500 or less. However, the technology involved in making a microcomputer into an "intelligent terminal" with AGNET is more complex. A high degree of interest in this type of software appears evident throughout the western region.

The important of operating a microcomputer in an intelligent mode with AGNET arises from the tremendous potential savings in telephone costs. A large part of the user's time during any terminal session is now spent typing in the needed information to respond to the AGNET questions. If such files could be developed on the microcomputer before the telephone call is actually made to AGNET, much of the telephone cost could be eliminated.

The authors have experienced substantial savings in telephone costs (50% or more) when using microcomputers as intelligent terminals. The capability to access AGNET as a central warehouse for information, download the information to the microcomputer, hang up the telephone, and work with the information that has been accessed results in even more cost savings.

In summary, it is a relatively easy procedure to turn a microcomputer into a dumb terminal for communicating with AGNET. It becomes a little more difficult to operate in the intelligent-terminal mode, but software has been developed in this study that makes this possible for most brands of microcomputers. The idea of interfacing farmer-owned microcomputers and a regional computer, such as AGNET, could be one of the most significant thrusts in extension service activities for computer applications to agriculture in the coming years.

Objective 2: Add current market-news information on AGNET for retrieval by producers and others.

This objective has been pursued heavily since the beginning of the pilot project. The system has been expanded so that 17 different market-price files are going onto AGNET daily. In addition, weekly and monthly analyses are going onto AGNET. During the six-month study period, 19,873 market-price files were retrieved by all AGNET users.

In addition to providing information for its current market value, most of the information included in the price files is captured by the computer and put into historical data files. By building such a data bank, files are in place for retrieval by the user in various programs for management and marketing decisions. Most of the captured information is already available for use in various retrieval and charting programs. However, retrieval programs for AGNET market information are still under development.

In an effort outside the scope of this study, the Foreign Agriculture Service (FAS) of USDA has begun a test using AGNET markets for distribution for much of their information. The response from users of the FAS information has been very enthusiastic, and several new users have subscribed to AGNET just to receive the FAS information.

The addition of market news and other information on AGNET will be a continuing process. Feedback from participating county agents and producers during this pilot study resulted in several files being added. Additional feedback on the final end-users' evaluations points to the need for even more types of files.

Market information that is currently being placed on AGNET is almost exclusively done by volunteer labor. Therefore, relatively little money is allocated to staffing explicitly for placing market information on AGNET. Several staff hours weekly are being devoted from numerous offices to place the information on AGNET.

By relying so heavily on manual labor to provide the information to the computer network, costs are magnified and the chances for errors arise. USDA Extension Service is working with Agricultural Marketing Service for direct electronic transfers of market information from AMS to the AGNET computer. If such a system could be put in place, cost savings for staff time would be tremendous and the timeliness of the availability of the information could be much improved.

Objective 3: Improve the documentation of marketing information and other pertinent information on AGNET and make it available to producers.

One of the developments coming out of this pilot study has been an AGNET Market Information Users Guide. The guide is intended to be just that, a guide to help the new user know what type of market information is available and how to access that information. In addition to documenting the market information, management decision tools are also referenced in the guide with a brief explanation of how to access and use those tools.

AGNET is a very "user-friendly" computer system. The major part of any documentation needed by the user is accessible directly from the computer with the use of "HELP" commands built into the system. Such HELP commands are unique to AGNET. Consequently, most of the needed documentation and aids are available at any time during a terminal session and preclude the necessity of having a manual available for reference while the user is online.

Objective 4: Evaluate the effectiveness and efficiency of this new delivery system in providing useful information to farmers.

The study tested two methods of market-information delivery. The first method was actual direct delivery to farmer-owned microcomputers or computer terminals located on the farm or ranch. The second method tested was "wholesaling" market information through county agents or trained agricultural professionals. These agricultural professionals used marketing bulletin boards in the county agent's office or in the financial institution center and then used frequent mailings of market information from these offices to selected producers in their area. This course of delivery was used mainly to acquaint producers with the type of information that could be obtained from the computer network and to test their responses to see if the information delivered was useful. The authors were also interested in seeing whether, after receiving the information in this manner, producers would be more interested in obtaining their own computer hardware for direct delivery of the information.

EVALUATION OF USERS DIRECTLY ACCESSING INFORMATION

An evaluation form was sent to the producers who were directly accessing the information from their own hardware. Evaluation forms were returned from 12 direct-access users. It should be noted that this group is a representative subset of farmers and not all AGNET users.

Users were asked to evaluate six general types of market information they could access from the computer. The results of that evaluation are listed in the following table:

Evaluation By Direct Users

Information	Very useful	Slightly useful	Not useful	No response	Total
Futures prices	3	4	1	4	12
Cash prices	5	3	1	3	12
Commentary & interpretation	6	3	0	3	12
News releases	2	5	1	4	12
Retrieval programs	3	2	1	6	12
Conferences	1	4	0	7	12

The files that contain commentary and interpretation of factors influencing the market were very well received by the users who were accessing them directly. Various comments received on this question included: "Good insights." "Comments really helped to get a feel for the market." "More of this type of information needed."

The files on various cash prices were also very well received by the users who were directly accessing the information. AGNET is very unusual in that several files contain localized information for various areas within a state that is not available anywhere else in a condensed, summarized form. Comments included such things as: "We need more local prices on the system." "Used these the most." "Often AGNET is the only source of this information." "Excellent."

The files on futures prices were not perceived by the end users to be as useful as the two previously discussed categories. One of the main reasons for this is that AGNET only offers each day's open and close of the futures. Producers who are active in the futures markets find that they need more current quotes, which they obtain from their farm radios or from their brokers. Also, not many producers use the futures market. Some comments on the futures included: "Would be better if we had a detailed report on weekly futures price movement." "Information didn't fit our area completely as there were no sugar futures." "Out of date by the time the producers really need this information." "Useful if picked off daily."

The retrieval programs were not used as heavily as the authors hoped they might be. One of the main reasons was that perhaps the users did not feel they were sufficiently versed in the correct technical aspects to use the program. Typical comments for this information included: "Did not use." "What are these?" "We need a lot more information on how to run these programs." "I liked these very much and accessed them regularly."

NEWSRELEASE items were also not rated very highly by the end users. This was not too surprising as the NEWSRELEASE program on AGNET is generally considered to be more consumer oriented, although there is much good useful information for livestock and grain producers. Typical comments included: "Very few used." "Checked only on occasional basis." "Some good, some bad." "Especially liked the ones on economic issues." "Some were excellent."

The lowest-rated information source by the end users was the electronic CONFERENCES. Again, this was not too surprising. CONFERENCES are of more use to people other than farmers. It is the responsibility of an individual AGNET user to link with the electronic CONFERENCES. Although the authors had sent out the procedure for doing this via U.S. mail, it is doubtful that many of the users took the time to go through the procedure to link up to the CONFERENCES. Typical responses included: "Did not use." "So what?" "Helped sometimes." "Need more information on how to use." "Not enough conferences sales or prices."

The users were asked to evaluate the timeliness of information delivered by AGNET. The response broke down as follows: very timely--5; average timeliness--5; too late to be useful--1; no response--1.

The users, who were all paying their own computer and telephone costs, had a very high expectation of when the information should be available on the computer. Many times the information for a given day would not be available until the following morning because of the manual transfer of the information onto the system. Once again, this points out the high desirability of automatic linkages with the AMS teletype system so that the information can be available much more quickly. Typical responses to the timeliness question included: "Most information was available from other sources at lower costs like newspapers and radio; however, this service shines in the fact that information is available on demand." "Many times it is hard to check information everyday. Why not put on a program that records daily futures-prices information and then on Friday evening we could pull them off for our records."

Users were asked to report costs. Very few had kept records of their costs, but those who did report indicated that $50 to $75 a month was a normal combined telephone and computer cost for accessing the AGNET information. A good share of the users responded that they did take advantage of nonprime-time telephone and computer costs by calling early in the morning or late in the evening.

Only three users indicated any problems from trying to access the information on AGNET. They also indicated that a workshop on operating technique would have been helpful. Most indicated that they felt it was quite easy to use the system. However, six respondents indicated that a workshop on how to apply the information being received from AGNET would be extremely useful.

Users were asked to give suggestions for improving AGNET delivery of market information and whether they felt it was worthwhile to continue providing information across the system. Most of the respondents who indicated that additional information would be desirable were looking for more localized cash prices and more commentary with specific projections for what the markets might do in the future. The overwhelming response was that the direct delivery of market information was extremely worthwhile and that the project should be continued. Only two users indicated that they did not intend to continue accessing AGNET information regularly.

Although very few respondents put a dollar value on the information received, the majority indicated that the cost-benefit ratios for accessing the information were highly favorable.

SUMMARY

The need for better information to be used by agricultural producers in making agricultural marketing decisions has been well documented. The thrust of this study has been to evaluate the feasibility of direct electronic delivery of this needed market information. The development of mechanisms for direct producer access to AGNET via farmer-owned microcomputers and computer terminals was one of the main objectives.

The best evaluation of this project lies in the large increase in retrievals of market information. During the time period of the study there was over a three-fold increase in the number of times that AGNET was accessed for market information.

Of the cooperators in this study, nearly three-fourths accessed AGNET through the use of microcomputers, while the remainder used dumb terminals for communication. It was found that operating a microcomputer in an intelligent mode with AGNET becomes increasingly desirable due to the tremendous potential in telephone savings. Savings of 50% or more resulted from using microcomputers in this manner.

It is relatively easy to turn a microcomputer into a dumb terminal for communicating with AGNET. It becomes much more difficult, however, to operate in the intelligent-terminal mode. This study uncovered software that makes this possible for most brands of microcomputers. The idea of interfacing farmer-owned microcomputers and a regional computer such as AGNET could be one of the most significant thrusts in extension service activities for agricultural computer applications in the future.

The pilot study was successful in providing current market news information to AGNET for retrieval by producers and others. A wide variety of new files has been made available in the MARKETS section on AGNET. Many of these files were the direct result of this pilot study. These files and others will continue to be available on AGNET.

Feedback from the final end user indicated the need for several more types of files, particularly of a regional type. The timeliness of the market information provided on AGNET was a concern to the producers involved in the study. Although the majority of the participants felt that the material was very helpful to them, they also expressed a desire for more timely information. This end-user evaluation points toward a critical need for direct electronic transfers of market information from AMS to the AGNET computer.

In summary, this pilot study provided much needed background information on the electronic delivery of market information. This study found that there is, indeed, a demand for the direct electronic delivery of marketing and management information to farm producers. There exits a distinct opportunity for the extension service to assume an even greater role in the delivery of this timely market information. In addition, the study provided the documentation of the need for increased development of computer applications to agriculture in information networking and evaluation of marketing alternatives. This information provides a base from which the extension service can evaluate and plan their activities in the computer arena for the future.

REFERENCES

Ainsle, et al. 1977. An evaluation of cooperative extension dairy programs. Specialist Report. Cornell University.

Bolen, Kenneth R. 1979. Economic information needs of farmers. Report of ESCS and SEA/Extension Study.

Brown, Thomas R. and Arthur Collins. 1978. Large commercial family farms information needs and sources. A Report of the National Extension Study Committee.

Skelton, Irvin. 1980. Wyoming agricultural extension service accomplishment report for FY-1980.

Watkins, Ed and Sharon Hoobler. 1980. Report of ECOP Subcommittee on agriculture forestry, and related industries extension marketing program and priorities survey. SEA/Extension.

USDA. 1968. A people and a spirit. Report of the joint USDA/NASULGC study committee on cooperative extension. Colorado State University.

WLMIP. 1977. Evaluation of the western livestock marketing information project. Report of WLMIP Technical Advisory Committee Survey of Users.

NAMES AND ADDRESSES OF THE LECTURERS AND STAFF

GEORGE AHLSCHWEDE
Sheep and Goat Specialist
Texas Agricultural Extension Service
Route 2, Box 950
San Angelo, TX 76901
--Sheep Specialist-Tour Coordinator

JOE B. ARMSTRONG
Associate Professor and
Extension Horse Specialist
Animal Science Department
New Mexico State University
Las Cruces, NM 88003
--Horse Geneticist

ROY. L. AX
Assistant Professor
Department of Dairy Science
University of Wisconsin
Madison, WI 53706
--Dairy Physiologist

HAROLD KENNETH BAKER
Meat and Livestock Commission
Box 44, Bletchley
Milton Keynes
MK2-2EF ENGLAND
--Beef Cattle Specialist & Geneticist

R. L. BAKER
Visiting Professor of Animal
Breeding & Genetics
Animal Science Department
University of Illinois
Urbana, IL 61801
--New Zealand Geneticist

R. A. BELLOWS
Location Research Leader
USDA-ARS
Route 1, Box 2021
Miles City, MT 59301
--Beef Cattle Physiologist

W. T. BERRY, JR.
Executive Vice-President
National Cattlemen's Association
P. O. Box 3469
Englewood, CO 80155
--Beef Cattle Organization Leader

HENRY C. BESUDEN
Vinewood Farm
Route 2
Winchester, KY 40391
--All-Time Great (Sheepman)

RONALD BLACKWELL
Executive Secretary-General Manager
American Quarter Horse Association
Amarillo, TX 79168
--Horse Organization Leader

BILL BORROR
Tehama Angus Ranch
Route 1, Box 359
Gerber, CA 96035
--Cattle Breeder

MELVIN BRADLEY
Professor of Animal Science and
State Extension Specialist
University of Missouri
Columbia, MO 65211
--Horse Specialist

B. C. BREIDENSTEIN
Director
Research and Nutrition Information
National Livestock and Meatboard
444 North Michigan Avenue
Chicago, IL 60611
--Meat Scientist

JENKS SWANN BRITT, D.V.M.
Veterinarian/Dairyman
Logan County Animal Clinic/J&W Dairy
Route 1
Russellville, KY 42276
--Veterinarian

HERB BROWN
Research Associate
Lilly Research Laboratories
P. O. Box 708
Greenfield, IN 46140
--Animal Scientist

O. D. BUTLER
Associate Deputy Chancellor
for Agriculture
Texas A&M University
College Station, TX 77843
--Animal Scientist

EVERT K. BYINGTON
Range Scientist
Winrock International
Route 3
Morrilton, AR 72110
--Range Scientist

B. P. CARDON
Dean
College of Agriculture
University of Arizona
Tucson, AZ 85721
--Animal Scientist

ARTHUR CHRISTENSEN
Manager
Christensen Ranch
P. O. Box 186
Dillon, MT 59725
--Sheep Producer

ROBERT L. COOK
Wildlife Biologist
Shelton Land and Ranch Company
P. O. Box 1107
Kerrville, TX 78028
--Wildlife Management Specialist

CARL E. COPPOCK
Professor
Animal Science Department
Texas A&M University
College Station, TX 77843
--Dairy Nutritionist

DICK CROW
Publisher
Western Livestock Journal
Crow Publications Inc.
P. O. Drawer 17F
Denver, CO 80217
--Journalist

STANLEY E. CURTIS
Professor of Animal Science
University of Illinois
Urbana, IL 61801
--Animal Scientist

A. JOHN DE BOER
Agricultural Economist
Winrock International
Route 3
Morrilton, AR 72110
--Agricultural Economist

WAYNE L. DECKER
Professor
Department of Atmospheric Science
University of Missouri
Columbia, MO 65211
--Meteorologist and Agri Weather Specialist

R. O. DRUMMOND
Laboratory Director
U.S. Livestock Insects Laboratory
P. O. Box 232
Kerrville, TX 78028
--Entomologist

ED DUREN
Extension Livestock Specialist
P. O. Box 29
Soda Springs, ID 83276
--Animal Scientist

WILLIAM EATON
Clear Dawn Angus Farm
R. R. 1
Huntsville, IL 62344
--Registered Cattle Breeder

WILLIAM D. FARR
Farr Farms Company
Box 878
Greeley, CO 80632
--All-Time Great (Cattleman)

H. A. FITZHUGH
Animal Scientist
Winrock International
Route 3
Morrilton, AR 72110
--Animal Scientist

MIGUEL A. GALINA
Professor
Universidad Nacional Autonoma de Mexico
A.P. 25, Cuautitlan Izcalli
Edo de Mexico, MEXICO
--Animal Scientist

HENRY GARDINER
Route
Ashland, KS 67831
--Cattleman

DONALD R. GILL
Extension Animal Nutritionist
Oklahoma State University
005 Animal Science
Stillwater, OK 74078
--Nutritionist

HUDSON A. GLIMP
Blue Meadows Farm, Inc.
Route 2, Box 407
Danville, KY 40422
--Sheep Producer

MARTIN H. GONZALEZ
President
ECO TERRA SA de CV
Fernando de Borja 208
Chihuahua, Chihuahua
MEXICO
--Range Scientist

TEMPLE GRANDIN
Livestock Handling Consultant
Department of Animal Science
University of Illinois
Urbana, IL 61801
--Livestock Facilities Specialist

SAMUEL B. GUSS, D.V.M.
Professor Emeritus
Veterinary Science Extension
Pennsylvania State University
2410 Shingletown Road
State College, PA 16801
--Veterinarian

JAMES C. HEIRD
Assistant Professor
Department of Animal Science
Texas Tech University
P. O. Box 4169
Lubbock, TX 79409
--Horse Specialist

A. L. HOERMAN
Extension Livestock Specialist
Texas A&M University Extension Center
P. O. Drawer 1849
Uvalde, TX 78801
--Beef Tour Coordinator

DOUGLAS HOUSEHOLDER
Extension Horse Specialist
Texas A&M University
College Station, TX 77843
--Tour and Clinic Moderator

HARLAN G. HUGHES
Agricultural Economist
University of Wyoming
Laramie, WY 82071
--Agricultural Economist

CLARENCE V. HULET
Research Leader
U.S. Sheep Experiment Station
USDA, ARS
Dubois, ID 83423
--Sheep Physiologist

HENRYK A. JASIOROWSKI
Director, Cattle Breeding Research
Warsaw Agricultural University
AGGW-AR 02-528 Warszawa
Rakowiecka Str. 26/30
POLAND
--Animal Scientist

DONALD M. KINSMAN
Professor, Animal Industries Department
University of Connecticut
Storrs, CT 06268
--Meat Scientist

JACK L. KREIDER
Associate Professor
Horse Program
Texas A&M University
College Station, TX 77843
--Physiologist

JAMES W. LAUDERDALE
Senior Scientist
The Upjohn Company
Performance Enhancement Research
Kalamazoo, MI 49001
--Physiologist

ROBERT A. LONG
Professor
Department of Animal Science
Texas Tech University
P. O. Box 4169
Lubbock, TX 79409
--Animal Scientist

CRAIG LUDWIG
Director of TPR
American Hereford Association
715 Hereford Drive
P. O. Box 4059
Kansas City, MO 64101
--Cattle Organization Leader

JAMES P. MC CALL
Director of Horse Program
Stallion Station
Louisiana Tech University
P. O. Box 10198
Ruston, LA 71272
--Horse Specialist

WILLIAM C. MC MULLEN, D.V.M.
Large Animal Medicine & Surgery
Texas A&M University
College Station, TX 77843
--Veterinarian

JOHN W. MC NEILL
Beef Cattle Specialist
Texas Agricultural Extension Service
6500 Amarillo Blvd., West
Amarillo, TX 79106
--Beef Cattle Specialist

DOYLE G. MEADOWS
Manager
Robinwood Farm
2822 East 2nd Street
Edmond, OK 73034
--Horse Specialist

JOHN L. MERRILL
XXX Ranch
Route 1, Box 54
Crowley, TX 76036
--Range Scientist and Rancher

BRET K. MIDDLETON
Animal Science Department
Iowa State University
Ames, Iowa 50011
--Computer Cow Game Coordinator

J. D. MORROW
Executive Vice-President
International Brangus Breeders Association
9500 Tioga Drive
San Antonio, TX 78230
--Cattle Organization Leader

HARRY C. MUSSMAN, D.V.M.
Administrator
APHIS/USDA
Washington, D.C. 20250
--Veterinarian

CHARLES W. NICHOLS
Manager
Davidson & Sons Cattle Company
Route 2, Box 15-0
Arnett, OK 73832
--Cattleman

J. DAVID NICHOLS
Anita, IA 50020
--Registered Cattle Breeder

MICHAEL J. NOLAN
Executive Secretary
Health and Regulatory Committee
American Horse Council, Inc.
1700 K Street, N.W., Suite 300
Washington, D.C. 20006
--Horse Specialist

JAY O'BRIEN
P. O. Box 9598
Amarillo, TX 79105
--Cattleman

JERRY O'SHEA
Principal Research Officer
The Agricultural Institute
Moorepark, Fermoy Company
Cork, IRELAND
--Dairy Equipment and Mastitis Expert

RICHARD O. PARKER
Assistant to the President
Agriservices Foundation
648 West Sierra Avenue
P. O. Box 429
Clovis, CA 93612
--Physiologist

BRENT PERRY, D.V.M.
President
Rio Vista International, Inc.
Route 9, Box 242
San Antonio, TX 78227
--Veterinarian

GUSTAV PERSON, JR.
Extension Agent
Guadalupe County
Ag Building
Seguin, TX 78155
--Horse Tour Coordinator

L. S. POPE
Dean
College of Agriculture
New Mexico State University
Las Cruces, NM 88003
--Animal Scientist

DOUGLAS PRESLEY
Extension Agent
Bexar County
Room 310
203 W. Nueva
San Antonio, TX 78207
--Horse Clinic Coordinator

NED S. RAUN
Vice-President, Programs
Winrock International
Route 3
Morrilton, AR 72110
--Animal Scientist

PATRICK O. REARDON
Assistant General Manager
Chaparrosa Ranch
P. O. Box 489
La Pryor, TX 78872
--Range Scientist

RUBY RINGSDORF
President
American Agri Women
28781 Bodenhamer Road
Eugene, OR 97402
--Farm Organization Leader

DON G. ROLLINS, D.V.M.
Technical Veterinary Advisor
Mid-America Dairymen, Inc.
800 West Tampa
Springfield, MO 65805
--Veterinarian

BUDDY ROULSTON
Professional Horse Trainer
Brenham, TX 77833
--Horse Trainer

MANUEL E. RUIZ
Head
Programa de Produccion Animal
CATIE
Turrialba, COSTA RICA
--Nutritionist

CHARLES G. SCRUGGS
Vice-President and Editor
Progressive Farmer
P. O. Box 2581
Birmingham, AL 35202
--Journalist

RICHARD S. SECHRIST
Executive Secretary
National Dairy Herd Improvement Association
3021 East Dublin-Granville Road
Columbus, OH 43229
--Dairy Organization Leader

MAURICE SHELTON
Professor
Texas Agricultural Experiment Station
Route 2, Box 950
San Angelo, TX 76901
--Sheep Specialist

PAT SHEPHERD
Manager
South Plains Feed Yard
Drawer C
Hale Center, TX 79041
--Feedlot Manager

JOHN STEWART-SMITH
President
Beefbooster Cattle Ltd.
P. O. Box 396
Cochrane, Alberta
CANADA TOL OWO
--Cattle Breeder

GEORGE STONE
President
National Farmers Union
12025 East 45th Avenue
Denver, CO 80251
--Farm Organization Leader

JACK D. STOUT
Associate Professor
Extension Dairy Specialist
Oklahoma State University
Stillwater, OK 74074
--Dairy Specialist

MRS. BAZY TANKERSLEY
Al-Marah Arabians
4101 North Bear Canyon Road
Tucson, AZ 85715
--All-Time Great (Horsewoman)

MAURICE TELLEEN
Editor and Publisher
The Draft Horse Journal
Box 670
Waverly, IA 50677
--Horseman and Journalist

THOMAS R. THEDFORD, D.V.M.
Extension Veterinarian
Oklahoma State University
Stillwater, OK 74078
--Veterinarian

TOPPER THORPE
General Manager
Cattle-Fax
5420 South Quebec Street
Englewood, CO 80155
--Agricultural Economist

ALLEN D. TILLMAN
Rockefeller Foundation, Emeritus
523 West Harned Place
Stillwater, OK 74074
--Animal Scientist

JAMES N. TRAPP
Associate Professor
Agricultural Economics Department
Oklahoma State University
Stillwater, OK 74078
--Agricultural Economist

ROBERT WALTON
President
American Breeders Service
P. O. Box 459
DeForest, WI 53532
--All-Time Great (Dairyman)

RODGER L. WASSON
Executive Director
American Sheep Producers Council, Inc.
200 Clayton Street
Denver, CO 80206
--Sheep Association Leader

DOYLE WAYBRIGHT
Mason Dixon Farms
RD 2
Gettsburg, PA 17325
--Dairyman

GARY W. WEBB
Stallion Manager
Winmunn Quarter Horses, Inc.
Route 1, Box 460
Brenham, TX 77833
--Horse Specialist

RICHARD O. WHEELER
President
Winrock International
Route 3
Morrilton, AR 72110
--Agricultural Economist

DICK WHETSELL
Vice-Chairman
Board of Directors
Oklahoma Land & Cattle Company
P. O. Box 1389
Pawhuska, OK 74056
--Range Scientist

R. GENE WHITE, D.V.M.
Coordinator of the Regional College
of Veterinary Medicine
University of Nebraska
Lincoln, NE 68583
--Veterinarian

RICHARD L. WILLHAM
Professor of Animal Science
Iowa State University
Kildee Hall
Ames, IA 50011
--Animal Breeding Specialist

DON WILLIAMS, D.V.M.
Henry C. Hitch Feedlot, Inc.
Box 1442
Guymon, OK 73942
--Feedlot Manager and Veterinarian

JAMES N. WILTBANK
Professor
Department of Animal Husbandry
Brigham Young University
Provo, UT 84602
--Beef Cattle Physiologist

CHRIS G. WOELFEL
Dairy Specialist
Texas Agricultural Extension Service
218 Kleberg Center
College Station, TX 77843
--Dairy Tour Coordinator

JAMES A. YAZMAN
Animal Scientist
Winrock International
Route 3
Morrilton, AR 72110
--Dairy Goat Specialist and Nutritionist

Other Winrock International Studies Published by Westview Press

Hair Sheep of West Africa and the Americas, edited by H. A. Fitzhugh and G. Eric Bradford

Future Dimensions of World Food and Population, edited by Richard G. Woods

Other Books of Interest from Westview Press

Carcase Evaluation in Livestock Breeding, Production and Marketing, A. J. Kempster, A. Cuthbertson, and G. Harrington

Energy Impacts Upon Future Livestock Production, Gerald M. Ward

Science, Agriculture, and the Politics of Research, Lawrence Busch and William B. Lacy

Developing Strategies for Rangeland Management, National Research Council

Proceedings of the XIV International Grassland Conference, edited by J. Allan Smith and Virgil W. Hays

Animal Agriculture: Research to Meet Human Needs in the 21st Century, edited by Wilson G. Pond, Robert A. Merkel, Lon D. McGilliard, and V. James Rhodes

Other Books of Interest from Winrock International*

Ruminant Products: More Than Meat and Milk, R. E. McDowell

The Role of Ruminants in Support of Man, H. A. Fitzhugh, H. J. Hodgson, O. J. Scoville, Thanh D. Nguyen, and T. C. Byerly

Potential of the World's Forages for Ruminant Animal Production, Second Edition, edited by R. Dennis Child and Evert K. Byington

Research on Crop-Animal Systems, edited by H. A. Fitzhugh, R. D. Hart, R. A. Moreno, P. O. Osuji, M. E. Ruiz, and L. Singh

Bibliography on Crop-Animal Systems, H. A. Fitzhugh and R. Hart

*Available directly from Winrock International, Petit Jean Mountain, Morrilton, Arkansas 72110